This series aims to report new developments in physical research and teaching – quickly, informally, and at a high level. The type of material considered for publication includes:

1. Preliminary drafts of original papers and monographs

2. Lectures on a new field, or presenting a new angle on a classical field

3. collections of seminar papers

4. Reports of meetings

Texts which are out of print but still in demand may also be considered if they fall within these categories.

The timeliness of a manuscript is more important than its form, which may be unfinished or tentative. Thus, in some instances, proofs may be merely outlined and results presented which have been or will later be published elsewhere.

Publication of *Lecture Notes* is intended as a service to the international physical community, in that a commercial publisher, Springer-Verlag, can offer a wider distribution to documents which would otherwise have a restricted readership. Once published and copyrighted, they can be documented in the scientific libraries.

Manuscripts

Manuscripts are reproduced by a photographic process; they must therefore be typed with extreme care. Symbols not on the typewriter should be inserted by hand in indelible black ink. Corrections to the typescript should be made by sticking the amended text over the old one, or by obliterating errors with white correcting fluid. The figures (in the original size) ready for reproduction should be inserted into the text. Should the text, or any part of it, have to be retyped, the author will be reimbursed upon publication of the volume. Authors receive 50 free copies.

The typescript is reduced slightly in size during reproduction, therefore a large size of type should be used; best results will not be obtained unless the text on any one page is kept within the overall limit of 18 x 26.5 cm (7 x 10½ inches). The publishers will be pleased to supply on request special stationery with the typing area outlined.

Manuscripts in English, German or French should be sent to Springer-Verlag, 6900 Heidelberg, Postfach 1780.

Die „*Lecture Notes*" sollen rasch und informell, aber auf hohem Niveau, über neue Entwicklungen in der Physik berichten. Zur Veröffentlichung kommen:

1. Vorläufige Fassungen von Originalarbeiten und Monographien.

2. Spezielle Vorlesungen über ein neues Gebiet oder ein klassisches Gebiet in neuer Betrachtungsweise.

3. Seminarausarbeitungen.

4. Vorträge von Tagungen.

Ferner kommen auch ältere vergriffene spezielle Vorlesungen, Seminare und Berichte in Frage, wenn nach ihnen eine anhaltende Nachfrage besteht.

Die Beiträge dürfen im Interesse einer größeren Aktualität durchaus den Charakter des Unfertigen und Vorläufigen haben. Sie brauchen Beweise unter Umständen nur zu skizzieren und dürfen auch Ergebnisse enthalten, die in ähnlicher Form schon erschienen sind oder später erscheinen sollen.

Die Herausgabe der „*Lecture Notes*" Serie durch den Springer-Verlag stellt eine Dienstleistung an die physikalischen Institute dar, indem der Springer-Verlag für ausreichende Lagerhaltung sorgt und einen großen internationalen Kreis von Interessenten erfassen kann. Durch Anzeigen in Fachzeitschriften, Aufnahme in Kataloge und durch Anmeldung zum Copyright sowie durch die Versendung von Besprechungsexemplaren wird eine lückenlose Dokumentation in den wissenschaftlichen Bibliotheken ermöglicht.

Lecture Notes in Physics

Edited by J. Ehlers, Austin, K. Hepp, Zürich and
H. A. Weidenmüller, Heidelberg
Managing Editor: W. Beiglböck, Heidelberg

6

G. A. Goldin, R. Hermann, B. Kostant,
L. Michel, C. C. Moore, L. O'Raifeartaigh, W. Rühl,
D. H. Sharp, I. T. Todorov

Group Representations in Mathematics and Physics

Battelle Seattle 1969 Rencontres

Edited by V. Bargmann, Princeton University

Springer-Verlag
Berlin Heidelberg GmbH 1970

ISBN 978-3-540-05310-1 ISBN 978-3-540-36441-2 (eBook)
DOI 10.1007/978-3-540-36441-2

 Library of Congress Catalog Card Number 75–146233

Originally published by Springer-Verlag Berlin Heidelberg New York in 1970.

Offsetdruck: Julius Beltz, Weinheim/Bergstr.

PREFACE

The Battelle Seattle Summer Rencontres in Mathematics and Physics have been established in order "to provide a channel of communication between mathematicians and physicists". Each year a topic of common interest was chosen which--it was hoped--would lead to fruitful discussions.

The 1969 Rencontres dealt with "Group Representations in Mathematics and Physics". In the view of most participants, I believe, these Rencontres lived up to expectation. The discussions were lively, mathematicians and physicists found many questions of common interest to talk about, and they learned from each other. On the mathematical side the emphasis was on the problems of infinite-dimensional representations, a subject which has attracted and continues to attract many workers in functional analysis. On the side of physics the applications to quantum theory of both finite- and infinite-dimensional representations were considered, with particular attention to the theory of elementary particles, including its most recent branch, "current algebra", which leads beyond the framework of Lie groups and calls for new methods.

As it should be, individual contacts and informal gatherings played a major role during these Rencontres. In addition, there were four series of lectures--two in Mathematics (by E. M. Stein and by C. C. Moore) and two in Physics (by L. Michel and by L. O'Raifeartaigh)--and a number of seminar talks by various participants.

The present volume contains three of the four lecture series (see the Note below) and five seminar talks, written up by their authors. The reader is expected to be familiar with the main facts of the theory of group representations. On the other hand, to help the mathematician, Michel and O'Raifeartaigh have carefully

---

Note. Due to the pressure of other work Professor Stein was, unfortunately, unable to prepare his lecture notes for publication. His course of lectures was entitled "Special Topics in the Representation Theory of Semi-Simple Groups" and dealt primarily with the analytic continuation of representations. Professor Stein has kindly supplied the following list of his publications on this subject.

(a) E. M. Stein, "Analytic Continuation of Group Representations", *Advances in Mathematics,* 4 (2), 172-207 (April, 1970). Academic Press, New York and London.

(b) A. W. Knapp and E. M. Stein, "Singular Integrals and the Principal Series I. and II.", I. *Proc. Nat. Acad. Sci. U.S.A.,* 63 (2), 281-284 (June, 1969); II. *Ibid.,* 66 (1), 13-17 (May, 1970).

(c) A. W. Knapp and E. M. Stein, "The Existence of Complementary Series". To appear in *Problems in Analysis,* Symposium in Honor of Solomon Bochner. Princeton University Press.

stated the quantum theoretical postulates and described the experimental evidence on which the group theorectical treatment of quantum physics is based.

The contributions to this volume range from systematic accounts of extensive fields to reports on current research on very specific questions. Only the papers by Moore and by Kostant may be called "purely mathematical". All other papers--although highly mathematical in content--are concerned with, or at least motivated by physical problems. A few remarks on the interrelation between these more physical papers may help some readers. The papers by Michel and O'Raifeartaigh give a very comprehensive account of the application of group theory to the most varied problems of quantum physics. While the two papers are, in the main, complementary to each other they overlap in some measure, but wherever they do, the discussion of the questions they both treat is sufficiently different to be highly interesting and illuminating. The contributions by Goldin-Sharp and by Hermann are concerned with current algebra. An introduction to the concepts involved here may be found in the last chapter of Michel's paper and, in greater detail, in the last chapter of O'Raifeartaigh's paper. Lastly, the papers by Todorov and by Rühl are independent of each other and of the remaining papers in this volume.

On behalf of all participants I take great pleasure in thanking the officers and the staff of the Battelle Memorial Institute, who did so much to provide an ideal setting for these Rencontres.

Special thanks are due to the technical typing staff at the Battelle Seattle Research Center, for their painstaking efforts in preparing this volume for publication.

April, 1970 V. Bargmann

TABLE OF CONTENTS

RESTRICTIONS OF UNITARY REPRESENTATIONS TO SUBGROUPS AND ERGODIC THEORY: GROUP EXTENSIONS AND GROUP COHOMOLOGY

by

Calvin C. Moore*

PREFACE

These notes are divided into two rather distinct parts, the first of which concerns the restriction of unitary representations of a group to one of its subgroups, and the connection of this with ergodic theory, while the second part concerns group extensions and the connection of this with unitary ray representations. Some background concerning representation theory is assumed and the reader should consult relevant portions of Mackey's notes [33], and survey article [34], Dixmier's book [9], and Chapter I of [3]. The square brackets refer to the common bibliography for both Part I and Part II at the end.

PART I. RESTRICTIONS OF UNITARY REPRESENTATIONS TO SUBGROUPS AND ERGODIC THEORY

1. INTRODUCTION

This first part concerns the general question of what happens when one takes a unitary representation of a locally compact group G, say an irreducible one, and restricts it to a subgroup H of G. One source of interest in this problem is ergodic theory as we will indicate below, but we believe a thorough study of this type of question will shed much light not only on representation theory as such but will produce much useful information concerning the structure of locally compact groups and their subgroups. Gelfand and Fomin [13] were perhaps the first to realize the relevance of this kind of problem concerning unitary representations for ergodic theory. They showed how one could study geodesic flows on surfaces of constant negative curvature by looking at unitary representations of the group $SL_2(R)$. This approach was extended by Parasyuk [43], Mautner [35], Green [1], Auslander and Green [2], and the author [39]. Part of these notes are an exposition of some of the results in [39] without proofs, and the reader is referred to this paper for further details. We shall also discuss some related results which will appear shortly. Chapter I of [3] contains an exposition of some aspects of the theory of unitary representations which we shall use as a general source both for Part I and Part II.

* Department of Mathematics, University of California, Berkeley, California 94720.

We shall suppose for the moment in order to illustrate our approach that G is a Lie group with Lie algebra $\mathcal{G}$. Let π be a continuous unitary representation of G and let X be an element of $\mathcal{G}$. Then $x(t) = \exp(tX)$ is a one-parameter subgroup of G where exp denotes the exponential map of $\mathcal{G}$ into G, and so $\pi(x(t))$ is a one-parameter group of unitary operators. It has an infinitesimal generator, or in other words, there is a unique (usually unbounded) self adjoint operator A such that $\pi(\exp(tX)) = \exp(itA)$ where the second exp is understood in the usual way for unbounded operators. We write $d\pi(X) = iA$, but we shall not enter into more discussion concerning the definition and properties of these operators since this is discussed in other lectures in this Rencontres.

The problem that concerns us specifically is to determine for a given group G and given $X \in \mathcal{G}$, the various possibilities for the unitary type of the operator $A = id\pi(X)$. The object is to get results that hold for fixed G and X, and for an arbitrary representation. One might hope to be able to say that the spectrum of A is limited to a very few possibilities or that one can put limitations on the possible eigenvalues of A. If G is for instance the real line R, so that $\mathcal{G}$ is one-dimensional, then any self adjoint operator A defines a unitary representation of R by $\pi(t) = \exp(itA)$. In this case we can extract no information concerning A, and in fact the same situation holds for any vector group G. Our results will concern exactly the opposite case, namely when G is semi-simple.

This same problem can be viewed slightly differently; suppose that $H = \{\exp(tX)\}$, $X \in \mathcal{G}$ is a one-parameter subgroup of G and suppose that $\pi(\exp(tX)) = \exp(itA)$ is a representation of H. We assume that this representation can be extended to a representation of G, and then ask what conclusions concerning the operator A can be drawn from this fact. Clearly whether one starts with a representation of G and restricts to H, or whether one starts with a representation of H and assumes that it extends to G comes to exactly the same thing. It is only a matter of emphasis.

Such problems are relevant in physics for if G is some postulated symmetry group of a quantum mechanical system, one has associated a unitary representation π of G on the Hilbert space associated to the system. (This is not quite true, but rather one has a ray representation of G; at our present heuristic level this doesn't matter, and in fact can be gotten around by well known methods to be treated in Part II of these lectures.) In any case, the operators $A = -id\pi(X)$ for various X in $\mathcal{G}$ have in many cases natural physical interpretations such as energy, momentum, angular momentum, and so on. It is an obvious question to ask what one can conclude about the spectrum or unitary type of these operators, based solely on the fact that they form part of the infinitesimal generators of a unitary representation of a larger group of some specified algebraic structure.

We have spoken about the restrictions of representations of a group G to one-parameter subgroups H. One can raise the same kinds of questions for larger subgroups of G. Some of the theorems below make sense in this generality and we shall state them in that form.

2. STATEMENT OF RESULTS

We turn now to the statement of our results, the first of which concerns as a special case the study of possible eigenvalues for infinitesimal generators as discussed above. In order to formulate the theorem it is convenient to introduce the following definition. Let G be a locally compact group and H a subgroup; we shall say that H has property E (in G) if for every representation π of G, and for every vector v in the Hilbert space of this representation such that $\pi(h)v = v$ for every $h \in H$, we have $\pi(g)v = v$ for every $g \in G$. In other words, if we have a representation π of G and are looking for invariant vectors for the restriction of the representation to the subgroup H, the condition says that we have only the obvious ones, namely the G-invariant vectors. This definition singles out a property which a subgroup H may or may not have which will be quite relevant for ergodic theory. If H is a one-parameter group $\{\exp(tX)\}$ corresponding to an element of the Lie algebra, then the condition essentially forbids the infinitesimal generator $A = -id\pi(X)$ from having 0 as an eigenvalue, unless of course the representation π of G has G-invariant vectors.

We want also to single out a slightly stronger property that a subgroup may have. More precisely, we say that H has property WM if for every representation π of G and every finite dimensional subspace V of the Hilbert space of the representation such that $\pi(h)V \subset V$ for every $h \in H$, we must have $\pi(g)v = v$ for every $g \in G$ and every $v \in V$. This condition, if satisfied, forbids the restriction of any representation π of G to H to have any finite dimensional subrepresentations other than the obvious ones. If again H is a one-parameter subgroup of a Lie group, the condition forbids the infinitesimal generator $A = id\pi(X)$ from having any eigenvalues. The terminology E and WM is motivated by ergodic theory and as we shall see later, whenever these conditions are satisfied, one may infer results asserting that certain group actions are ergodic (E) or weakly mixing (WM).

We note that a subgroup H has property E or WM (in G) if and only if its closure $\overline{H}$ does, and so it would suffice to consider closed subgroups. It is easy to see that a proper compact subgroup H of a group G can never have property E be examining the representation π of G induced by the trivial one-dimensional representation of K. Furthermore, if G is abelian, no proper closed subgroup can have property E. Our goal is at least in some cases to characterize those subgroups of a given group which have one or both of these properties, and from the two examples above, we see that we are going to have to assume that G is sufficiently non-commutative, and that H is sufficiently non-compact.

Semi-simple Lie groups are certainly one of the most important classes of groups, and they are in a sense as non-commutative as possible. The main result below will characterize those subgroups of such a group which have properties E and WM. This result is contained in [39] and we refer the reader to this paper for a more detailed discussion.

If G is a semi-simple Lie group, let G^* be its adjoint group, that is, G/Z, where Z is the center of G. It is well known that G^* is the product $\Pi_{i=1}^n G_i^*$ of a finite number of simple Lie groups, each having center reduced to the identity element. This is the global version of the decomposition of the Lie algebra of G into a sum of simple ideals. Let p_i denote the projection of G onto G_i^*. We shall say that a subgroup H of G is <u>totally non-compact</u> if $p_i(H)$ has non-compact closure in G_i^* for each i. Intuitively this says that H sticks out non-compactly in the adjoint group of each simple factor of G. If G is simple with finite center, the condition is simply that the closure of H be non-compact.

<u>Theorem 1</u>

For a semi-simple group G and a subgroup H of G, the following are equivalent

(a) H is totally non-compact.

(b) H has property E.

(c) H has property WM.

Thus for a totally non-compact subgroup H of G and any representation π of G which has no G-invariant vectors, we can conclude that the restriction of π to H has no invariant vectors, or for that matter no finite dimensional invariant subspaces. Even if H only partially satisfies the non-compactness conditions we can still extract information. For instance if $H \subset G_1 \times G_2$ and if the projection of H into the first factor G_1 is totally non-compact, then one can conclude that any finite dimensional subspace for H is left fixed pointwise by G_1. This follows by a detailed analysis of the proof in [39]. If we specialize the theorem to the case of a one-parameter subgroup $H = \{\exp(tX)\}$, $X \in \mathcal{G}$, we can conclude that the operator $A = -id\pi(X)$ has no eigenvalues provided H is totally non-compact.

This result overlaps with the O'Raifeartaigh theorem [42], and gives a stronger conclusion under much stronger hypotheses. To check that a one-parameter group is totally non-compact is in any given situation, a rather routine matter. The result above contains as special cases the results of Gelfand and Fomin, Parasyuk and Mautner mentioned above.

For one-parameter groups we can in fact get much more information concerning the infinitesimal generator $A = -id\pi(X)$ when $X \in \mathcal{G}$ generates a totally non-compact one-parameter subgroup. The unitary type of A is in fact limited to a rather small number of possibilities. We introduce the Hilbert space $H^+(n)$ which is to consist of all square integrable functions on the interval $(0,\infty)$, with Lebesgue measure, taking values in a standard n-dimensional Hilbert space H^n. Here n is an integer or $+\infty$. Let $H^-(n)$ and $H(n)$ denote the similar spaces of functions on $(-\infty,0)$, and on $(-\infty,\infty)$, and let $M^{\pm}(n^{\pm})(M(n))$ denote the unbounded self adjoint operator on $H^{\pm}(n)(H(n))$ which is multiplication by the scalar function $f(x) = x$. We note that $H^{\pm}(n^{\pm})$ is a non-negative (respectively a non-positive) operator.

Theorem 2

Let G be semi-simple and let $H = \{\exp(tX)\}$ be a totally non-compact one-parameter subgroup, and π be a representation of G with no G-invariant vectors, and let $A = -id\pi(X)$. Then if we write $A = A^+ + A^-$ where A^+ and A^- are the positive and negative parts of A, there exists $n^{\pm}$ such that $A^{\pm}$ is unitarily equivalent to $M^{\pm}(n^{\pm})$.

The proof of this is contained in [39] and the reader is referred to that paper for the details. We also note that if we have a single element g of G such that its powers g^n form a totally non-compact subgroup, we can obtain an entirely analogous result for the unitary type of the operator $\pi(g)$ (see [39]).

The result above for one-parameter subgroups is best possible in that all choices of n^+ and n^- occur, and if one considers a subgroup for which the hypothesis fails, then one can find a representation for which the conclusion fails. In fact, for $G = SL_2(R)$ and for a one-parameter subgroup generated by a nilpotent matrix all possible choices of n^+ and n^- occur. The situation for irreducible representations of $SL_2(R)$ is quite interesting; for the principal and complementary series, $n^+ = n^- = 1$, and for one discrete series, $n^+ = 1$, and $n^- = 0$, while $n^+ = 0$, and $n^- = 1$ holds for the other discrete series. The one representation of the principal series which is not irreducible decomposes into two irreducible summands which behave like discrete series for $n^{\pm}$.

For higher dimensional semi-simple groups, the situation becomes a bit simpler. More precisely, if we exclude any group G which has a simple factor locally isomorphic to $SL_2(R)$, the only possible choices for n^+ and n^- are either 0 or ∞. This fact is implicit in the argument contained in [39]. Thus in this case $A = -id\pi(X)$ for a totally non-compact one-parameter subgroup is up to unitary equivalence, one of three types, $M^+(\infty)$, $M^-(\infty)$ or $M(\infty)$. Since changing X into $-X$ or replacing π by its contragradient representation will interchange M^+ and M^-, we really have only two distinct cases, which we can classify as one sided spectrum or two sided spectrum.

We can raise the question of when every totally non-compact one-parameter group has two sided spectrum for every representation. S. Scull in a dissertation in progress has shown that this is true for $SL_n(R)$ for $n \geq 3$. B. Kostant has proved that this is also true whenever the Weyl group of the maximal compact subgroup

K of G contains the element -1. In this connection we should also remark that not every element of $Sp_n(R)$ has one sided spectrum as we shall see in a later section of these notes.

Finally we should like to indicate one application of these results to qauntum physics. Let P denote the Poincaré group, and let us assume that P is a subgroup a larger symmetry group G of unknown origin. We shall assume that G is semi-simple, and that π is a representation of G on a Hilbert space. To extract physical information one would restrict π to P and decompose it, and Theorems 1 and 2 above supply information about what this decomposition can look like. Indeed let G_1 be the largest normal subgroup of G containing P and let us assume that there are no G_1 invariant vectors since such vectors cannot be of interest. Now let X be the element of the Lie algebra of P corresponding to translation in time so that $A = -id\pi(X)$ is the energy. The spectrum of A controls to some extent the representations of P that can occur since in an irreducible representation of mass m, the energy operator has spectrum $[m,\infty)$.

Theorems 1 and 2 and the remarks following them give us the following result.

Theorem 3

Under the above hypotheses, A is unitarily equivalent either to $M^+(\infty)$, $M^-(\infty)$ or $M(\infty)$, as defined in Theorem 2.

Proof. It is a simple algebraic matter to verify that the one-parameter subgroup generated by X is totally non-compact in G_1, and that G_1 has no factors locally isomorphic to $SL_2(R)$ so the result follows. (The result concerning multiplicities it should be noted, is obvious on other grounds once one has the spectrum of A.)

This result says that in the decomposition of π on P, we must find representations corresponding to arbitrarily small mass or zero mass or imaginary mass. For a survey of the representations of P, see the article of O'Raifeartaigh in this volume.

As we have said before, we shall not enter into the details of the proofs of Theorems 1 and 2. There is, however, one important fact, Lemma 4.2 of [39] used in the proof of Theorem 1, for which we now have an alternate argument. This lemma says that a one-parameter subgroup of the universal covering group of $SL_2(R)$ has property E. The argument in [39] is based on infinitesimal methods due to T. Sherman. The alternate argument is global in nature and has in addition the property that it works for $SL_2(k)$, where k is a p-adic field, and also for the covering groups of this group defined in [40]. To carry this out however, one needs the algebraic analysis of the covering groups of $SL_2(R)$ and $SL_2(k)$ contained in

[40]. We will carry out the proof for $SL_2(R)$ with the understanding that if one redefines the meaning of the symbols that we will introduce, the argument will carry over word for word to the general case.

We may clearly assume that the one-parameter subgroup under consideration is $x(t) = \begin{pmatrix} 1 & t \\ 0 & 1 \end{pmatrix}$. We let $y(t)$ be the transpose of the matrix $x(t)$, and we define $w(t) = x(t)y(-t^{-1})x(t)$ and $h(t) = w(t)w(-1)$. It may be verified that $h(t)$ is a diagonal matrix with entries t and t^{-1}. Let us suppose that π is a unitary representation of $SL_2(R)$ and that v is a vector of unit length such that $\pi(x(t))v = v$ for all t. An easy calculation shows that $(\pi(w(t))v,v)$ $= (\pi(y(-t^{-1}))v,v)$ and if we let $|t| \to \infty$, we see that $\lim(\pi(w(t))v,v) = 1$. We let $u = \pi(w(-1))^{-1}v$ so that $(u,u) = 1$, and note that $(\pi(h(t))u,v) \to (v,v) = 1$ as $|t| \to \infty$. We then write $\pi(h(t))u = a(t)v + s(t)$ where $s(t)$ is orthogonal to v, and note that $|a(t)|^2 + |s(t)|^2 = 1$. Since $(\pi(h(t))u,v) = a(t)$ we see that $|s(t)| \to 1$ as $|t| \to \infty$, or equivalently, $\pi(h(t))u \to v$. It follows immediately that $\pi(h(s^{-1}t))u \to v$ for any $s \neq 0$ and then that $\pi(h(s))v = v$. Since the one-parameter group $h(t)$, $t > 0$, has property E (this is Mautner's lemma [35]), it follows that $\pi(g)v = v$ for $g \in SL_2(R)$. This completes the proof of the lemma in question.

3. APPLICATIONS TO ERGODIC THEORY

We shall now turn our attention to the applications of these results to ergodic theory and defer for a later section a treatment of some more questions concerning the restriction of representations to subgroups. These final results have little direct connection with ergodic theory whereas Theorems 1 and 2 have a very direct connection.

Let us first introduce the setting in which we are going to study ergodic theory. Let M be a Borel space, that is, a set equipped with σ-field of subsets, called the Borel sets, and let G be a locally compact group, separable in the sense of the second axiom of countability. We shall suppose that G acts on M as a transformation group so that we have specified a map f of $G \times M \to M$, written $f(g,m) = g \cdot m$ such that for fixed $g \in G$, the map $m \to g \cdot m$ is a bijective map of M onto itself and such that the function that associates to each g in G, this bijective map, is a homomorphism of G into the group of all such maps of M into itself. We say that G is a Borel transformation group if the map $f(g,m) = g \cdot m$ is a Borel map from $G \times M$ into M where G is given the σ-field of sets generated by the open sets and where $G \times M$ is given the product Borel structure. A function f is a Borel function if $f^{-1}(E)$ is a Borel set in the domain for every Borel set in the range. Thus not only is $m \to g \cdot m$ a Borel automorphism of M, but this Borel automorphism varies "smoothly" with g. The reader should consult Chapter I of [3] for further details.

For any Borel set E of M, $g \cdot E = \{g \cdot m | m \in E\}$ is a Borel set called the transform of E by g. If μ is a measure on M, we define the transform $g \cdot \mu$ of μ by g using the formula $g \cdot \mu(E) = \mu(g^{-1} \cdot E)$. We say that μ is G-invariant (or simply invariant) if $g \cdot \mu = \mu$ for all $g \in G$. Unfortunately, many interesting measures arising in practice fail to have this property, but possess instead the weaker property of quasi-invariance. A measure μ is quasi-invariant if μ and $g \cdot \mu$ are equivalent in the sense of absolute continuity, or more concretely, $\mu(E) = 0 \Leftrightarrow \mu(g \cdot E)$ for all $g \in G$.

If G is the additive group of the integers, then specifying an action of G on M is the same thing as specifying the Borel automorphism h or M corresponding to the group element one. If $h \cdot \mu = \mu$ for some measure or if $h \cdot \mu$ is equivalent to μ then μ is invariant (or respectively quasi-invariant). The case of a single measure preserving automorphism of a finite measure space is the classical setting for ergodic theory (see [15]). If G is the real line, an action of G consists in giving a one-parameter group $h(t)$ of Borel automorphisms of M subject to the joint measurability condition. This condition is readily verifiable in cases of interest and indeed in general it is a condition that permits us to work with actions of groups that are not discrete groups.

If M is a compact manifold and if X is a C^∞ vector field on M, then the usual existence theorems for ordinary differential equations provide us with a one-parameter group $h(t)$ of diffeomorphisms of M such that $\dot{h}(t) = X(h(t))$ and such that $(t,m) \to h(t)(m)$ is a C^∞ map and hence certainly Borel. Such a flow may or may not leave invariant a measure, but if for instance X is of Hamiltonian type, then Liouville's theorem provides an invariant measure. Since this subject is discussed in Kostant's article in this volume we will not go into more details here (see also [4]).

Before proceeding in our general context of which we have seen several examples above, we must impose a regularity condition on M of a technical nature; more precisely, we shall assume that M is an analytic Borel space. The reader is referred to [3] for further exposition concerning this condition; in any case it is a condition that is satisfied in all reasonable examples. Suppose that μ is a quasi-invariant measure on M for G. (In fact this is not just a property of μ, but rather a property of the set of all measures equivalent to μ, so we may speak of a quasi-invariant measure class.) One says that G acts <u>ergodically</u> on M, with respect to the measure μ, or that μ is an ergodic measure if whenever we have $g \cdot E = E$ for all $g \in G$, and some Borel set E, then $\mu(E) = 0$ or $\mu(M - E) = 0$. In other words, the only invariant Borel sets under the action are null sets or their complements. It also says that the action is indecomposable in that we cannot write $M = M_1 \cup M_2$ where M_1 and M_2 are disjoint invariant Borel sets of positive measure.

A rather natural modification of this definition consists in assuming that whenever $\mu(g \cdot E \Delta E) = 0$ for all $g \in G$ and some Borel set E, then $\mu(E) = 0$ or $\mu(M - E) = 0$. Here $g \cdot E \Delta E$ denotes the symmetric difference of the two sets, that is, the points in one but not the other of the two sets. A set with $\mu(g \cdot E \Delta E) = 0$ might be called almost-invariant, and one would be asserting that any such set is a null set or the complement of a null set. The second definition of ergodicity is clearly more restrictive than the first, and if G is countable they can easily be seen to be equivalent. For a general locally compact group it is a non-trivial result of Mackey (see [32]) that the two conditions are equivalent.

Suppose now that μ is a finite invariant measure for an action of the integers. This as we have seen is specified by a single measure preserving transformation, and if the action is ergodic one has the Birkhoff ergodic theorem [15]. For $f \in L_1(\mu)$,

$$\lim_{n\to\infty} \frac{1}{n+1} \Sigma_{i=0}^{n} f(u^i(x)) = \int f d\mu$$

for almost all x. (There is a similar statement for ergodic actions of the real line.) If we interpret this formula in its classical context of statistical mechanics where $u(x)$ is the evolution of a state x after unit time, then the left-hand side is the time average of a function f (a dynamical variable) and the right-hand side is the phase average of the same dynamical variable. The equality of these two averages is to hold for almost all initial states of the system. In fact it is not difficult to see that the validity of such a formula is equivalent to ergodicity. The question of equality of time averages and phase averages has a long history in statistical mechanics, and the ergodic theorem just reduces the question to the problem of showing that certain actions are ergodic (see [24]).

Not only for this reason, but for many others, one of the fundamental questions in ergodic theory is to supply sufficient conditions for an action or a class of actions to be ergodic. Our object here is to review a general method one has available for doing this by means of unitary representations and to apply the results of the previous section. We have remarked before that the present technique was initiated by Gelfand and Fomin for geodesic flows, although the observation that ergodic theory and unitary representations are closely connected goes back to Koopman.

We shall assume now that we are dealing with actions of a group G on a space M with a finite invariant measure μ, and we may assume without loss of generality that $\mu(M) = 1$. One can then define an associated unitary representation π of G on the Hilbert space $L_2(M,\mu)$. More precisely for $f \in L_2$, and $g \in G$, we define $(\pi(g)f)(x) = f(g^{-1} \cdot x)$. It is easy to verify that for each $g \in G$, $\pi(g)$ is a unitary operator, and using joint measurability of the action, one can show that π is a continuous unitary representation of G. The key observation is that one can detect ergodicity of an action merely by looking at π. Heuristically, the non-existence of invariant sets is equivalent to the non-existence of invariant measurable functions, and since the space has finite measure this is equivalent to the non-existence of square summable invariant functions.

Lemma 3.1

The action of G on M is ergodic if and only if $\pi(g)f = f$ for all $g \in G$ implies that f is a constant almost everywhere, and hence a constant in $L_2(M,\mu)$.

There is a somewhat stronger notion than ergodicity which is often useful, namely weak mixing. For this one must define the Cartesian square of an action of G on M. We notice that G acts on $M \times M$ by $g \cdot (m,n) = (g \cdot m, g \cdot n)$ and that the product measure $\mu \times \mu$ is invariant. One says that the action of G on M is weakly mixing [15] if the action of G on $M \times M$ is ergodic. This condition trivially implies ergodicity, and if it is satisfied, every Cartesian power of the action is weakly mixing and ergodic. Weak mixing can also be detected by looking at the unitary representation π (see [15] and [39]).

Lemma 3.2

The action of G on M is weakly mixing if and only if any finite dimensional subspace V of $L_2(M,\mu)$ invariant under π consists of constant functions (and hence is one-dimensional).

These lemmas serve to motivate the definitions of properties E and WM in Section 2 above since these definitions concern invariant vectors and finite dimensional subspaces of representations of a group. Finally, if G is the real line R or the integers Z, an action may possess the yet stronger property of strong mixing. To motivate this we note that the ergodic theorem implies that for any pair of measurable sets A and B

$$\lim_{t\to\infty} \mu(h(t)A \cap B) = \mu(A)\mu(B)$$

in the sense of Cesaro limits. Here $h(t)$ is the action defined for $t \in R$ or $t \in Z$. One says that the action is strongly mixing if the above limit exists in the usual sense [15]. One can find a sufficient condition in terms of the representation π for this to be the case; indeed by Stone's theorem [25], there exists a projection valued measure P on the Borel sets of the dual of G (R or the circle group T) corresponding to π. We shall say that P is absolutely continuous if $P(E) = 0$ if E is a Lebesgue null set. See [15] for the following.

Lemma 3.3

If G is as above, then an action of G is strongly mixing provided that the subrepresentation of π on the orthogonal complement of the constant function has an absolutely continuous projection valued measure.

It is clear that strong mixing implies weak mixing and that Theorem 2 of Section 2 is exactly the sort of result that will enable us to establish that certain actions are strongly mixing.

More specifically the theorems from Section 2 will be applied in the following manner. Suppose that we have an action of a group H on M with a finite invariant measure, and suppose that G is a larger group containing H which acts on M preserving μ such that the action of H on M determined by the fact that H is a subgroup of G coincides with the given action of H on M. In other words, we are assuming that the given action of H on M may be "embedded" in the action of a larger group G. We shall assume that the larger group G acts ergodically, and then ask if we can conclude from this that H also necessarily acts ergodically. Equivalently we can start with an action of G on M known to be ergodic by some method, and pick a subgroup H of G, and ask if H also acts ergodically. Properties E and WM from Section 2 are immediately relevant to this situation.

Theorem 4

Suppose that H has property E (respectively WM) in G, and that G acts ergodically on M with a finite invariant measure. Then H is also ergodic (respectively weakly mixing).

Proof. We consider the representation π of G on $L^2(M,\mu)$. If f is an invariant function for H then by property E, it is an invariant function for G, and by ergodicity of G, f is a constant. Hence H acts ergodically; by the same argument we can conclude weak mixing if H has property WM.

The following follows immediately using Theorem 1.

Corollary. If G is semi-simple and acts ergodically on M as above, and if H is totally non-compact in G, then H is ergodic and weakly mixing.

If H is a subgroup of G isomorphic to the real line (or the integers) we can also obtain results concerning strong mixing using Theorem 2.

Theorem 5

Let G be semi-simple and let $H \subset G$ be totally non-compact, and isomorphic to the real line (or the integers). If G acts ergodically on M as above, then the action of H is strongly mixing.

Although the hypothesis of the above theorems, as far as H is concerned, may seem rather special, their interest lies in the observation that this hypothesis of embedability in a larger group G is satisfied in many cases. Indeed it is satisfied for some classical flows, that seemingly have no connection with group theory. One way that an action of a group G can be seen to be ergodic is if it is transitive; that is if m and n are given points of M, there is an element $g \in G$ such that $g \cdot m = n$. If we put $\Gamma = \{g | g \cdot m = m\}$ for a fixed m, one can identify the coset space G/Γ with M by means of the map $g\Gamma \to g \cdot m$. With our hypotheses on M, one can conclude that this is a Borel isomorphism and that Γ is closed so that we may as well assume that M is G/Γ. (See [30] and [3] for an exposition of the details of this reasoning.) The action of G on G/Γ is then given concretely by $g \cdot (h\Gamma) = gh\Gamma$. Our assumption that M has a finite G-invariant measure means that G/Γ also possesses such a measure, and this places rather severe restrictions on what Γ can be (see [7] for instance). A transitive action is immediately seen to be ergodic by the first of the two definitions of ergodicity above.

One of the simplest examples of an action of the real line which is embedable in a transitive action is that of a rotation on a torus. Let $T^n = \{(z_1,\cdots,z_n),\ z_i$ complex numbers with $|z_i| = 1\}$ be an n-torus. We pick real numbers $a_1,\cdots, a_n$ and let R act on T^n by $t \cdot (z_1,\cdots,z_n) = (\exp(ita_n)z_1,\cdots, \exp(ita_n)z_n)$. This action is well known to be ergodic if and only if the a_i are rationally independent [15]. We observe, however, that this action can be embedded in a transitive action of R^n since $T^n = R^n/Z^n$ where Z^n is a lattice. The classical proof of the result quoted above is based in its essence on this observation. Leon Green [1] has established a beautiful generalization of this result to nil manifolds which again uses exactly the same ideas (see also [2]). Gelfand and Fomin [13] observed that geodesic flows on surfaces of constant negative curvature are embedded naturally in transitive actions of the group $G = SL_2(R)$; Parasyuk [43] observed the same for horocycle flow, and Mautner did the same for geodesic flow on certain higher dimensional manifolds [35]. The proofs of ergodicity are all based on the same idea and the following general result subsumes all such results concerning semi-simple groups, and follows immediately from Theorems 4 and 5.

Theorem 6

Let G be a semi-simple Lie group and let Γ be a subgroup such that G/Γ has a finite invariant measure, and let H be totally non-compact. Then the natural action of H on G/Γ is ergodic and weakly mixing. If H is the line (or the integers), then the action is strongly mixing.

Many examples of subgroups Γ satisfying this condition are known; for instance $SL_n(Z)$, the subgroup of $SL_n(R)$ consisting of matrices with integral entries (see [8]).

We shall close this brief discussion of one aspect of ergodic theory with a duality theorem of sorts which was noticed independently by G. W. Mackey and the author (see [39]). Above we were dealing with two subgroups H and Γ of G but there was an assumed asymmetry since we let H act on G/Γ. We can just as well let Γ operate on $H\backslash G = \{Hg | g \in G\}$ by $\gamma \cdot Hg = Hg\gamma^{-1}$. The following fact holds for any pair of closed subgroups Γ and H of G.

Lemma 3.4

The action of H on G/Γ is ergodic if and only if the action of Γ on $H\backslash G$ is ergodic.

In general a coset space G/Γ or $H\backslash G$ has no invariant measure, much less a finite one, but it always has a unique quasi-invariant measure class [28], and it is with respect to this measure class that the above lemma applies.

As an application of this, let $G = SL_n(R)$ and let $\Gamma = SL_n(Z)$, the subgroup of matrices with integral entries, and let H denote the subgroup of G consisting of matrices with first column $(1,0,\cdots,0)$. Then G/Γ has finite volume and H is totally non-compact if $n \geq 2$. Since H is ergodic on G/Γ, the duality principle says that Γ is ergodic on $H\backslash G$. The space $H\backslash G$ is easy to identify and is in fact R^n minus the origin with Lebesgue measure, and the action of G on this space is the natural linear action. Since a single point is a Lebesgue null set, $H\backslash G$ is measure theoretically the same as R^n and we have the following result.

Theorem 7

The linear action of $\Gamma = SL_n(Z)$ on R^n with Lebesgue measure is ergodic. Moreover, the same is true for any Γ such that $SL_n(R)/\Gamma$ has a finite invariant measure.

This easily stated result does not appear to be amenable to any direct approach.

4. MORE ON RESTRICTIONS

In Section 2 we studied certain aspects of the general problem of restricting a representation of a semi-simple Lie group to a subgroup and examining how it decomposes. These results are of immediate interest in ergodic theory and the connection was discussed in Section 3. In this final section we want to discuss some additional questions concerning the restriction of representations to subgroups. Theorem 2 for instance concerned the restriction of representations to one-parameter subgroups, and it is natural to raise similar questions concerning the restriction of representations to more general subgroups H, particularly general abelian subgroups. From the general version of Stone's theorem for abelian groups H [25], we know that any unitary representation of H leads to a projection valued measure on the Borel subsets of the dual group $\hat{H}$. Together with an appropriate multiplicity function (see [33]) this projection valued measure determines the representation. We are interested in the equivalence class of this projection valued measure P and in particular we would like to compare it to Haar measure on $\hat{H}$. We will say that P is absolutely continuous (with respect to Haar measure) if $P(E) = 0$ for any Haar null set $E \subset \hat{H}$, and that P is equivalent to Haar measure if $P(E) = 0$ if and only if E is a Haar null set. Theorem 2 above says in particular that if G is semi-simple and if H is a totally non-compact one-parameter subgroup of G and if π is a representation of G with no G-invariant vectors, then the projection valued measure on $\hat{H}$ associated with the restriction of π to H is absolutely continuous. It is natural to raise the question of when other abelian subgroups have this property. We note that non-compactness is not an issue since the conclusion above is trivially satisfied when H is a compact subgroup of any group G. If one knows such results about the projection valued measure for a vector subgroup, one may immediately conclude results concerning eigenvalues of any operator corresponding to an element of the enveloping algebra of the Lie algebra of H, a question of some interest in physics.

Let us now suppose that G is simple, and we write the Iwasawa decomposition $G = KAN$. Any abelian subgroup is in some vague sense made up of a part from K, a part from A, and a part from N, and we shall consider the three cases separately. We have already noted that the question posed is trivial for compact subgroups, and we turn to subgroups of A. Since A is abelian, we may as well consider the case $H = A$. T. Sherman [48] has observed at least in a special case that the answer is affirmative, and the same holds in general one can easily see.

Theorem 8

If G is simple and if π is a unitary representation of G with no G-invariant vectors, then the projection valued measure associated with the restriction of π to A is equivalent to Haar measure on $\hat{A}$. Moreover, the multiplicity is uniform.

We now turn to the consideration of subgroups of N, and here all we have at present is a counterexample. We consider the symplectic group $Sp_n(R)$ of real $2n \times 2n$ matrices preserving a non-degenerate skew bilinear form. Since the fundamental group of this group is the integers, there is a unique double covering group G of $Sp_n(R)$. (This is Weil's metaplectic group [51].) It is easy to verify that N for this group contains a normal abelian subgroup V isomorphic to the vector space of real $n \times n$ symmetric matrices. In fact N is the semi-direct product of V with a group T isomorphic to all strictly triangular $n \times n$ matrices (that is, all entries above the diagonal are 0 and all diagonal entries are one). Weil [51] has constructed a representation π of G called the metaplectic representation (see also Shale [46]). One may compute the projection valued measure on $\hat{V}$ associated with the restriction of π to V quite easily. In fact if one identifies V with $\hat{V}$ by means of the bilinear form $(a,b) = \mathrm{Tr}(ab)$, this projection valued measure on $V = \hat{V}$ is concentrated on the set of positive definite matrices of rank one. Thus, if $n > 1$, this is a Haar null set, and gives an example where P is singular with respect to Haar measure. This example also shows that many one-parameter subgroups of V have one sided spectrum and hence that the phenomenon noticed for $SL_2(R)$ persists in higher dimensions.

There is another point worth noticing about this situation; let $\pi(k)$ denote the kth tensor power of the representation π with itself. Then it is quite easy to check that the projection valued measure associated with the restriction of $\pi(k)$ to V is concentrated on the set of positive definite matrices of rank equal to the minimum of k and n. Thus $\pi(n)$ is the first tensor power to have absolutely continuous spectrum. Since it is virtually obvious that any discrete series representation of G has a projection valued measure which is absolutely continuous, any connection between discrete series and the tensor powers of π analogous to the situation for $n = 1$, is likely to involve large tensor powers.

We notice that the condition above for abelian groups H that the projection valued measure on the dual group $\hat{H}$ associated to a representation be absolutely continuous, can be rephrased so as to make sense for any subgroup H. The condition can be readily seen to be equivalent to the condition that the representation of H should be unitarily equivalent to a subrepresentation of the direct sum of the regular representation with itself infinitely many times. This makes sense for any H, and we shall say following [33], that a representation of H satisfying this condition is quasi-contained in the regular representation. (If as often happens, the regular representation is equivalent to the infinite direct sum of itself, the condition is simply that the given representation is a subrepresentation of the regular representation.)

We do not have any general theorems concerning this situation, but there is one case of special interest. If G is semi-simple with Iwasawa decomposition $G = KAN$, we let M be the centralizer of A in K, and define $B = MAN$. One knows that B is a group [6], and in some sense it is one of the most important subgroups of G. The principal series representations of G consist simply of the representations of G induced by the finite dimensional representations of B [6], and B plays a key role in the structure of G. When G is $SL_n(R)$, then B is simply the subgroup of triangular matrices.

Our interest here will be the study of restrictions of representations of G to B in the spirit indicated above, and for this we shall need to know something about the structure of the regular representation of B. When $G = SL_2(R)$, the regular representation of B is rather striking. It is known that B has in addition to its obvious one-dimensional representations, exactly four other irreducible representations, all infinite dimensional [33], say π_i, $i = 1, 2, 3, 4$. A simple calculation will show that the regular representation π of B is the direct sum of the π_i each taken infinitely often, $\pi = \infty(\pi_1 \oplus \pi_2 \oplus \pi_3 \oplus \pi_4)$. Thus π is the discrete direct sum of irreducible representations, with in fact only a finite number of distinct types entering into the decomposition. It is easy to see that the same is true for $G = SL_2(C)$.

Is this an accident or do we find the same phenomenon for other semi-simple groups? A calculation for $G = SL_n(R)$ for $n \geq 3$ reveals that the regular representation of B decomposes as a continuous direct integral and contains no irreducible summands. On the other hand for $G = Sp_n(R)$, one can find exactly 4^n irreducible infinite dimensional representations of B such that the regular representation is the discrete direct sum of these representations, each taken infinitely often. We shall now state a general criterion which will appear in a forthcoming paper.

As usual we consider the normalizer M_1 of A in K, and note that $W = M_1/M$ is a finite group, called the Weyl group, which acts as a group of automorphisms of A, and hence also on the Lie algebra of A. The group W may or may not contain the automorphism $a \to a^{-1}$ of A (or equivalently the map $Y \to -Y$ of the Lie algebra of A). If it does we shall say that -1 belongs to W.

Theorem 9

If G is semi-simple, then the regular representation of its subgroup B either decomposes as a discrete direct sum of irreducible representations of B (with a finite or countable number of inequivalent summands) or it decomposes as a continuous direct integral with no irreducible summands. The first possibility occurs if and only if -1 belongs to W.

If $-1 \in W$, we may think of the finite or countably infinite number of irreducible representations which are summands of the regular representation as "discrete series", but we prefer to call them generic representations since they are in a sense those irreducible representations of B which are in "general position". One may not conclude as in the semi-simple case that the matrix entries are square integrable functions on B since B is not unimodular. Finally we note that Harish Chandra [21] has given a necessary and sufficient condition that the group G have a discrete series, that is, there are irreducible representations which are summands of the regular representation. It is interesting to note that his condition implies that $-1 \in W$ (and in fact is almost equivalent to it) and hence if G has a discrete series so does B.

Once we have this kind of control over at least some of the representations of B, it is natural to raise the question of what the restriction of a representation π of G to B looks like. This technique is exceedingly fruitful for $G = SL_2$, and Stein in his lectures in this volume uses a similar technique except with B replaced by an even larger subgroup.

One may ask if it is true that any representation π of a simple group G which has no G-invariant vectors has its restriction to B quasi-contained in the regular representation of B. This is true for all the series of representations constructed by Harish-Chandra [20] which are used to obtain the Plancherel formula. For discrete series this follows from the observation that any representation contained in the regular representation of G, has its restriction to any subgroup H quasi-contained in the regular representation of H. The general case follows from known facts concerning induced representations. Unfortunately the answer to the above question is negative in general, and the counterexample is our friend the metaplectic representation π of the double covering G of $Sp_n(R)$, $n \geq 2$. We have $B = MAN$ and N contains a normal subgroup isomorphic to the vector space of symmetric $n \times n$ matrices. If the restriction of π to B is quasi-contained in the regular representation of B, it follows by the comment above that its further restriction of V is quasi-contained in the regular representation of V which we know is false. Again we do not know what the best possible theorems are in the general case.

PART II. GROUP EXTENSIONS AND GROUP COHOMOLOGY

5. STRUCTURE OF LOCALLY COMPACT GROUPS

In this second part we shall take up a rather different aspect of group representations, and indeed here the major considerations will concern more the structure of locally compact groups. The motivation for the study of group extensions comes from the phenomenon of ray or projective representations of groups; however, to treat these questions properly, we feel it is better to first widen the problem, and then come back to the original questions using the general techniques which we shall develop.

We shall suppose that G and A are topological groups with A abelian, and that G operates on A is a topological transformation group of automorphisms. More precisely, we are given a continuous map of $G \times A \to A$ written $(g,a) \to g \cdot a$ such that for fixed $g \in G$, the map $a \to g \cdot a$ is an automorphism of A, which we denote by $p(g)$, and further that p is a homomorphism of G into the group of automorphisms of A. The hypothesis of joint continuity of $g \cdot a$ assures not only that $p(g)$ is an automorphism of the topological group A, but that $p(g)$ varies "smoothly" with g. If G and A satisfy the above, we say that A is a topological G-module or simply that A is a G-module [37].

This definition includes a wide variety of examples. If for instance A is a Hilbert space with its norm topology, and if π is a continuous unitary representation of G on A in the usual sense, then one may verify that $(g \cdot a) = \pi(g)(a)$ defines A as a G-module. If π is again a representation of G on a Hilbert space, and if A is some group of unitary operators on this Hilbert space such that $\pi(g)a\pi(g)^{-1} \in A \ \forall a \in A$, then $g \cdot a = \pi(g)a\pi(g)^{-1}$ defines for each $g \in G$ an automorphism of A. It may be verified that A, equipped with the strong operator topology, is a G-module. Finally, if A is any topological group $g \cdot a = a$ defines A as a G-module. Such modules will be called trivial topological G-modules.

A group extension of a given G by a given topological G-module will be first of all an exact sequence of groups

$$1 \to A \overset{i}{\to} E \overset{\pi}{\to} G \to 1$$

where i is an injection of A into E, and π is a surjection of E onto G, and where the kernel of π is exactly the range of i. We assume not only that i is continuous, but also that it is a homeomorphism onto its range, and we assume that π is continuous and open. This means that A and $i(A)$ can be identified not only as groups, but as topological groups where $i(A)$ has the relative topology from E, and that $E/i(A)$ and G may be identified as topological groups, the first of these having the quotient topology. Finally, we impose an algebraic assumption to take account of the action of G on A. We note that if $g \in G$ and if g' is an element of E with $\pi(g') = g$, then $a \to i^{-1}(g'i(a)(g')^{-1})$ is an automorphism of A which depends only on g, and not on the choice of g'. We demand that this automorphism be the given automorphism $a \to g \cdot a$ in the definition of A as a G-module. We note that whenever we have an extension of G by A, then by the above A becomes a G-module, the joint continuity of the map $G \times A \to A$ following from the axioms for a topological group. This observation is one of the main motivations for defining G-modules as we did by imposing the condition of joint continuity. The reader is referred to [37] and [38] for more details.

One of the simplest examples of a topological group extension is the extension of the circle T by the integers Z (viewed as a trivial module), defined by the real line R, namely

$$1 \to Z \to R \to T \to 1 .$$

Another example which is of more significance particularly in quantum physics is as follows. Let H be a Hilbert space, and let $U(H)$ be the group of all unitary operators on H with the strong operator topology. Then the circle group T viewed as scalar multiples of the identity operator is a normal subgroup, and

$U(H)/T = PU(H)$ is called the projective unitary group. Then

$$1 \to T \to U(H) \to PU(H) \to 1$$

is an extension of $PU(H)$ by T, T being trivial module. Indeed if $P(H)$ is the set of one-dimensional subspaces of H, then any $a \in U(H)$ defines a collineation on $P(H)$ by $r \to a \cdot r$ where $a \cdot r$ is the transform of $r \in P(H)$ under a. A classic theorem of Wigner (see [52] or [5]) says that except for anti-unitary transformations, these are the only maps of $P(H)$ onto itself which preserve the function $f(r_1,r_2) = |(u_1,u_2)|$ which is defined for $r_i \in P(H)$ by picking unit vectors $u_i \in r_i$. We observe that the projective transformations corresponding to a and $b \in U(H)$ agree if and only if $a = tb$ with $t \in T$. Thus $U(H)/T = PU(H)$ is isomorphic naturally to a group of projective transformations.

If H is the Hilbert space associated with a quantum mechanical system, and if G is a symmetry group of this physical system, the axioms of quantum mechanics say that we have a homomorphism of G into $PU(H)$ (except for those symmetries which we would want to be anti-unitary, but this will not change anything essential in this heuristic discussion). Such a homomorphism is precisely what is known as a projective or ray unitary representation [5]. For the moment let us assume that $G \subset PU(H)$; then if p is the projection from $U(H)$ onto $PU(H)$, we let $G' = p^{-1}(G)$, and then

$$1 \to T \to G' \to G \to 1$$

becomes a group extension of G by T. Even when we do not want to identify G as a subgroup of $PU(H)$, we shall see that we can still construct a group extension

$$1 \to T \to E \to G \to 1$$

where E has a homomorphism into $U(H)$, or in other words, a unitary representation. The fact that projective representations can be viewed as ordinary representations of a suitable group extension is a well known and fundamental fact.

In these notes we want to present a brief outline of a systematic theory of group extensions and more generally of a theory of group cohomology which is intimately related to the initial problem. We refer the reader to [37] and [38] for more details and to the references cited there, in particular the pioneering work of G. W. Mackey [29], [31] who originated this point of view concerning group extensions. A large part of the contents of these notes will be the subject of a forthcoming paper of the author, and we will try to summarize the major new points involved. These results extend and generalize those in [38] and [39].

One of the most important problems is to classify the set of all extensions of a given group G by a given topological G-module A. Two extensions are said to be equivalent if there is a commutative diagram

$$\begin{array}{ccccccccc} 1 & \to & A & \to & E & \to & G & \to & 1 \\ & & \downarrow & & \downarrow & & \downarrow & & \\ 1 & \to & A & \to & E' & \to & G & \to & 1 \end{array}$$

of continuous maps where the end vertical maps are the identity maps and where the middle vertical is an isomorphism of topological groups. It should be noted that it is not sufficient to assume that E and E' are isomorphic as topological groups to have equivalent extensions, but rather there must be a particular isomorphism which respects the data of a group extension. One of the first facts is that the set of equivalence classes of extensions of G by A forms a group $\mathrm{Ext}(G,A)$ by means of the Baer product (cf. [16]), and at least in many cases this group is given as a two-dimensional cohomology group $H^2(G,A)$. It turns out to be useful to study the other cohomology groups $H^n(G,A)$, both to gain a better understanding of extensions, and also to have at hand general methods of computing $\mathrm{Ext}(G,A)$ in many specific situations.

6. G-MODULES

After this introduction we shall now proceed to some of the details. We will henceforth assume that G is locally compact and separable in the sense of the second axiom of countability. (Local compactness seems to be essential for this treatment, although we hope in the future to be able to dispense with it; separability is an assumption of a more technical nature used to avoid certain pathologies.) We shall also assume that A is separable, metrizable and moreover metrizable by some complete metric. Following Bourbaki, one might call such groups polonais, and we denote the family of all such groups by P. Since we will always be dealing with G-modules, we consider all polonais G-modules which we denote by $P(G)$. We note that $P(G)$ contains all separable locally compact G-modules A. Group extensions were studied in the case of locally compact A in [38] and [39], and one of the key points in the present treatment is that we now enlarge the category of modules to $P(G)$. In addition to including many important and interesting examples which were excluded before, we also achieve more technical versatility in that the larger category will contain cohomologically trivial modules, will enable us to define induced modules in a natural way, and will allow us to construct resolutions without going outside the category.

If $A,B \in P(G)$ a G-homomorphism f of A into B is simply a continuous intertwining homomorphism, that is, one satisfying $f(g \cdot a) = g \cdot f(a)$. We note that P (resp. $P(G)$) is closed under the operations of countable Cartesian products, closed subgroups (closed submodules), and quotient groups (quotient modules). In addition, if we have a sequence of elements of P

$$1 \to A' \xrightarrow{i} A \xrightarrow{\pi} A'' \to 1$$

which is exact in the sense of Section 5, then one can show that $A \in P$ if and only if A' and A'' are in P. The same is clearly true for $P(G)$ if the homomorphisms in the sequence above are G-equivariant. Morever, if A, A', A'' are in P, and it is only assumed that i and π are continuous, it follows by classical closed graph theorems (cf. [3]) that i is a homeomorphism and that π is open, and hence that the sequence is exact in the sense of Section 5.

In addition P is closed under the following construction which might be described as a sort of direct integral. This construction will be of paramount importance to us. Let (M,u) be a σ-finite measure space such that the measure algebra of (M,μ) is separable [14]. This means that we may as well assume that M is $[0,1]$ with Lebesgue measure together with a countable number of atoms. Now let $A \in P$, and define $U(M,A)$ to be the group of all measurable functions from M to A modulo the group of functions equal to 1 (the identity in A) almost everywhere. An element of $U(M,A)$ is then an equivalence class of measurable functions, all of which are equal to each other almost everywhere. (A function f is measurable if $f^{-1}(0)$ is measurable in M for every open 0 in A.) It is clear that $U(M,A)$ is a group under pointwise multiplication.

We topologize $U(M,A)$ by the topology of convergence in measure; more precisely let ρ_1 be a bounded metric on A, which always exists, and let ν be a finite measure on M equivalent to μ. We define a metric on $U(M,A)$ by

$$\rho(f,g) = \int \rho_1(f(x),g(x))d\nu(x)$$

which is always finite since ν is finite and ρ_1 is bounded.

<u>Lemma 6.1</u>

$U(M,A)$ with ρ as defined above is in P, and the topology is independent of the choice of ρ_1 and ν.

If $A = T$ is the circle group, $U(M,T)$ has a natural interpretation; namely let H be $L_2(M)$, the space of square integrable functions on M, and let $f \in U(M,T)$. Then f defines a unitary operator on H by multiplication by F, $(U(f)h)(x) = f(x)h(x)$. Clearly $U(f) = U(g)$ if and only if $f = g$ in $U(M,T)$ and so $U(M,T)$ may be viewed as a group of unitary operators on H. It may be verified that the topology on $U(M,T)$ introduced above is exactly the strong operator topology when we view $U(M,T)$ as operators. If $A = R$ is the real line, $U(M,R)$ is a topological vector space; in fact a Frechet space, although it is not locally convex. Finally, if $A \in P(G)$, we can define an action of G on $U(M,A)$ by means of the formula $(g \cdot F)(m) = g \cdot (F(m))$. This may be thought of as a direct integral of copies of A. If M is an atomic measure space, the construction does give the Cartesian product of copies of A. In analogy with direct integrals of representations [33] one might hope to find a reasonable definition of a measurable map of M into $P(G)$, and then define a direct integral where the "fiber" $A(m)$ over $m \in M$ is allowed to vary instead of remaining constant as above. Since we have found no use for this kind of construction as yet, we shall not proceed any further.

The group $U(M,A)$ for $A \in P$ has many interesting properties, and one of the most important for our purposes is a "law of exponents". We let $M = M_1 \times M_2$, and then intuitively a function of two variables on M into A can be thought of as a function of one variable (say m_1) into the space of functions of the second variable m_2 into A. Such a correspondence holds exactly and indeed follows from a version of the Fubini theorem.

Lemma 6.2

There is a canonical isomorphism of $U(M_1 \times M_2,A)$ onto $U(M_1,U(M_2,A))$ as topological groups.

A most important special case of the construction of $U(M,A)$ is when $M = G$ is a locally compact (separable) group with Haar measure. In this case $U(M,A)$ will be denoted by $I(A)$ and we note that $I(A)$ is itself a G-module for any $A \in P$. In fact we simply let G act by translation: $(g \cdot F)(s) = F(g^{-1}s)$. If in addition $A \in P(G)$ so that G also operates on A we can embed A into $I(A)$ by the map f defined by $(f(a))(s) = s^{-1} \cdot a$.

Lemma 6.3

If $A \in P$, then $I(A)$ is in $P(G)$. Moreover, f is an equivariant isomorphism of A onto a closed submodule of $I(A)$ so that

$$1 \to A \to I(A) \to U(A) \to 1$$

is exact where $U(A)$ is the quotient module.

It is clear that $I(A)$ is in some sense the regular representation of G with coefficients in A. In the case of a finite group, it is known that $I(A)$ is cohomologically trivial in that $H^n(G,A) = 0$ for $n \geq 1$ [45]. The fact that this can also be proved in the present context will be of vital importance. Lemma 6.3 above would then assert that any A may be embedded in a cohomologically trivial module, and this fact will allow us to use many techniques from homological algebra. It should also be noted that $I(A)$ is almost never locally compact. Finally, once we have defined the regular representation, it is but a short step to the notion of induced representations. If H is a closed subgroup of G, and if $A \in P(H)$, we define $I_H^G(A)$, the induced module, as a submodule of $I(A)$. More precisely, $I_H^G(A) = \{f | f \in I(A),\ f(gs) = s^{-1}f(g)$ for almost all pairs (s,g) in $H \times G$ where Haar measure is understood$\}$. We have engaged in the usual abuse of notation and have regarded elements of $I(A)$ as functions instead of equivalence classes of functions, but this poses no problem. It is easy to show that $I_H^G(A)$ is a closed submodule of $I(A)$ and hence is in $P(G)$. All of the expected properties of induced representations such as inducing in stages hold in our context, but we shall defer these details to our forthcoming paper.

7. GROUP EXTENSIONS

Having discussed the G-modules which will enter into our theory, we turn to a more explicit discussion of group extensions and group cohomology. In complete analogy with the case of discrete groups [11], we shall introduce cohomology groups $H^n(G,A)$, $n \geq 0$ for $A \in P(G)$. These groups have simple interpretations in low dimensions; namely $H^0(G,A) = A^G$, the G-fixed points in $A = \{a | g \cdot a = a$ for all $g \in G\}$. For $n = 1$, and a trivial G-module, $H^1(G,A)$ will be the continuous homomorphisms of G into A (while for a general module we will have equivalence classes of continuous crossed homomorphisms). For $n = 2$, $H^2(G,A)$ will be $\text{Ext}(G,A)$, the group of topological group extensions of G by A.

By way of introduction to the cohomology we shall begin with a discussion of how one may parameterize the group extensions of G by A using cocycles. Let

$$1 \to A \to E \overset{\pi}{\to} G \to 1$$

be a given extension. The identity element of the group $\text{Ext}(G,A)$ is the semi-direct product of G and A, and in the special case when A is a trivial G-module, the direct product of G and A. This extension is characterized by the property that one may find a continuous homomorphism f of G back into E such that $(\pi \circ f)(g) = g$. The idea behind the following is to compare a general extension of G by A to the semi-direct product. It is natural to consider a map f of G to E such that $\pi \circ f$ is the identity map, and compute the defect of f from being a homomorphism. Since we are dealing not with abstract groups, but with topological groups, it would not be sensible to choose any arbitrary map f. Ideally one would want to look for a continuous map f of G into E satisfying the above, however, it is simply a fact of life that such a continuous map does not always exist. Indeed even in the case

$$1 \to Z \to R \to Z \to 1$$

such a map does not exist, and in general the existence of such a continuous map for a general extension would imply that E viewed as a principal fiber bundle with base G and fiber A would be a trivial bundle and so in particular $E = G \times A$ as topological space. Mackey has shown how to resolve this, and following him we observe that one may always find a Borel map f of G into E satisfying $(\pi \circ f)(s) = s$ for $s \in G$, (see [9]). Other choices of an appropriate map may be considered such as those continuous at the identity element of G or those continuous in a neighborhood of the identity element of G, but we believe that the choice of a Borel map f leads to a theory which is in general more satisfactory.

Once we have selected such a Borel function f (or cross section as it is sometimes called) we note that $a(g,h) = f(g)f(h)f(gh)^{-1}$ is a Borel function from $G \times G$ into the subgroup $i(A)$ of E. We view it as a function from $G \times G$ into A and we notice that it is a Borel function, and as a consequence of the associative law in E satisfied the "cocycle identity",

$$a(s,t)a(st,r) = (s \cdot a(t,r))a(st,r)$$

for all $s, t, r, G \times G \times G$. We denote by $Z^2(G,A)$ the group of all Borel maps of $G \times G$ into A satisfying this identity, and call such functions 2-cocycles. The group structure is understood to be multiplication of such functions pointwise. We have associated now to each element of $\mathrm{Ext}(G,A)$ an element of $Z^2(G,A)$, but this depends on the selection of a Borel cross section f of G into E. If we replace f by any other Borel cross section f', the cocycle a changes, but it changes only by multiplication by a 2-cocycle of the form $(s \cdot b(t))b(s)b(st)^{-1}$ for some Borel function b of G into A. We call such functions 2-coboundaries, and denote the group of such by $B^2(G,A)$, and notice the very important fact that to each extension in $\mathrm{Ext}(G,A)$ we can associate a unique element of the quotient group $Z^2(G,A)/B^2(G,A)$ which is independent of any choices. This quotient group is denoted by $H^2(G,A)$, the two dimensional cohomology group of G with coefficients in the topological G-module A.

The map of $\mathrm{Ext}(G,A)$ into $H^2(G,A)$ may be verified to be a homomorphism of groups, and moreover may be seen to be injective. If A is locally compact Mackey [30] has shown that this map is surjective as well. We are able to show (see below) that this is also true for any $A \in P(G)$. This construction gives a parameterization of $\mathrm{Ext}(G,A)$ in terms of a cohomology group and also motivates the introduction of the general cohomology groups $H^n(G,A)$.

If $A \in P(G)$, we define a complex of groups $C^n(G,A)$, $n \geq 0$, where $C^n(G,A)$ is the set of all Borel functions from $G \times \cdots \times G$ (n factors) into A, and we define a coboundary operator δ_n from $C^n(G,A)$ into $C^{n+1}(G,A)$ by the classical formula [11],

$$\begin{aligned}(\delta_n f)(s_1,\cdots,s_{n+1}) &= s_1 \cdot f(s_2,s_3,\cdots,s_{n+1}) \\ &- f(s_1 s_2,s_3,\cdots,s_{n+1}) \cdots \pm f(s_1,\cdots,s_n s_{n+1}) \\ &\mp f(s_1,\cdots,s_n)\end{aligned}$$

where we are writing A additively. The verification that $\delta_n f$ is a Borel function if f is a Borel function is routine [38], as is also the formula $\delta_{n+1}\delta_n = 0$. We define $Z^n(G,A)$ to be the kernel of δ_n and $B^n(G,A)$ to be the range of δ_{n-1}, and $H^n(G,A)$ to be the quotient group Z^n/B^n. For $n = 2$, this gives the group $H^2(G,A)$ as defined above, so everything is compatible. For $n = 0$, a function of zero variables is by convention an element of A, and δ_0 is given by $\delta_0(a)(s) = s \cdot a - a$. Thus $B^0 = 0$ and $Z^0 = H^0(G,A) = A^G$, the G-fixed points in A. For $n = 1$, and a trivial G-module, $B^1 = 0$, and $Z^1 = H^1(G,A) = \{f \mid f(st) = f(s) + f(t),$ f Borel$\}$. By a classical theorem of Banach, every such Borel homomorphism is automatically continuous (cf. [3]), so $H^1(G,A)$ is the group of continuous homomorphisms of G into A. If G acts on A, $B^1 \neq 0$, and Z^1 consists of Borel crossed homomorphisms of G into A or functions satisfying $f(st) = s \cdot f(t) + f(s)$. Such a function is by the same theorem of Banach continuous so $H^1(G,A)$ consists of classes of continuous crossed homomorphisms of G into A where a class consists of a coset of $B^1(G,A) = \{f(s) = s \cdot a - a \text{ for some } a \in A\}$.

Elements of $Z^2(G,T)$ arise naturally in the study of unitary ray representations [31]. Let p be a continuous homomorphism of G into PU(H), the projective unitary group of a Hilbert space H as defined previously. We can find a Borel cross section f of PU(H) back into U(H) by general theorems as above, and then $f(p(s))f(p(t))f(p(st))^{-1} = a(s,t)$ can be seen to define an element of $Z^2(G,T)$, and hence an element of $H^2(G,T)$. It is clear that the element of $H^2(G,T)$ is zero if and only if we may find a continuous unitary represnetation π of G on H which "induces" p [31]. Thus an analysis of $H^2(G,T)$ is crucial for an understanding of when a ray representation "is" in fact an honest unitary representation. Even if the element of $H^2(G,T)$ is non zero we can still construct according to Mackey's theorem a group extension of G by T

$$1 \to T \to E \to G \to 1$$

and one may verify that E possesses an "honest" unitary representation on H which is of the form $t \to t \cdot 1$ on T and which "induces" the given projective or ray representation of G. This makes explicit our earlier comment that ray representations may be interpreted as ordinary representations of a group extension.

8. COHOMOLOGY GROUPS

The introduction of cohomology groups $H^n(G,A)$ is grantedly very *ad hoc*. First of all we selected a particular class of functions (Borel functions) which happened to give us what we wanted in low dimensions, and moreover we selected a perhaps somewhat artificial definition of δ_n. One's doubts are further compounded by the observation that the constructions of Section 6 suggest a somewhat different definition of the groups $H^n(G,A)$.

We defined $C^n(G,A)$ to be all Borel functions from $G\times\cdots\times G = G^n$ into A, but one is led to consider the possibility of replacing $C^n(G,A)$ by $U(G^n,A)$, the group of equivalence classes modulo null functions of measurable functions from G^n (Haar measure) into A, and we denote this group by $\underline{C}^n(G,A)$. It is not difficult to verify that δ_n as above is a well defined map from $\underline{C}^n$ to $\underline{C}^{n+1}$, and hence that we get cohomology groups $\underline{H}^n(G,A) = \underline{Z}^n(G,A)/\underline{B}^n(G,A)$ where $\underline{Z}^n$ is the kernel of δ_n and $\underline{B}^n$ is the range of δ_{n-1}. The cocycles in dimension zero consist of the kernel of δ_0 or the elements a of A such that $s \cdot a = a$ for almost all s in G. It is not hard to see that this implies that $s \cdot a = a$ for all $s \in G$, and hence $\underline{H}^0(G,A) = A^G$. If A is a trivial G-module, then the cocycles in dimension one are exactly the equivalence classes of functions f from G to A such that $f(st) = f(s) + f(t)$ for almost all pairs s and t. Similarly, in dimension two we look at functions which satisfy the cocycle identity above for almost all triples (s,t,r). A result of Mackey in [36] suggests that such an approach is not as outlandish as it first appears.

Motivated by the above, together with the possibility of a wide variety of other choices of cohomology groups we ask if we can somehow find a set of reasonable axioms which any cohomology theory should in principle satisfy, and then prove that there is up to isomorphism only one way of satisfying these axioms. We shall show that this is the case, and moreover that the groups H^n and $\underline{H}^n$ defined above by cocycles do satisfy these axioms. We then will know not only that these two definitions of cohomology groups agree, but also that any other attempt to define cohomology groups satisfying the axioms below must necessarily lead to the same groups.

(a) Our first axiom is of a general algebraic nature. We assume given for each $A \in P(G)$, G fixed, and for each $n \geq 0$, an abelian group denoted by $H^n(G,A)$ such that these are "functors of cohomological type". More precisely, we assume that for any G-homomorphism f of A into B, we have induced homomorphisms f^n of $H^n(G,A)$ into $H^n(G,B)$ such that the law of composition is satisfied: $(gf)^n = g^n f^n$ when g is a G-homomorphism of B into C. Moreover $1^n = 1$ where 1 denotes the identity homomorphism of A into A, and we assume that for any short exact sequence

$$1 \to A' \to A \to A'' \to 1$$

in $P(G)$, we have natural coboundary operators $\partial_n: H^n(G,A'') \to H^{n+1}(G,A')$ such that the infinite long sequence

$$0 \to H^0(G,A') \to H^0(G,A) \to H^0(G,A'') \to H^1(G,A') \to \cdots \to H^n(G,A)$$
$$\to H^n(G,A'') \to H^{n+1}(G,A'') \to H^{n+1}(G,A) \to \cdots$$

is exact (see [38] and [45]).

(b) The second axiom demands $H^0(G,A) = A^G$ for any $A \in P(G)$.

(c) The third axiom is a vanishing axiom which is motivated by the cohomology of finite groups; namely we demand $H^n(G,I(A)) = (0)$ for $n \geq 1$, and every polonais group, where $I(A)$ is the "regular representation" as defined in Section 6.

Axiom (c) is of course the really crucial one; it asserts that certain modules are cohomologically trivial and although there is a great deal of motivation for it from the cohomology of abstract groups, it does represent a definite choice. One could conceivably select some other class of modules and assume them to be cohomologically trivial, and this would lead to a unicity theorem for some possibly different cohomology theory. Our defense here is that the groups defined by cochains above do satisfy this vanishing axiom, and that the groups $I(A)$ do seem to play a natural role in analysis and group representations.

The following unicity theorem follows immediately from Lemma 6.3 and standard methods of homological algebra.

Theorem 10

If $H_i^n(G,A)$, $i = 1, 2$ are two assignments of cohomology groups defined for $A \in P(G)$ for a fixed G which satisfy Axioms (a), (b), and (c) above, there are canonical isomorphisms of $H_1^n(G,A)$ onto $H_2^n(G,A)$ for all n and all A.

One of our major results is that the groups $H^n(G,A)$ and $\underline{H}^n(G,A)$ defined above by Borel cochains, and equivalence classes of measurable cochains do satisfy these axioms.

Theorem 11

The groups $H^n(G,A)$ and $\underline{H}^n(G,A)$ satisfy Axioms (a), (b), and (c) and hence are isomorphic. More precisely, the map which attaches to each Borel cochain in $C^n(G,A)$, its equivalence class in $\underline{C}^n(G,A)$, induces this isomorphism on cohomology.

The verification of Axioms (a) and (b) is routine in both cases (see [38]); however, the verification of Axiom (c) is non-trivial. In fact for $n = 1$, this verification is for all intents and purposes equivalent to Mackey's general version of the Stone-von Neumann theorem in [26]. A close examination of Mackey's argument in [26] reveals that what is essentially being proved is that $H^1(G,I(T)) = 0$. (Actually one wants to replace T by a unitary group $U(H)$ on a Hilbert space, and this would lead us into non-abelian cohomology (see [45]). The essential analytic details however are the same as when H is one-dimensional so that $U(H) = T$.) Theorem 11 is proved first for $n = 1$, and then the general case is reduced to this case by an induction argument. The argument follows in spirit the argument for abstract groups where in fact the result is trivial; however, there are non-trivial analytical complications concerning null sets in our case.

In view of Theorem 11 we shall henceforth use the notations $H^n(G,A)$ and $\underline{H}^n(G,A)$ interchangeably; our choice of notation will serve to emphasize that we are interested in a particular facet of these groups which may be evident from one of the definitions, but not the other. We note in particular that such results as the above are not approachable if one stays within the category of locally compact G-modules, and that essential use is made of non-locally compact modules.

We have remarked before that we have a natural notion of induced modules which gives us for each $A \in P(H)$, a module $I_H^G(A) \in P(G)$ where H is a closed subgroup of G. A very useful tool for finite groups is Shapiro's lemma [45] which relates the cohomology of A with that of the induced module.

Theorem 12

There are canonical isomorphisms $H^n(H,A) \simeq H^n(G,I_H^G(A))$ for all $A \in P(H)$ and all n.

The proof is obtained by noting that both sides of the above as functors on $P(H)$ satisfy Axioms (a), (b), and (c), and then one applies Theorem 10. We note that for $H = (e)$, this is simply the vanishing theorem. Also we note for $n = 1$, that this theorem is essentially Mackey's imprimitivity theorem [27].

9. ADDITIONAL PROPERTIES

We shall now discuss some additional properties of these cohomology groups, and in particular nail down the connection with group extensions. For $n = 0$, we have already seen that $H^0(G,A) = A^G$ and that $n = 1$, $H^1(G,A)$ is the group of continuous crossed homomorphisms of G into A modulo principal ones, and if A is a trivial G-module, it is simply the group of all continuous homomorphisms of G into A. In Section 7 we constructed an injective homomorphism of the group $\mathrm{Ext}(G,A)$ of equivalence classes of topological group extensions into $H^2(G,A)$. For A locally compact, Mackey has shown that this map is onto, but his argument [30] does not extend since it makes essential use of the Haar measure on A. We have an alternate argument which works in general and which we outline below.

If $a \in H^2(G,A)$, we embed A into $I(A)$ by Lemma 6.3 and let a' be the image of the class a in $H^2(G,I(A))$ under the map given in Axiom (a). Since $H^2(G,I(A)) = 0$, $a' = 0$, and so there is clearly an extension of G by $I(A)$ corresponding to a', namely the semi-direct product $I(A) \cdot G$. We wish to construct an extension of G by A corresponding to $a \in H^2(G,A)$ and on general principles we would expect this extension, if it exists, as a subgroup of $I(A) \cdot G$. In fact if we pick a cocycle in the class a, we can immediately construct a subgroup E' of $I(A) \cdot G$ and then prove that it has all the required properties. (This particular construction is virtually forced on us, again by general principles.) Thus $\mathrm{Ext}(G,A) \simeq H^2(G,A)$.

The higher cohomology groups have as yet no direct interpretation, however, we certainly do expect $H^3(G,A)$ to contain obstructions to the construction of non-abelian extensions as in [3], Chapter IV.

When the cohomology groups are constructed via equivalence classes of measurable cochains, another interesting and significant property emerges. Namely, since $\underline{C}^n(G,A)$ is a polonais group, and since it may be readily checked that the coboundary operators δ_n are continuous, it follows that $\underline{Z}^n(G,A)$ is closed and hence in P. Thus $\underline{H}^n(G,A)$ is the quotient of a group in P by a subgroup and, hence when given the quotient topology, is itself a topological group. There is no *a priori* reason for $\underline{B}^n(G,A)$ to be closed, and it is an unpleasant fact of life that it is not always closed so that $\underline{H}^n(G,A)$ may not even be Hausdorff. The closure of the identity element in such a group is a closed subgroup, and upon dividing by it, we obtain a Hausdorff group which in the case of $\underline{H}^n(G,A)$ is simply

$\underline{Z}^n(G,A)$ divided by the closure of $\underline{B}^n(G,A)$. This quotient group will again be polonais, and $\underline{H}^n(G,A)$ will satisfy all the axioms of a polonais group except with "metric" replaced by "pseudo-metric". Thus $\underline{H}^n(G,A)$ is in a class of groups one might reluctantly call pseudo-polonais.

In any case, the fact that $\underline{H}^n(G,A)$ and hence $H^n(G,A)$ have a natural and more or less reasonable topology will be quite important for us. In fact we can strengthen Axiom (a) above and prove that the groups $\underline{H}^n(G,A)$ are functors of cohomological type taking values in the category of topological groups. Moreover, if $n = 1$, and if A is a trivial G module, $H^1(G,A)$ being continuous homomorphisms of G into A has a natural Hausdorff topology, namely that of convergence on compact sets. It may be verified that the topology on $\underline{H}^1(G,A)$ coincides with this topology. In [39] a great deal of effort was devoted to constructing a topology for $H^2(G,A)$ for various G and A by rather *ad hoc* methods. It is not hard to show that this topology coincides with the one above on $\underline{H}^2(G,A)$ whenever the former exists. Details of this will appear in our subsequent paper.

One reason for seeking a topology on $H^n(G,A)$ (aside from the esthetic one of expecting a topological object when one starts with topological data) is so that one can hope to make sense out of the spectral sequence for the cohomology of a group extension (cf. [19]). If H is a closed normal subgroup of G, the analogy with finite groups leads us to hope for a spectral sequence $E_r^{p,q}$ converging to $H^*(G,A)$ with $E_2^{p,q} = H^p(G/H,H^q(H,A))$ (see [19]). We observe that for this to begin to make sense, we must have $H^q(H,A) \in P(G/H)$, and in particular it must have a topology. We can show that there is always a spectral sequence of this type, and moreover that if $H^q(H,A)$ happens to be Hausdorff then the $E_2^{p,q}$ term is given by the expected formula. The existence of such a spectral sequence is quite important since it is an almost indispensable tool in making all but the simplest calculations of our cohomology groups. The reader is referred to [38] and [39] for examples in the case when A is locally compact.

We shall close this section with one final result concerning direct integrals of G-modules. Recall from Section 6 that if $A \in P(G)$, the group $U(X,A)$ had a natural structure as G-module which we called the direct integral. Since Cartesian products are a special case of this, and since cohomology commutes with products, we may ask if the same is true for integrals and we have the following result.

Theorem 13

If $\underline{H}^n(G,A)$ is Hausdorff, we have an isomorphism of topological groups

$$\underline{H}^n(G,U(X,A)) \simeq U(X,\underline{H}^n(G,A)) .$$

The content of this result is that a cocycle with values in a direct integral module $U(X,A)$ may be represented as a direct integral of cocycles. If $n = 1$

with trivial action, the side condition is satisfied and since one cocycles are homomorphisms, this result essentially gives us a new proof of the existence of direct integral decompositions of unitary representations.

10. EXAMPLES AND APPLICATIONS

We want to conclude with some examples, some computations, and some applications of the general theory above.

Suppose that $G = G_1 \times G_2$ and suppose for simplicity that A is a trivial G-module. Then either as a consequence of the spectral sequence above, or as a result of explicit computations (cf. [31]), we may obtain a structure theorem for $H^2(G,A)$ as follows:

$$H^2(G,A) \simeq H^2(G_1,A) \oplus H^2(G_2,A) \oplus H^1(G_1,H^1(G_2,A)) \ .$$

The first two terms are easy enough to understand and represent the contributions of the factors G_1 and G_2 to the cohomology of G, while the final term is a cross-term representing the interaction of G_1 and G_2. This enables us for instance to immediately compute $H^2(R^n,T)$, $H^2(Z^n,T)$, and $H^2(T^n,T)$ by induction on n. Indeed it is easy to verify that $H^2(R,T) = H^2(Z,T) = H^2(T,T) = 0$ by looking at the possible group extensions in these three cases. Since $R^n = R^{n-1} \times R$, and so on, it follows readily by induction that $H^2(R^n,T)$ is isomorphic to a vector space V of dimension $n(n - 1)/2$, and that $H^2(Z^n,T)$ is isomorphic to a torus S again of dimension $n(n - 1)/2$, and that $H^2(T^n,T) = 0$. Moreover the topology defined above on the groups $\underline{H}^2(R^n,T)$ and $\underline{H}^2(Z^n,T)$ coincides with the usual topology on the vector V and torus S. The isomorphism can also be implemented quite explicitly since one may show that each class in $H^2(R^n,T)$ contains a unique skew symmetric continuous bilinear function, and one may identify $H^2(R^n,T)$ with the group of such functions which is a vector group of dimension exactly $n(n - 1)/2$. A similar but slightly more involved statement holds for $H^2(Z^n,T)$.

If G is a semi-simple Lie group and if A is a trivial locally compact G-module, it is classical [47] that $H^2(G,A) \simeq H^1(\pi_1(G),A)$ where $\pi_1(G)$ is the usual fundamental group of G. Furthermore $\underline{H}^2(G,A)$ is Hausdorff in its natural topology and this topology coincides with the compact open topology on $H^1(\pi_1(G),A)$ $= \mathrm{Hom}(\pi_1(G),A)$ which in this simple case is simply the topology of pointwise convergence. This result also holds for any trivial G-module in $P(G)$ and moreover a similar result holds for a much broader class of groups G if one is willing to suitably redefine and generalize the notion of the fundamental group $\pi_1(G)$ of G (see [40]).

Using the spectral sequence of the previous section one may compute $H^2(G,T)$ when G is a semi-direct product of a semi-simple group and say a vector group. One may verify in this case known results for the inhomogeneous Lorentz group, and similar kinds of groups. We refer the reader to [38] and [39] for more details.

Another application of this material, and especially of our results concerning non-locally compact G-modules concerns the following situation. Let $A = T^n$ be a finite or infinite dimensional torus where $n = 1, 2, \cdots, \infty$, and suppose that

$$1 \to A \to E \to G \to 1$$

is a group extension of G by A where E is locally compact and abelian. It is a trivial and well known consequence of the duality theory of locally compact abelian groups that such an extension splits (that is, represents the identity element of Ext(G,A)) and so $E \simeq A \oplus G$ is a direct sum of A and G as topological groups.

With this result in mind for Cartesian products of circles, it is natural to ask if a similar result holds for direct integrals of the circle group and the answer is affirmative.

Theorem 14

If $1 \to U(X,T) \to E \to G \to 1$ is an extension of G by U(X,T) with E abelian and G locally compact, then the extension is split so that $E = U(X,T) + G$ as topological groups.

The idea of the proof is quite simple; one may verify that $\underline{H}^2(G,T)$ is Hausdorff and so Theorem 13 is applicable. After some extra argument using the fact that E is assumed to be abelian, the problem is thrown back using Theorem 13 to case when $U(X,T) = T$ where the result is known. Theorem 14 is found to be quite useful in settling certain questions concerning the structure of non-locally compact topological groups. Moreover exactly the same technique allows us to establish the following result.

Theorem 15

If $H^2(G,T) = (0)$ then any extension $1 \to U(X,T) \to E \to G \to 1$ with G locally compact and U(X,T) a trivial G-module splits.

This final result leads to a very useful theorem concerning automorphism groups of von Neumann algebras which will have some applications in quantum field theory. Suppose that B is a von Neumann algebra of operators on a separable Hilbert space and that G is a locally compact group. We suppose given a homomorphism f of G into the group of inner *-automorphisms of B satisfying the continuity requirements set down in [22]. Thus for each $g \in G$, we have a unitary operator u(g) in B such that $f(g)(b) = u(g)b\, u(g)^{-1}$ for all $b \in B$. The question we raise is whether one can choose the operators u(g) so that they form a continuous unitary representation of G. This question is relevant in quantum field theory when for instance B is some algebra of observables and G is some symmetry group of the physical system. If B is the algebra of all bounded operators on Hilbert space, a moment's reflection will show that we are raising exactly the question of when a projective or ray representation of G can be converted into an ordinary representation since the group of *-inner automorphisms of B is PU(H). It follows from our general discussion of group extensions that we can do this for projective representations if $H^2(G,T) = 0$, or equivalently if every group extension of G by the circle group splits as a product. The theorem to follow asserts that the same is true in the general context described above.

Theorem 16

If f is any homomorphism of G into the group of *-inner automorphisms of a von Neumann algebra B on a separable Hilbert space, continuous in the sense described in [22], and if $H^2(G,T) = 0$, then there is a unitary representation π of G with $\pi(g) \in B$ such that $f(g)(b) = \pi(g)b\pi(g)^{-1}$ for $b \in B$.

The proof is almost immediate for the map f immediately gives rise to a cohomology class a in $H^2(G,W)$ where W is the group of unitary operators in the center of B, such that $a = 0$ if and only if a representation π as described in the theorem exists. However, by the structure theory of von Neumann algebras W is of the form U(X,T) and the result follows by Theorem 15.

When $G = R$, Kadison in [22] established a special case of this. Recently R. Kallman [23] has obtained a far more general result. For the Poincaré group, another case of physical interest, L. Michel has already obtained the above result by rather different methods [36].

REFERENCES

[1] Auslander, L., *et al.* "Flows on Homogeneous Spaces", *Annals of Mathematics Studies, No. 53,* Princeton (1963).

[2] Auslander, L., and Green, L. "G-induced Flows", *Am. J. Math.,* 88, 43-60 (1966).

[3] Auslander, L., and Moore, C. C. "Unitary Representations of Solvable Lie Groups", *Mem. Am. Math. Soc.,* No. 62 (1966).

[4] Avez, A. "Ergodic Theory of Dynamical Systems", *Notes, University of Minnesota* (1966).

[5] Bargmann, V. "On Unitary Ray Representations of Continuous Groups", *Ann. Math.,* 59, 1-46 (1954).

[6] Bruhat, F. "Sur les Representations Induites des Groupes de Lie", *Bull. Soc. Math. France,* 84, 97-205 (1956).

[7] Borel, A. "Density Properties for Certain Subgroups of Semi-simple Groups Without Compact Components", *Ann. Math.,* 72, 179-188 (1960).

[8] Borel, A., and Harish-Chandra. "Arithmetic Subgroups of Algebraic Groups", *Ann. Math.,* 75, 485-535 (1962).

[9] Dixmier, J. "Dual et Quasi-dual d'une Algebre de Banach Involutiv", *Trans. Am. Math. Soc.,* 104, 278-283 (1962).

[10] Dixmier, J. *Les C* Algebres et Leur Representations,* Gauthier-Villars, Paris (1964).

[11] Eilenberg, S., and MacLane, S. "Cohomology Theory in Abstract Groups, I", *Ann. Math.,* 48, 51-78 (1947).

[12] Eilenberg, S. "Cohomology Theory in Abstract Groups, II", *Ann. Math.,* 48, 326-341 (1947).

[13] Gelfand, I., and Fomin, S. "Geodesic Flows on Manifolds of Constant Negative Curvature", *Uspehi Mat. Nauk,* 7, 118-137 (1952).

[14] Halmos, P. *Measure Theory,* Van Nostrand, New York (1950).

[15] Halmos, P. *Lectures on Ergodic Theory,* Publications of the Mathematical Society of Japan, No. 3 (1956).

[16] Hochschild, G. "Group Extensions of Lie Groups I", *Ann. Math.,* 54, 96-109 (1951).

[17] Hochschild, G. *The Structure of Lie Groups,* Holden Day, San Francisco (1965).

[18] Hochschild, G., and Mostow, G. D. "Cohomology of Lie Groups", *Illinois J. Math.,* 6, 367-401 (1962).

[19] Hochschild, G., and Serre, J. P. "Cohomology of Group Extensions", *Trans. Am. Math. Soc.,* 74, 110-134 (1953).

[20] Harish-Chandra. "Representations of Semi-simple Lie Groups V", *Proc. Nat. Acad. Sci.,* 40, 1076-1077 (1954).

[21] Harish-Chandra. "Discrete Series for Semi-simple Lie Groups", *Acta Math.*, 113, 242-318 (1965).

[22] Kadison, R. V. "Transformations of States in Operator Theory and Dynamics", *Topology*, 3, 177-198 (1965).

[23] Kallman, R. "Spatially Induced Groups of Automorphisms of Certain von Neumann Algebras" (to appear).

[24] Khinchin, A. *Mathematical Foundations of Statistical Mechanics*, Dover, New York (1949).

[25] Loomis, L. *An Introduction to Abstract Harmonic Analysis*, Van Nostrand, New York (1953).

[26] Mackey, G. W. "A Theorem of Stone and von Neumann", *Duke Math. J.*, 16, 313-326 (1949).

[27] Mackey, G. W. "On Induced Representations of Groups", *Am. J. Math.*, 73, 576-593 (1951).

[28] Mackey, G. W. "Induced Representations of Locally Compact Groups I", *Ann. of Math.*, 55, 101-139 (1952).

[29] Mackey, G. W. Les Ensembles Boreliens et Les Extensions des Groupes", *J. Math. Pures Appl.*, 36, 171-178 (1957).

[30] Mackey, G. W. "Borel Structures in Groups and Their Duals", *Trans. Am. Math. Soc.*, 85, 134-165 (1957).

[31] Mackey, G. W. "Unitary Representations of Group Extensions", *Acta Math.*, 99, 265-311 (1958).

[32] Mackey, G. W. "Point Realizations of Transformation Groups", *Illinois J. Math.*, 6, 327-335 (1962).

[33] Mackey, G. W. "The Theory of Group Representations", mimeographed notes, University of Chicago (1955).

[34] Mackey, G. W. "Infinite Dimensional Group Representations", *Bull. Amer. Math. Soc.*, 69, 628-686 (1963).

[35] Mautner, F. I. "Geodesic Flows on Symmetric Riemann Spaces", *Ann. Math.*, 65, 416-431 (1957).

[36] Michel, L. "Sur les Extensions Centrales du Groupe de Lorentz Inhomogene Connexe", *Nucl. Phys.*, 57, 356-385 (1964).

[37] Moore, C. C. "Extensions and Low Dimensional Cohomology Theory of Locally Compact Groups, I", *Trans. Am. Math. Soc.*, 113, 40-63 (1964).

[38] Moore, C. C. "Extensions and Low Dimensional Cohomology Theory of Locally Compact Groups, II", *Trans. Am. Math. Soc.*, 113, 63-86 (1964).

[39] Moore, C. C. "Ergodicity of Flows on Homogeneous Spaces", *Am. J. Math.*, 88, 154-178 (1966).

[40] Moore, C. C. "Group Extensions of P-adic and Adelic Linear Groups", *Inst. Hautes Études Sci. Publ. Math.*, (35), 5-74 (1968).

[41] Mostow, G. D. "Cohomology of Topological Groups and Solvmanifolds", *Ann. Math.*, 73, 20-48 (1961).

[42] O'Raifeartaigh, L. "Mass Differences and Lie Algebras of Finite Order", *Phys. Rev. Lett.*, 14, 575-577 (1965).

[43] Parasyuk, O. "Horocycle Flows on Surfaces of Negative Curvature", *Uspehi Mat. Nauk*, 8, 125-26 (1953).

[44] Segal, I. "An Extension of a Theorem of L. O'Raifeartaigh", *J. Functional Analysis*, 1, 1-21 (1967).

[45] Serre, J. P. *Cohomologie Galoisienne*, Berlin, Springer (1964).

[46] Shale, D. "Linear Isometries of Free Boson Fields", *Trans. Am. Math. Soc.*, 103, 149-167 (1962).

[47] Shapiro, A. "Group Extensions of Compact Lie Groups", *Ann. Math.*, 50, 581-585 (1949).

[48] Sherman, T. "A Weight Theory for Unitary Representations", *Canad. J. Math.*, 18, 159-168 (1966).

[49] Seminaire "Sophus Lie", Paris (1954).

[50] Weil, W. *L'Integration dans les Groupes Topologiques et ses Applications*, Hermann, Paris (1940).

[51] Weil, A. "Sur Certains Groupes D'Operateurs Unitaires", *Acta Math.*, 111, 143-211 (1964).

[52] Wigner, E. *Group Theory*, Academic Press, New York (1959).

APPLICATIONS OF GROUP THEORY TO QUANTUM PHYSICS

ALGEBRAIC ASPECTS

by

Louis Michel†

TABLE OF CONTENTS

† Institut des Hautes Etudes Scientifiques, 91 - BURES-SUR-YVETTE - France.

0. INTRODUCTION

Since you mathematicians and we physicists came here to meet together, there is no need to emphasize that we both believe that the progress of physics requires for its theoretical formulation more and more advanced mathematics. I thought fit however to give you the opportunity to read what Dirac wrote on this subject, 38 years ago, as an introduction to the very paper where he predicted the existence of the "antielectron", which we now call positron.

Notwithstanding Dirac's prediction, when positrons were observed one year later by Blackett and Occhialini, and by the Joliot-Curies, they were not immediately recognized. And Anderson who was the first to identify a positron (in cosmic rays) did not know Dirac's paper. This illustrates the communication difficulties which existed and still exist between theoretical and experimental physicists. You should also expect them between mathematicians and physicists. (Not to speak of the difficulties due to my use of English.) You and I are here determined to overcome them, but I beg your patience in advance.

Extract from *Proc. Roy. Soc.*, Ser. A, 130, 60 (1930):

Quantised Singularities in the Electromagnetic Field

By P. A. M. DIRAC, F. R. S., St. John's College, Cambridge

§ 1. Introduction

The steady progress of physics requires for its theoretical formulation a mathematics that gets continually more advanced. This is only natural and to be expected. What, however, was not expected by the scientific workers of the last century was the particular form that the line of advancement of the mathematics would take, namely, it was expected that the mathematics would get more and more complicated, but would rest on a permanent basis of axioms and definitions, while actually the modern physical developments have required a mathematics that continually shifts its foundations and gets more abstract. Non-euclidean geometry and non-commutative algebra, which were at one time considered to be purely fictions of the mind and pastimes for logical thinkers, have now been found to be very necessary for the description of general facts of the physical world. It seems likely that this process of increasing abstraction will continue in the future and that advance in physics is to be associated with a continual modification and generalization of the axioms at the base of the mathematics rather than with a logical development of any one mathematical scheme on a fixed foundation.

There are at present fundamental problems in theoretical physics awaiting solution, e.g., the relativistic formulation of quantum mechanics and the nature of atomic nuclei (to be followed by more difficult ones such as the problem of life), the solution of which problems will presumably require a more drastic revision of our fundamental concepts than any that have gone before. Quite likely these changes will be so great that it will be beyond the power of human intelligence to get the necessary new ideas by direct attempts to formulate the experimental data in mathematical terms. The theoretical worker in the future will therefore have to proceed in a more indirect way. The most powerful method of advance that can be suggested at present is to employ all the resources of pure mathematics in attempts to perfect and generalize the mathematical formalism that forms the existing basis of theoretical physics and after each success in this direction, to try to interpret the new mathematical features in terms of physical entities (by a process like Eddington's Principle of Identification).

A recent paper by the author† may possibly be regarded as a small step according to this general scheme of advance. The mathematical formalism at that time involved a serious difficulty through its prediction of negative kinetic energy values for an electron. It was proposed to get over this difficulty, making use of Pauli's Exclusion Principle which does not allow more than one electron in any state, by saying that in the physical world almost all the negative-energy states are already occupied, so that our ordinary electrons of positive energy cannot fall into them. The question then arises as to the physical interpretation of the negative-energy states, which on this view really exist. We should expect the uniformly filled distribution of negative-energy states to be completely unobservable to us, but an unoccupied one of these states, being something exceptional, would make its presence felt as a kind of hole. It was shown that one of these holes would appear to us as a particle with a positive energy and a positive charge and it was suggested that this particle should be identified with a proton. Subsequent investigations, however, have shown that this particle necessarily has the same mass as an electron†† and also that, if it collides with an electron, the two will have a chance of annihilating one another much too great to be consistent with the known stability of matter.†††

It thus appears that we must abandon the identification of the holes with protons and must find some other interpretation for them. Following Oppenheimer,†††† we can assume that in the world as we know it, all, and not merely nearly all, of the negative-energy states for electrons are occupied. A hole, if there were one, would be a new kind of particle, unknown to experimental physics, having the same mass and opposite charge to an electron. We may call such a particle an antielectron. We should not expect to find any of them in nature, on account of their rapid rate of recombination with electrons, but if they could be produced experimentally in high vacuum they would be quite stable and amenable to observation. An encounter between two hard γ-rays (of energy at least half a million volts) could lead to the creation simultaneously of an electron and antielectron, the probability of occurence of this process being of the same order of magnitude as that of the collision of the two γ-rays on the assumption that they are spheres of the same size as classical electrons. This probability is negligible, however, with the intensities of γ-rays at present available.

The protons on the above view are quite unconnected with electrons. Presumably the protons will have their own negative-energy states, all of which normally are occupied, an unoccupied one appearing as an antiproton.

Let me just remind you that antiprotons were first observed twenty-four years later (1955).

There will be many advanced seminars on the applications of group theory to quantum physics. So I believe that these lectures must be introductory, and that I have to present concepts that will be used by all physicists here. That will be Part 1.

---

† *Proc. Roy. Soc., Ser. A,* 126, 360 (1930).

†† H. Weyl, *Gruppentheorie und Quantenmechanik,* 2nd ed., p. 234 (1931).

††† I. Tamm, *Z Physik,* 62, 545 (1930); J. R. Oppenheimer, *Phys. Rev.,* 35, 939 (1930); P. Dirac, *Proc. Camb. Philos. Soc.,* 26, 361 (1930).

†††† J. R. Oppenheimer, *Phys. Rev.,* 35, 562 (1930).

The ultimate goal of these lectures will be to bring you to the present problems on the subject, mainly in the field of the fundamental particle physics. Then there might be some overlap with Professor O'Raifeartaigh's lectures, but there should be no inconvenience to see some aspects of physics from probably two different points of view. We have to face the fact that fruitful discussions of frontier problems of physics between mathematicians and physicists are difficult, because these problems often cannot be presented in a formalized language, but only through some physical analogy. So, obviously, to understand what is the problem, one must know some physics!

In these lectures I will therefore present a quick survey of applications of group theory to atomic, molecular and nuclear physics. Often, I will even follow an historical approach. Indeed, physicist minds are partly conditioned by the recent history of physics. But I also hope to use the power of your language, mathematics, to convey to you a maximum of physics in a minimum of time. Of course I shall have succeeded only if I have also been able to convince you that physics is fascinating!

It is fit to end this introduction by the history of the birth of our subject. Less than three years after the first paper on quantum mechanics (W. Heisenberg, *Z. Phys.*, 33, 879 (1925), there appeared the first two papers devoted to the application of group theory to quantum mechanics:

- E. P. Wigner, "Einige Folgerungen aus der Schrödingerschen Theorie für die Termstrukturen", *Z. Phys.*, 43, 624 (1927).
- F. Hund, "Symmetriecharaktere von Termen bei Systemen mit gleichen Partikeln in der Quantenmechanik", *Z. Phys.*, 43, 788 (1927).

Wigner will surely be the most quoted author on our subject. Let us just say that, with J. von Neumann, he applied group theory to atomic spectra ("Zur Erklärung einiger Eigenshaften der Spektren aus der Quantenmechanik des Drehelektrons I., II., III., *Z. Phys.*, 47, 203; 49, 73; 51, 844 (1928)), and published a self-contained book on this question: E. P. Wigner, *Gruppentheorie und ihre Anwendung auf die Quantenmechanik der Atomspektren,* Vieweg, Braunschweig (1931).

It is remarkable that two famous mathematicians, Hermann Weyl and Van der Waerden, also published very early books on our subject:

- H. Weyl, *Gruppentheorie und Quantenmechanik,* Hirzel, Leipzig (1928).
- Van der Waerden, *Die Gruppentheoretische Methode in der Quantenmechanik,* Springer, Berlin (1932).

Then the excellent, but more elementary book, by E. Bauer, *Introduction à la Théorie des Groupes et ses Application à la Physique Quantique,* Presses Universitaires de France, Paris (1933), continued a list of books which, today, may have reached several dozen.

For the interested mathematicians I would still recommend the two very first books, but in their second, revised and enlarged edition: H. Weyl, *The Theory of Groups and Quantum Mechanics*, Methuen, London (1931); Paper Back reprint, Dover, New York (1949); the translation by J. J. Griffin of Wigner's book, *Group Theory and Its Application to the Quantum Mechanics of Atomic Spectra*, Academic Press, New York (1959).

For The Mathematician Readers

Physics will be injected in these notes as needed. However, it seems convenient to gather here some information on physical constants which might be useful at any time.

We will study quantum phenomena. In atomic, nuclear, fundamental particle physics, the key number to pass from macroscopic scale is the Avogadro number:

$$a = 6.0228 \times 10^{23} \sim 6.03 \times 10^{23} \quad . \tag{0.1}$$

It is the number of atoms in a mass of one gram of hydrogen.

A hydrogen atom is made of one proton (mass m_p) and one electron (mass m_ε).

$$\frac{m_p}{m_\varepsilon} = 1836.5 \quad .$$

These two particles are electrically charged, p^+, ε^-, the absolute value of this charge is

$$e = \frac{1 \text{ Faraday}}{a} = \frac{96{,}600}{6.03 \times 10^{23}} \text{ Coulombs} \quad .$$

The most convenient unit systems, for us, will use

$$\hbar = (\text{Planck constant}) \times (2\pi)^{-1} = 1$$
$$c = (\text{velocity of light}) = 1 \quad .$$

In this system $e = (137.04)^{-1/2}$, indeed

$$\alpha = \frac{e^2}{\hbar c} = \frac{1}{137.04}$$

is a dimensionless number. Atoms of the other elements are made of Z electrons and a nucleus which contains Z protons and N neutrons; Z is the atomic number, $A = Z + N$ the atomic mass number, e.g., for hydrogen $Z = 1$; hydrogen has 3 isotopes $N = 0$, $A = 1$; $A = 2$, deuterium; $A = 3$, tritium (unstable, lifetime 12 years). For uranium $Z = 92$; the most abundant isotope is $A = 238$. The neutron has no electric charge and its mass is nearly equal to that of the proton,

$$\frac{m_n}{m_\varepsilon} = 1839.0 \quad .$$

So the atom mass is practically concentrated in its nucleus.

Many more particles will be introduced, e.g., the photon γ, and the four different neutrinos all with zero rest-mass and zero charge, etc... see 3.5 and 3.6.

We remind the reader that in relativistic physics mass is not conserved; mass is a form of energy. The energy of a particle of mass m, velocity v is $E = mc^2(1-(v/c)^2)^{-1/2}$. The rest energy $(v = 0)$ is mc^2.

We need to choose another unit to complete our unit systems. The best choice for atomic physics is the electron mass m_ε. Then the other units are:

momentum	$m_\varepsilon c$
energy	$m_\varepsilon c^2$
length	$\hbar/m_\varepsilon c = 3.86 \times 10^{-11}$ cm
time	$\hbar/m_\varepsilon c^2 = 1.28 \times 10^{-21}$ sec.

However, due to the nature of its measurement, the most common energy unit used for particles is the electron volt (eV). It is the energy that a particle with the universal electric charge e gains by traversing an electric field of potential difference one volt.

The conversion with the preceeding unit system is

$$m_\varepsilon c^2 = 0.511 \times 10^6 \text{ eV} \sim 1/2 \text{ MeV}$$

$$m_p c^2 = 938.256 \text{ MeV} \sim 1 \text{ GeV} = 10^9 \text{ eV} \quad .$$

Note that $1 = \hbar \sim (10^{-13}\text{ cm}) \times (200\text{ MeV})$

Before 1932, only two kinds of interactions were known, gravitation and electromagnetism. In the static approximation the two interactions can be described by proportional potentials K/r where r is the distance. So the absolute ratio of the (attractive) gravitational energy to the (repulsive) electrostatic energy between two protons is independent of their distance.
It is

$$\frac{G' m_p^2}{e^2} = \frac{\alpha'}{\alpha} = \frac{137}{175} \times 10^{-36} \quad .$$

Thus, gravitation will be completely neglected in these lectures.†

---

† The gravitational energy of a system increases roughly as the square of the number N of nucleons while, in neutral matter, the electrostatic energy is roughly proportional to N. So gravitation becomes important only for masses as large as that of asteroids, planets (we know it on the earth!) or stars. It is not a coincidence that most stars have a number of nucleons $\sim \alpha'^{-3/2} = 10^{57}$ (see for instance E. E. Salpeter, "Dimensionless Ratio and Stellar Structure", in *Frontier in Physics, Bethe Festschrift*, p. 463, R. Marshak Editor).

The binding energy of atoms, molecules, solids, etc., is of electromagnetic origin. This energy can be released in chemical form, with an order of magnitude:

$$a \times 1 \text{ eV} = 23 \text{ cal/mole}$$

which is $\sim 10^{-9}$ to 10^{-11} the rest-mass energy. If the energy we receive from the sun came from chemical reactions, the sun would produce it for less than 10^5 years!

There are two other known kinds of interactions: the nuclear interaction (see part 3 and 5), stronger than the electromagnetic interaction at distance smaller than 10^{-13} cm; the Fermi or "weak interaction" (see 3.6 and 5), which is very short range. Both interactions are important in stars and nuclear reactions and can yield an energy up to 10^{-3} the rest mass energy.

For the Physicist Reader

All mathematical terms used here are not defined. Of course many of them are known to physicists (e.g., for the notion of root vectors of Lie algebra, see Salam's lectures in *High Energy Physics and Elementary Particles, Trieste Seminar 1962* (International Atomic Energy Agency, Vienna (1963)). Some terms (used mainly in I) come from a modern mathematical terminology. They were not absolutely necessary and they are used explicitly as synonyms of other terms generally used by physicists. Physicists should know the proper mathematical terms of the mathematical concepts they need: indeed their students, and even their young children know them and physicists want to communicate with their students and their children!

An excellent and elementary exposition of this modern mathematical language is given in the text *Algebra* by S. Mac Lane and G. Birkhoff, Macmillan, New York, (1967), particularly Chapter I; note also the list of symbols, p. XVII to XIX.

1. COVARIANCE IN QUANTUM THEORY AND ITS MATHEMATICAL TOOLS

1.1. What Is Quantum Mechanics

Less than two years after the first paper (quoted above) of Heisenberg on quantum mechanics, J. von Neumann answered this question in three successive papers in *Göttingen Nachrichten*, (1927), (pp. 1, 245, 273) expanded in a book: *Mathematische Grundlagen der Quanten Mechanik*, (1930) (English translation, Princeton University Press (1955)).†

† He later published with G. Birkhoff, "The Logic of Quantum Mechanics", *Ann. of Math.*, 37, 935 (1936). This subject is still controversial and lively.

Two early books on quantum mechanics by physicists are reedited and still very advisable reading: P. A. M. Dirac, *The Principles of Quantum Mechanics*, Clarendon Press, Oxford 1st ed. (1930) 4th ed. (1958). W. Pauli, "Prinzipien der Quanten Theorie", *Handbuch der Physik*, 1, Springer (1958) 1st ed. (1933).†

If you have not read these books it is not too late to do it, but today let us just give a mini-description of quantum mechanics.

a) To each physical system corresponds a separable complex Hilbert space $\mathcal{H}$. A physical state is represented by a vector $x> \in \mathcal{H}$. (Normed to 1 for convenience: $<x,x> = 1$.)

b) Each physical observable $\underline{a}$ (e.g., energy, electric charge, etc.) is represented by a self adjoint operator A on $\mathcal{H}$. The spectrum of A is the set of possible values of $\underline{a}$.

c) Quantum mechanics does not predict, in general, the value of $\underline{a}$ for the state $x>$, it gives only its expectation value:

$$<x,Ax> = \mathrm{Tr}\, A\, P_x \tag{1.1}$$

where P_x is the Hermitian projector $(P_x = P_x^*)$ onto the one dimensional space spanned by $x>$. Note that unit eigen vectors of P_x (with eigen-value 1) differ only by a scalar phase factor and describe the same state since they yield the same physical predictions. The projectors P_x are themselves observables. Indeed

$$\mathrm{Tr} P_x P_y = |<x,y>|^2 \tag{1.2}$$

is the probability to observe in the state $x>$ (respectively $y>$) the physical system which is known to be in the state $y>$ (respectively $x>$). Part of the art of the quantum physicist is to code what he sees in nature into vectors of Hilbert space! This always requires "physical approximations".

When we can describe a state by a rank one projector (or a vector up to a phase) we say that we have a pure state and that we have a complete information on it.

More often our information on the state is only partial. In the simplest case we know only a set of probabilities c_i (with $\Sigma c_i = 1$) for the system to be in the set of orthogonal pure states P_i (i.e., $P_i P_j = \delta_{ij} P_j$) so the expectation value is

$$\Sigma_i c_i\, \mathrm{Tr}\, A\, P_i = \mathrm{Tr}\, A\, R \tag{1.3}$$

with

$$R = \Sigma_i c_i P_i \qquad \mathrm{Tr}\, R = \Sigma_i c_i = 1 \quad . \tag{1.3'}$$

Since $0 \leq c_i$, the self adjoint operator R is positive and it is called the

---

† There are also books on the mathematical foundation of quantum mechanics by mathematicians: G. Mackey, L. Schwartz.

density matrix† of the mixture (= not pure) state of the system. Pure states are extremal points of the convex domain of states.

This leads us to a natural generalization. One defines a Banach*-algebra $\mathcal{B}$ with unit 1, generated by the observables.†† (More specifically it is usually a C*-algebra). Then a state is a linear functional ϕ on $\mathcal{B}$ which is positive, that is $\forall A \in \mathcal{B}$, $\phi(A^*A) \geq 0$. For systems with a finite number of degrees of freedom this is not an essential generalization. It becomes so for infinite degrees of freedom as in quantum field theory and statistical mechanics. Classical statistical mechanics can also be put in the same mathematical mould with an abelian algebra.†††

1.2 Group Invariance

We assume that there is a relativity group G for every physical theory considered here. That is G acts on a physical system S, and there is an isomorphism between the physics of S (its Hilbert space of states $\mathcal{H}$, its algebra of observables $\mathcal{B} \subset L(\mathcal{H})$††††, etc...) and the physics of $\underline{g}(S)$, the transform of S, by $\underline{g} \in G$ (e.g., $\underline{g}$ can be a rotation). This will be called the "active" point of view of G-invariance. The "passive" point of view for a transformation group is simply the isomorphism between the physical description of the same system S by two observers choosing different coordinate frames, G-transforms of each other.

For any $g \in G$, we denote by P_{gx_i} the transformed of the state P_{x_i}. To say that G is an invariance group is equivalent to saying that all probabilities of Equation (1.2) are invariant

$$x> \in \mathcal{H}, \quad g \in G, \quad \mathrm{Tr} P_{gx_1} P_{gx_2} = \mathrm{Tr} P_{x_1} P_{x_2} \tag{1.4}$$

---

† J. von Neumann introduced the density-matrix in 1927 in the papers quoted above.

†† Quite early physicists also considered non-associative algebras formed by the observables and introduced Jordan algebras. The first fundamental paper on those algebras is by P. Jordan, J. von Neumann and E. Wigner; "On an Algebraic Generalization of the Quantum Mechanical Formalism", *Ann. of Math.*, <u>35</u>, 29 (1934).

††† I. E. Segal advocated twenty years ago the use of C*-algebra for quantum physics. The fundamental paper showing the benefits from this choice (physical approximation and Feld's ε-equivalence; introduction of super-selection rules) is that of R. Haag and D. Kastler, "An Algebraic Approach to Quantum Field Theory", *J. Math. Phys., Supplement* 848 (1964). Most of the C*-algebra physics is published in the journal: *Communications in Mathematical Physics*, and is written in a rigorous mathematical style. For statistical mechanics, see D. Ruelle, *Statistical Mechanics*, Benjamin, New York (1969). Soon there will appear in the collection of C. N. R. S. Colloquia (France) "Rigorous Results on Interacting Systems with Infinite Degrees of Freedom".

†††† We denote by $L(\mathcal{H})$ the space of linear operators on $\mathcal{H}$.

or

$$|\langle gx_1, gx_2\rangle|^2 = |\langle x_1, x_2\rangle|^2 \tag{1.4$'$}$$

This means that G acts on $\mathcal{H}$ by isometries.

Wigner proved in his book (Appendix to Chapter 20)[†] that $x\rangle \rightsquigarrow gx\rangle$ is either a unitary operator U(g) or an antiunitary operator V(g) on $\mathcal{H}$. We recall that an antiunitary operator V has the characteristic properties

$$x\rangle,\ y\rangle \in \mathcal{H},\ V(\alpha x\rangle + \beta y\rangle) = \bar{\alpha}Vx\rangle + \bar{\beta}Vy\rangle \tag{1.5}$$

$$\langle Vx, Vy\rangle = \overline{\langle x,y\rangle} = \langle y,x\rangle \tag{1.5$'$}$$

it has an inverse (1.5″)

Given an isometry on $\mathcal{H}$, there is a simple criterion[††] for deciding whether it is realized by a unitary operator U or an antiunitary operator V. In either case U or V is defined up to a scalar phase factor. The product of two antiunitary operators is a unitary operator.

Let $\mathcal{V}(\mathcal{H})$ be the group of unitary and antiunitary operators on $\mathcal{H}$ and $\mathcal{U}(\mathcal{H})$ the subgroup of unitary operators. $\mathcal{U}(\mathcal{H})$ is an invariant subgroup of $\mathcal{V}(\mathcal{H})$ since it is a subgroup of index two. We assume that G acts effectively on $\mathcal{H}$, i.e., no other element than $1 \in G$ acts trivially on $\mathcal{H}$. The U(g)'s or V(g)'s for $g \in G$ generate a subgroup $\mathcal{E}(G)$ of $\mathcal{V}(\mathcal{H})$ which is an extension of G by the group U_1 (phase multiplication of the vectors of $\mathcal{H}$, leaving invariant the states) with the action

$$G \xrightarrow{f} \text{Aut}\ U_1$$

where Ker f is the invariant subgroup of index two $G_+ \subset G$ which acts by unitary transformations and the non-trivial element of Im f is the complex conjugation $\alpha \rightsquigarrow \bar{\alpha} = \alpha^{-1} \in U(1)$.

We can also say that G_+ acts by a linear unitary projective representation and Wigner has coined the word projective "corepresentation" for the action of G (when G is strictly larger than G_+).

Wigner also showed from physical arguments that antiunitary operators are to be used with transformations which reverse the direction of time, this in order that energy be positive: indeed, the time translation t is represented by the operator e^{iHt}; if $t \to -t$, i has to go to $-i$ in order that both H and e^{iHt} be invariant.

---

† A more explicit proof of Wigner's theorem has been given by V. Bargmann, *J. Math. Phys.*, 5, 862 (1964). See also proofs of slight generalizations by U. Uhlhorn, *Arkiv för Fysik*, 23, 307 (1963). In the framework of Birkhoff and von Neumann axiomatics, the equivalent theorem has been proven by G. Emch and C. Piron, *J. Math. Phys.*, 4, 469 (1963).

†† See Bargmann:† for any triplet of vectors $x\rangle, y\rangle, z\rangle$, $\langle x,y\rangle\langle y,z\rangle\langle z,x\rangle$ is invariant under a unitary transformation U and is transformed into its complex conjugate under the antiunitary transformation V.

Continuous projective linear unitary representation of finite groups or Lie groups are well known. For instance, for the three-dimensional rotation group SO(3,R) these projective representations are in a one to one correspondence with the "linear irreducible unitary representations" (= irrep through all these lectures) of SU(2) the universal covering of SO(3,R). This justifies the introduction of spinors in quantum physics.

In Part 4 we will study invariance under the relativity groups of non-relativistic (= Newtonian) mechanics and of special relativity theory. But there are other invariance groups in physics. For instance the permutation group $S(n)$ acting on n identical particles (as the electrons of an atom). In nuclear physics and fundamental particle physics we shall meet many "approximate invariances". The corresponding invariance group is most often a U(n) or SU(n) group (group of unitary $n \times n$ matrices, with determinant 1 for SU(n)) with $n = 1, 2, 3, 4, 6$. We shall have more to say for the word "approximate" symmetry.

We will also have to study invariance under a group G when G is a symmetry group for a physical system, e.g., the symmetry group (one of the crystallographic group) of a crystal. This example raises a fascinating question about group invariance in physics. Surely the interaction between atoms are translation invariant (and may be invariant under a larger transformation group). How is it possible that atoms aggregate to form a crystal whose lattice is invariant only under a subgroup of a translation group? When such a phenomenon occurs, i.e., when a stable state has a lesser symmetry than that of the physical laws we will say that we have a broken symmetry.†

We will continue this Part 1 by introducing some mathematical tools that we will use quite frequently.

1.3 G-Vector Spaces

Let G be a given group. If you like you can say that we consider a category whose objects are vector spaces E (over a given field K) with a linear action of G on E (i.e., $G \ni x \rightsquigarrow g(x) \in L(E)$, where $L(E)$ is the algebra of endomorphisms of E, with $xy \rightsquigarrow g(x)g(y) = g(xy)$.

The morphisms of the category are the vector space homomorphisms $E \xrightarrow{f} E'$ compatible with the group action, i.e., they are the commutative diagrams for every $x \in G$, of vector space homomorphisms. We will call these morphisms G-homomorphisms of G-vector spaces.

---

† This short section on group invariance is too sketchy. Much more should be said of the symmetry of physical laws (e.g., E. P. Wigner, "Symmetry and Conversation Laws", *Proc. Nat. Acad. Sci., U.S.A.*, 51, 956 (1964)) withouc which symmetries of states, that we have considered, would not last. Of course much more will be said in these lectures.

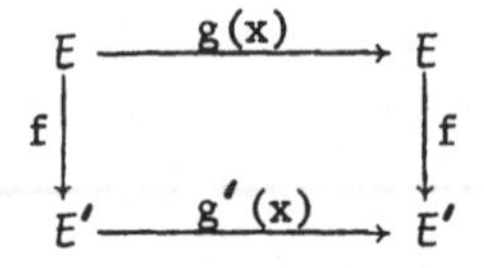

Diagram 1.

Of course we could have also said that we study bimodules (G- and K-modules) or even more simply that we are interested in the linear representations of G; and G-homomorphisms are also called "intertwining" operators. Note that the G-morphisms from E_1 to E_2 form a vector space that we denote $\mathrm{Hom}(E_1, E_2)^G$. Indeed, it is the subspace of the invariant vectors of $\mathrm{Hom}(E_1, E_2)$; they are the intertwining operators for the two representations of G on E_1 and E_2.

Given G-vector spaces, $E_1, E_2 \ldots$ all vector spaces which can be formed functorially from them are also G-vector spaces, e.g., $E_1 \otimes E_2$, $\mathrm{Hom}(E_1, E_2)$, $L(E) = \mathrm{Hom}(E, E)$, the vector space of the tensor algebra on E: $T(E) = \oplus_{n=0}^{\infty} E^{(n)}$ when $E^{(n)} = E \otimes E \otimes \ldots \otimes E$, n factors, (and $E^{(0)} = K$), etc.

Given a physical system, let $\mathcal{H}$ be the Hilbert space of its state vectors. Assume that $\mathcal{H}$ is a G-vector space. So is $L(\mathcal{H})$. We are then led to the study of the objects (of the category of G-vector spaces) "above" $L(\mathcal{H})$. They are called in the physical literature "Tensor operators on $\mathcal{H}$". (A notable exception is the book devoted to this subject, *Irreducible Tensorial Sets*, by V. Fano and G. Racah, Academic Press, New York (1959)). *By definition*, for physicists, an "E_i*-tensor operator*" *is a* G*-morphism* (*or intertwining operator*) *from* E_i *to* $L(\mathcal{H})$. If the representation of G on E_i is irreducible, then the corresponding G-morphism is called in physics an "irreducible tensor operator". If G acts trivially on E_i then we have "scalar tensor operator". (Just try to remember that tensor operators on $\mathcal{H}$ are not operators on $\mathcal{H}$!)†

It is time to specify the field K. Generally, of course, it is the field of complex numbers since $\mathcal{H}$ is a complex Hilbert space. However, reality is also essential in physics. So often E is a real vector space and the "E-tensor operator" is a G-homomorphism T of real vector spaces from E to the real vector space of self-adjoint operators on $\mathcal{H}$. Of course it is always possible later to enlarge the field from $\mathbb{R}$ to $\mathbb{C}$.

When G is a Lie group we consider, of course, only continuous differentiable representations so a G-vector space is also a $\underline{g}$-module for the Lie algebra $\underline{g}$ of G. We denote G the vector space of $\underline{g}$. Among the G-tensor operators on $L(\mathcal{H})$ there is a particular one F' which is also the Lie algebra representation of $\underline{g}$ on $\mathcal{H}$. When the representation of G on $\mathcal{H}$ is unitary, then $F = iF'$ has self-adjoint operators for images which satisfy

---

† "Scalar" is often used by physicists for "invariant"!

$$[F(a),F(b)] = (F(a)F(b) - F(b)F(a)) = iF(a \wedge b) \quad . \tag{1.6}$$

When G is respectively the group of rotations, space translations, time-translations, etc., F corresponds respectively to the observables; angular-momentum, momentum, energy, ... In the technical sense of 1.1, what we called observables there, are the elements of the image of F, i.e., for instance, the component of the angular momentum or of the momentum, in a given direction. But I hope it is by now clear that G-morphisms on $L(\mathcal{H})$ are what correspond to the physical concepts with a tensorial character with respect to a group G (other examples: velocity, magnetic moment, electric quadrupole moment, energy-momentum tensor, tensor of inertia, etc.).

Let R and U (unitary) be the representations of G respectively on E and on $\mathcal{H}$. By definition of the E-tensor operator T

$$\forall x \in E,\ \forall g \in G,\quad U(g)T(x)U^{-1}(g) = T(R(g)x) \quad . \tag{1.7}$$

If D and $F' = iF$ are the corresponding representations of the Lie algebra

$$D(a) = \frac{d}{d\alpha} R(e^{\alpha a})\Big|_{\alpha=0} \ ; \ iF = \frac{d}{d\alpha} U(e^{\alpha a})\Big|_{\alpha=0} \tag{1.8}$$

then an equivalent definition of the E-tensor operator T is

$$\forall x \in E,\ \forall a \in \underline{g},\ [F(a),T(x)] = iT(D(a)x) \quad . \tag{1.9}$$

In a nutshell, I would say that much of the application of group theory to quantum physics consist in the study of the "tensor-operators" on the G-vector (Hilbert) space $\mathcal{H}$ of a physical system. They form a ring† (and an algebra). Let T_1 and T_2 be respectively E_1 and E_2-tensor operators on $\mathcal{H}$, then

$$E_1 \oplus E_2 \ni x \oplus y \leadsto T_1(x) + T_2(y) \quad \text{defines a } E_1 \oplus E_2\text{-tensor operator}$$

$$E_1 \otimes E_2 \ni x \otimes y \leadsto T_1(x)T_2(y) \quad \text{defines a } E_1 \otimes E_2\text{-tensor operator}$$

that we denote respectively $T_1 \oplus T_2$ and $T_1 \otimes T_2$. The latter is generally reducible and can be decomposed into a direct sum of irreducible "tensor operators".

I believe that many problems arise which have not been systematically studied by physicists although they work very much with this ring (for fixed G, $\mathcal{H}$ and action of G on $\mathcal{H}$).

For instance, if G is simple, and T is a G-tensor operator and $\forall x, y \in G$, $[T(x),T(y)] = 0$. I believe this implies $\dim \mathrm{Hom}(G,\mathcal{H})^G$ is infinite.††

Of course the subalgebra generated by an element is well known; given an E-tensor operator T there is a functorial G-morphism $\hat{T}$ from the tensor algebra

† For infinite dimensional $\mathcal{H}$, the operators T(x) are not bounded so their product is not always well defined. I will forget here this difficulty which has to be faced in quantum mechanics and is considered in O'Raifeartaigh's lectures.

†† C. Moore proved it during the Rencontres.

$T(E)$ on E to $L(\mathcal{H})$, which is moreover an algebra homomorphism. If i is the canonical injection of E into $T(E)(\mathrm{Im}\ i = E^{(1)})$, then the Diagram 2 is commutative.

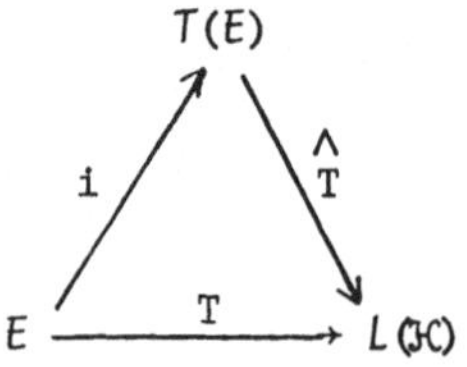

Diagram 2.

In the particular case where T is the representation F (up to a factor i, (see Equation (1.6)) of $\underline{g}$ on $\mathcal{H}$, then it appears also in the representation of $U(G)$, the universal enveloping algebra of $\underline{g}$

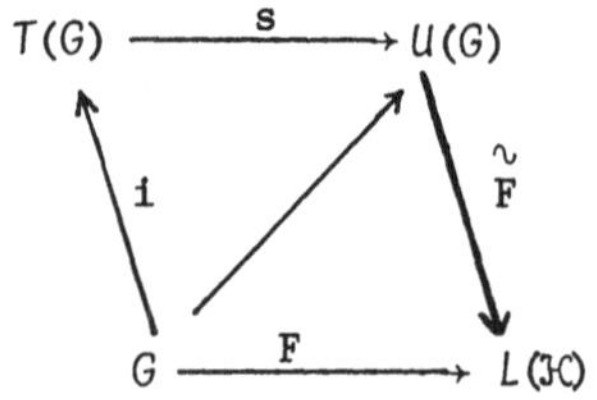

Diagram 3.

$$\hat{F} \text{ (in diagram 2) } = \tilde{F} \cdot s \quad .$$

A remarkable "scalar tensor operator" is the Casimir† operator.

Let G be a semi-simple Lie group. Let $a \rightsquigarrow D(a)$ the adjoint representation of the Lie algebra $\underline{g}$ on its vector space G

$$D(a)b = a \wedge b, \; [D(a),D(b)] = D(a \wedge b) \quad . \tag{1.10}$$

The symmetrical bilinear Cartan-Killing form

$$\beta(a,b) = \mathrm{Tr} D(a) D(b) \tag{1.11}$$

is non-degenerate. Therefore, it defines a G-isomorphism i' between G and its dual G'. This also defines the isomorphism $i' \otimes I$, (I = identity)

$$G \otimes G \xleftarrow{i' \otimes I} G' \otimes G \xleftarrow{j} \mathrm{Hom}(G,G) \quad .$$

The well known canonical homomorphism j is also a G-homomorphism. The identity operator 1 on G is an invariant G-vector $\in \mathrm{Hom}\,(G,G)^G$. So

$$c = (i' \otimes I) \cdot j(1) \tag{1.12}$$

is an invariant vector of $E \otimes E \subset T(E)$ with a fixed normalisation and $\hat{F}(c)$ is the Casimir operator on $\mathcal{H}$.

† Casimir is a physicist with a high position in Philips.

It occurs that neither physicists, nor some mathematicians (cf., Bourbaki, *Groupes et Algèbres de Lie*, Chapter I §3 No. 7) use this canonical normalization for c. In the physics literature nowadays, the images by $\tilde{F}$ of a set of algebraically independant elements of the center of $U(\underline{g})$ are called "the Casimir operators".

In order to induce physicists to use the more canonical point of view exposed here, let us end this section by a very simple theorem proven elaborately in particular cases in the physics literature.

Theorem

If G has no non-trivial one-dimensional representation, and if T is for G a non-invariant irreducible E-tensor operator on $\mathcal{H}$, a finite dimensional space, then $\forall a \in E$, tr T(a) = 0. Indeed, the field ($\mathbb{C}$ for instance) is a trivial one-dimensional G-vector space, and "trace" $\in \mathrm{Hom}(L(\mathcal{H}),\mathbb{C})^G$ since $T \in \mathrm{Hom}(E,L(\mathcal{H}))^G$, then "trace T" = "trace" $\circ$ T $\in \mathrm{Hom}(E,\mathbb{C})^G = 0$ by our hypothesis.

1.4. Unitary Groups U(n) and Permutation Groups $S(n)$†

We have to survey briefly some results on irreps of U(n) and $S(n)$ that we shall use very much in these lectures. The irreps of $S(n)$ can be labeled by integer partitions of n

$$[\lambda_1^{\alpha_1} \ldots \lambda_i^{\alpha_i} \ldots \lambda_k^{\alpha_k}] \quad \text{with} \quad \lambda_1 > \lambda_2 \ldots \lambda_k > 0$$

and

$$\Sigma_{i=1}^k \alpha_i \lambda_i = n \quad .$$

There is a more picturesque notation of $[\lambda_1^{\alpha_1} \ldots \lambda_k^{\alpha_k}]$ which is an ideogram made with n small squares, α_1 lines of λ_1 squares, α_2 lines of λ_2 squares, etc. and called a Young diagram.

Example of $[\lambda_i^{\alpha_i}]$

$\lambda_1 = 9,\ \alpha_1 = 1,\ \lambda_2 = 5,\ \alpha_2 = 3$

$\lambda_3 = 3,\ \alpha_3 = 2,\ \lambda_4 = 1,\ \alpha_4 = 1$

$n = 9 + (3 \times 5) + (2 \times 3) + 1 = 31$

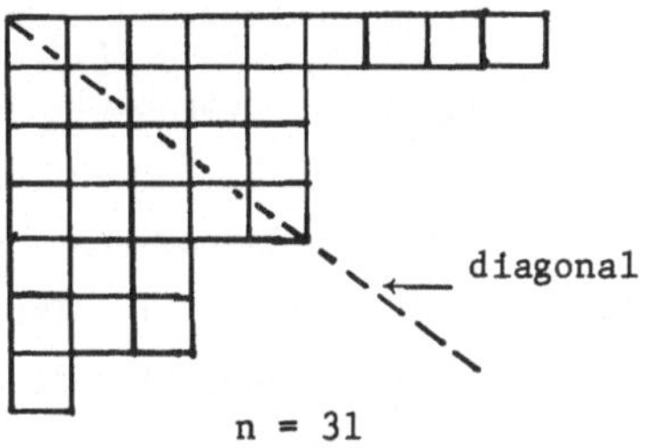

n = 31

The Young diagrams contain a qualitative information, the more horizontal (vertical) the diagram the more symmetrical (antisymmetrical) are the vectors of the representations.

---

† For an exposition of the representations of U(n) and $S(n)$ see Weyl's book, Chapter V. A survey for the needs of physicists has been made by C. Itzykson and M. Nauenberg, *Rev. Mod. Phys.*, 38, 95 (1966).

There are only two one-dimensional irreps of $S(n)$

$[n]$ =  completely symmetrical
n squares

$[1^n]$ =  n squares completely antisymmetrical.

The representations of $S(n)$ are self contragredient. To each irrep of $S(n)$ we can associate a complementary representation

$$[\lambda_i^{\alpha_i}]^c = [\hat{\lambda}_i^{\hat{\alpha}_i}] \quad \text{with} \quad \hat{\lambda}_i = \Sigma_{j=1}^{k+1-i}\alpha_j, \; \hat{\alpha}_i = \lambda_{k-i+1} - \lambda_{k-i+2} \; . \tag{1.13}$$

Its Young diagram is simply obtained by a symmetry through the diagonal.

We recall that the tensor product of two irrep contains $[n]$ (resp., $[1^n]$) only if the two irreps are equivalent (resp. complementary) and then it contains $[n]$ (resp., $[1^n]$) once only.

We will also use a shorter symbol $[\;]_\lambda$ for a linear unitary representation of $S(n)$.

We call factorial a group representation which is direct sum of equivalent irreps.

Let $\mathcal{H}^{(1)}$ be a Hilbert space and

$$\mathcal{H}^{(n)} = \overset{n}{\otimes}\, \mathcal{H}^{(1)} = \mathcal{H}^{(1)} \otimes \ldots \otimes \mathcal{H}^{(1)} \otimes \mathcal{H}^{(1)} \quad (n \text{ factors}). \tag{1.14}$$

By permutation of the factors, $S(n)$ acts linearly on $\mathcal{H}^{(n)}$ through a representation that we denote $[\;]_{\mathcal{H}^{(n)}}$ and which can be decomposed canonically into factorial representation. We denote $\mathcal{H}^{(n)}_{[\;]_\lambda}$ the subspace of $\mathcal{H}^{(n)}$ on which acts the factorial representation $\oplus\, [\;]_\lambda$.

For instance $\mathcal{H}^{(n)}_{[1^n]}$, also denoted $\overset{n}{\wedge}\, \mathcal{H}^{(1)}$, and $\mathcal{H}^{(n)}_{[n]}$, also denoted $\overset{n}{\vee}\, \mathcal{H}^{(1)}$, are the spaces of completely antisymmetrical and symmetrical rank-n tensors on $\mathcal{H}^{(1)}$.

Let us assume that $\dim \mathcal{H}^{(1)} = k$ is finite. Then $U(k)$ acts on $\mathcal{H}^{(1)}$ and on $\mathcal{H}^{(n)}$ by $\overset{n}{\otimes}\, U(k)$. The decomposition of this linear representation of $U(k)$ on $\mathcal{H}^{(n)}$ into factorial representations yields the <u>same</u> subspaces $\mathcal{H}^{(n)}_{[\;]_\lambda}$. One can therefore denote by the same symbols $[\;]_\lambda$ the corresponding irreps of $U(k)$.

To summarize:

$$\text{for } S(n),\; [\;]_{\mathcal{H}^{(n)}} \sim \oplus_\lambda u_\lambda [\;]_\lambda \tag{1.15}$$

$$\text{for } U(k),\; [\;]_{\mathcal{H}^{(n)}} \sim \oplus_\lambda s_\lambda [\;]_\lambda \tag{1.15'}$$

where u_λ = dimension of the irrep $[\ ]_\lambda$ of $U(k)$, s_λ = dimension of the irrep $[\ ]_\lambda$ of $S(n)$.

The above is a theorem which is a leit-motiv of Weyl's book quoted in the introduction and is implicit in the two other books. When $n > k$, irreps of $U(k)$ are labeled only by those partitions of n such that $\Sigma\alpha_i \leq k$, i.e., the Young diagrams of irreps of $U(k)$ have at most k lines, but an arbitrary number n of squares; $n = 1$, $\square$ corresponds to the k-dimensional (= fundamental) representation of $U(k)$ and $n = 0$, "•" to the trivial (= one-dimensional) representation. For example, the representations of $U(2)$ are labeled $[\lambda_1,\lambda_2]$ with the integers $\lambda_1 \geq \lambda_2 \geq 0$.

The diagram of the contragredient representation of $[\lambda_1^{\alpha_1}]$ is $[\lambda_i^{\alpha_i'}]$ with $\Sigma\alpha_i < k$, $\alpha_1' = k - \Sigma_1^p\alpha_i$, $\lambda_1' = \lambda_1 \alpha_j' = \alpha_{p+2-j}$, $\lambda_i' = \lambda_i - \lambda_{p+1-i}$ if $i,j > 1$. Of course it is shorter to say that the Young diagram of $[\lambda_i'^{\alpha_i'}]$ is the complement (upside down) of that of $[\lambda_i^{\alpha_i}]$ in the rectangle of k lines of λ_1 squares.

<u>SU(n) Representations</u>. The restriction of an irrep of $U(k)$ to the subgroup $SU(k)$ of $U(k)$ is an irrep of $SU(k)$. Irreps of $U(k)$ whose Young diagrams differ only on the left by a rectangular block of columns of length k yield by restriction equivalent irreps of $SU(k)$. Taking into account this remark, one unambiguously label irreps of $SU(k)$ by Young diagrams. Moreover, this yields all inequivalent irreps of $SU(k)$.

<u>Example</u>. The equivalence classes of irrep of $SU(2)$ obtained by restriction of the irreps $[\lambda_1,\lambda_2]$ of $U(2)$ are given by the value of the integer $\lambda_1 - \lambda_2$. So their Young diagram can be written as a horizontal line of $\lambda_1 - \lambda_2$ squares. For $SU(2)$ irreps it is customary to use the symbol

$$D_j \quad \text{with} \quad j = \frac{1}{2}(\lambda_1 - \lambda_2) \tag{1.16}$$

where j is called the spin of the representation. $2j + 1$ is its dimension. The Casimir operator of D_j, as normalized by physicists, is $j(j + 1)\mathbf{1}$, which is twice that defined by (1.9). (Indeed, physicists take as Cartan-Killing form $1/2\ \mathrm{Tr}D(a)D(b)$).

We also recall the well known decomposition

$$D_{j_1} \otimes D_{j_2} = \bigoplus_{j=|j_1-j_2|}^{j_1+j_2} D_j \tag{1.17}$$

Note that all representations of $SU(2)$ are self contragredient.

<u>Representations of the Adjoint Groups $SU(n)/Z_n$</u>. The center of $SU(k)$ is Z_k, the cyclic group of k elements, so the adjoint group of $SU(k)$ is $SU(k)/Z_k$.

The representations of this group are those of SU(k) whose Young diagram has a number of squares multiple of k. For example, $SU(2)/Z_2 = SO(3)$. Its representations are D_j with integral j. Their Young diagrams contain only one line of an even number ($= 2_j$) of squares.

The adjoint representation of SU(n) or its adjoint group is that on the space of its Lie algebra, it has dimension $n^2 - 1$ and label $[2, 1^{n-2}]$; it is equivalent to its contragredient.

Remark For Any Group. For any group G, let $\mathcal{H}^{(1)}$ be the space of a linear unitary representation (it may be reducible and $\dim \mathcal{H}^{(1)}$ may be infinite). As we saw $S(n)$ and G acts on $\mathcal{H}^{(n)} = \overset{n}{\otimes} \mathcal{H}^{(1)}$. Subspaces $\mathcal{H}^{(n)}_{[\ ]_\lambda}$ of primary representations of $S(n)$ are not in general subspaces of primary representation of G. Methods for knowing the nature of the G-representation of the different $\mathcal{H}^{(n)}_{[\ ]_\lambda}$ would be interesting for the physicists, especially in some case, for $\mathcal{H}^{(n)}_{[n]}$ (bosons) and $\mathcal{H}^{(n)}_{[1^n]}$ (fermions). Here is an example of a result, proven by A. Bohr, *Mat. Fys. Medd. Dan. Vid. Selsk*, 26 (No. 14), 16.

G is SO(3), $\mathcal{H}^{(1)}$ is the five-dimensional Hilbert-space of the representation D_2. For any n the representation of SO(3) on $\mathcal{H}^{(n)}_{[n]} = \overset{n}{\vee} \mathcal{H}^{(1)}$ does not contain D_1 in its reduction into direct sum of irreps. (Physically, a nucleus with spin 0 ground state has no spin 1 state corresponding to collective excitations.)

Of course we also can add that if an irrep of G appears on $\mathcal{H}^{(n)}$ only once, then it acts either on $\mathcal{H}^{(n)}_{[n]}$ or on $\mathcal{H}^{(n)}_{[1^n]}$.

1.5. More Algebras and More Tensor Operators. Pseudo Roots of SU(n)

An algebra on the vector space E is an element of $\mathrm{Hom}(E \otimes E, E)$. The algebra is symmetrical, (respectively, antisymmetrical) if it is an element of $\mathrm{Hom}(E \vee E, E)$, (resp., $\mathrm{Hom}(E \wedge E, E)$). Similarly we can define a co-algebra, symmetrical, antisymmetrical co-algebra as an element of $\mathrm{Hom}(E, E \otimes E)$, $\mathrm{Hom}(E, E \vee E)$ or $\mathrm{Hom}(E, E \wedge E)$.

If E is the space of a linear representation of G, elements of $\mathrm{Hom}(E \otimes E, E)^G$, resp., $\mathrm{Hom}(E, E \otimes E)^G$ are algebras, resp., co-algebras, whose group of automorphisms contains G.

When G is a semi-simple compact Lie group, a necessary condition for $\dim \mathrm{Hom}(E \otimes E, E)^G > 0$ is that the representation on E has a null weight. For example, for the space G of the adjoint representation, $\dim \mathrm{Hom}(G \wedge G, G)^G = 1$ for all simple compact Lie groups and the corresponding antisymmetrical algebra is the Lie algebra itself.

In V we shall see two examples of symmetrical algebras uniquely defined on a real irrep space E of $G = SU(3) \times SU(3)$, with $\dim \mathrm{Hom}(E \vee E, E)^G = 1$. For the adjoint representation of a simple compact Lie algebra $\dim \mathrm{Hom}(G \vee G, G)^G = 0$ or 1. It has the latter value for the $SU(n)$, $n > 2$. Let us give some properties of this symmetrical algebra of $SU(n)$, since it has been used in the physics literature of elementary particles, after its introduction by Gell-Mann. What follows is extracted from a preprint written in collaboration with L. A. Radicati.

Let G_n be the $n^2 - 1$ real vector space of the $n \times n$ traceless hermitian matrices x. The action of $u \in SU(n)$ on G_n (vector space of the Lie algebra) is $x \overset{u}{\leadsto} uxu^{-1} = uxu^*$. The euclidean scalar product

$$(x,y) = \frac{1}{2} \text{ trace } xy \tag{1.18}$$

is invariant (= $1/n$ the Cartan-Killing bilinear form). The SU(n) Lie algebra law is

$$x \wedge y = -\frac{i}{2}(xy - yx) \equiv -\frac{i}{2}[x,y] \tag{1.19}$$

and the symmetrical algebra law is†

$$x \vee y = \frac{1}{2}\{x,y\} - \frac{2}{n}(x,y)\mathbb{1} \quad \text{where} \quad \{x,y\} = xy + yx \quad . \tag{1.19'}$$

Note that for $n = 2$ it is trivial: $x \vee y = 0$.

In the physics literature (mainly for $n = 3$) one introduces an orthonormal basis $(e_i,e_j) = \delta_{ij} (i,j = 1,\dots,n^2 - 1)$ and uses traditionally the notation f_{ijk}, d_{ijk} for the structure constants

$$e_i \wedge e_j = \Sigma_k f_{ijk} e_k, \; e_i \vee e_j = \Sigma_k d_{ijk} e_k \quad .$$

Let us use F(a), D(a) for the linear mappings of G

$$F(a)x = a \wedge x, \; D(a)x = a \vee x \tag{1.20}$$

(the matrices are $F(e_j)_{ik} = f_{ijk}$, $D(e_j)_{ik} = d_{ijk}$).

With the scalar product (1.18), F(a) is antisymmetric and D(a) is symmetric. D and F are tensor-operators $\in \mathrm{Hom}(G, L(G))^{SU(n)}$ so from the theorem at the end of 1.3, trace $D(a) = 0$. As is well known, in the Lie algebra SU(n), the centralizer of an element x, i.e., the set $\{y, y \wedge x = 0\}$ is a Lie subalgebra of dimension $\geq n - 1$. When its dimension is $n - 1$ it is abelian and it is called the Cartan subalgebra C_x of x. (All Cartan subalgebras are transformed into each other by the group.) C_x is spanned by the $n - 1$ linearly independent vectors x, $x \vee x$, $(x \vee x) \vee x = x \vee (x \vee x)$, $((x \vee x) \vee x) \vee x$, etc., up to $n - 1$ factors and C_x is also a subalgebra for the law " $\vee$ ". The <u>roots</u> of SU(n) are solutions of the equation $r^n - (r,r)r^{n-2} = 0$. We shall normalize them by $(r,r) = 1$. In a

---

† This is not a Jordan algebra. However, one could have started from the n^2 dimensional representation realized by the $n \times n$ hermitian matrices. The corresponding symmetrical algebra is a Jordan algebra.

Cartan algebra C, there are $n(n-1)$ normalized roots r_k, (if r is a root, $-r$ is also one), for every $a \in C$, the spectrum of $F(a)$ has $n-1$ zeros for the eigen space C and on the orthogonal space $C^{\perp}$ the spectrum is the set

$$\text{Spectrum } F(a)\big|_{C^{\perp}} = \{i(a,r_k)\} \quad . \tag{1.21}$$

Define (for $n > 2$)

$$\sqrt{\frac{n-2}{n}}\, q_k = r_k \vee r_k = (-r_k) \vee (-r_k) \tag{1.22}$$

then

$$(q_k,q_k) = 1 \tag{1.22'}$$

and they are idempotents of the $\vee$-algebra

$$q_k \vee q_k = \frac{n-4}{\sqrt{n(n-2)}}\, q_k \quad . \tag{1.23}$$

We will call them "pseudo roots" (they are weights of $SU(n)$) for they satisfy for every $a \in C$

$$\text{Spectrum } D(a)\big|_{C^{\perp}} = \{\frac{n-2}{n}(q_k,a) = (a,r_k \vee r_k)\} \tag{1.24}$$

(all the eigen values have at least multiplicity 2).

Let us denote by $\lambda \in \mathrm{Hom}(G \wedge G,G)^{SU(n)}$, $\nu \in \mathrm{Hom}(G \vee G,G)^{SU(n)}$ the vector space homomorphisms

$$\lambda(x \otimes y) = x \wedge y, \quad \nu(x \otimes y) = x \vee y$$

and consider their right inverse

$$\lambda \circ \lambda' = \text{Identity on } G, \quad \nu \circ \nu' = \text{Identity in } G \quad . \tag{1.25}$$

Note that λ and λ' can be defined for any semi-simple Lie algebra. As we said λ' and ν' define co-algebras on G. If T is a G-tensor operator, using the mapping $\hat{T}$ of Diagram 2, one can define the G-tensor operators

$$T \wedge T = \hat{T} \circ \lambda' \quad \text{and} \quad T \vee T = \hat{T} \circ \nu' \tag{1.26}$$

and by recursion

$$(\ldots(T \,\tau_1\, T)\,\tau_2\, T)\ldots)\,\tau_k\, T \tag{1.26'}$$

where "τ_i" is either " $\wedge$ " or " $\vee$ ". For physicists who need to see coordinates, in $SU(3)$ octet space

$$\lambda'(e_i) = \Sigma_{j,k} -\frac{1}{3} f_{ijk} e_j \otimes e_k, \quad \nu'(e_i) = \Sigma_{jk} \frac{3}{5} d_{ijk} e_j \otimes e_k \quad .\dagger$$

† The f_{ijk} and d_{ijk} are the structure constants introduced by Gell-Mann.

If we set $T(e_i) = T_i$, then

$$(T \wedge T)_i = \Sigma_{j,k} - \frac{1}{3} f_{ijk} T_j T_k, \quad (T \vee T)_i = \Sigma_{jk} \frac{3}{5} d_{ijk} T_j T_k \quad .$$

Note of course that we can define $T \wedge T$, $T \vee T$ for any real irreducible E-tensor operator when G is a compact group when $\dim \mathrm{Hom}(E \top E, E)^G = 1$ ($\top$ is $\wedge$ or $\vee$) since there is the irrep of G on E is orthogonal and leaves invariant a euclidean scalar product. Indeed, λ or ν are then surjective and are isomorphisms between $(\mathrm{Ker}\ \lambda)^{\perp}$ and E (resp., $(\mathrm{Ker}\ \nu)^{\perp}$ and E) so we can define their right inverse.

Let us consider the more particular case when the G-morphism T is F itself (see Equation (1.6)), i.e., the representation (up to the factor i) of the Lie algebra on $\mathcal{H}$. Then $F \wedge F = iF$. When SU(3) is used for elementary particles, $F \vee F$ is often called the D-coupling operator (see 5.1b). For SU(2), in order to follow the tradition started in elementary school, we denote by $\times$ the Lie algebra law (= vector product)

$$[F(\vec{a}), F(\vec{b})] = iF(\vec{a} \times \vec{b}) \tag{1.27}$$

and by ε_{ijk} the structure constants

$$e_i \times e_j = \Sigma_k \varepsilon_{ijk} e_k \quad . \tag{1.28}$$

So if $\vec{A}$ is a vector operator (with $A(e_i) = A_i$

$$(\vec{A} \times \vec{A})_i = \sum_{jk} \varepsilon_{ijk} A_j A_k = \frac{1}{2} \sum_{jk} \varepsilon_{ijk} [A_j, A_k] \tag{1.29}$$

<u>Remark</u>. Given two G-tensor operators A and B, we can also define

$$A \vee B = A \otimes B \circ \nu' \, , \quad A \wedge B = A \otimes B \circ \lambda'$$

and in particular $\vec{A} \times \vec{B}$. This reduces to Equation (1.26) when $A = B$.

<u>1.6. More on SU(2) and its Tensor Operators</u>

For SU(2) the symmetrical algebra $\vee$ on the adjoint representation G is trivial, $\dim(G \otimes G, G)^{SU(2)} = \dim(G \wedge G, G)^{SU(2)} = 1$

Much more generally, given any three irrep on E_{j_1}, E_{j_2}, E_{j_3},

$$\dim \mathrm{Hom}(E_{j_1} \otimes E_{j_2}, E_{j_3})^G = \Delta(j_1, j_2, j_3) = 0 \text{ or } 1 \tag{1.30}$$

where $\Delta(j_1, j_2, j_3) = 1$ if $|j_1 - j_2| \le j_3 \le j_1 + j_2$ (triangular relation), 0 otherwise. This property, under an equivalent formulation, is called the <u>Wigner-Eckart theorem</u> by physicists, and groups with the property (1.30) have been called simply reducible by Wigner.

Let us give here two references that we shall quote often in this section.

A. *Quantum Theory of Angular Momentum* - a collection of reprints and original papers edited by L. C. Biedenharn and H. Van Dam, Academic Press, New York (1965).

B. *Spectroscopic and Group Theoretical Methods in Physics, Racah Memorial Volume,* North Holland, Amsterdam (1968).

In B p. 131-136, Wigner proves the following theorem for finite groups.

Theorem

Let G be a finite group and H a subgroup. The following conditions are equivalent

a) The restriction to H of any irrep of G is multiplicity free when decomposed into irrep of H;

b) The ring of conjugation classes by H of elements of G is abelian.

Let us explain a) and b) in more detail.

a) Given an irrep of G on $\mathcal{H}$, its restriction as a representation of H is generally reducible. To say that it is multiplicity free means that in its decomposition into irreps of H no such irrep appears more than once. Equivalently, one can say that the commutant of the representation of H (i.e., the set of all bounded elements of $\mathcal{L}(\mathcal{H})$ which commute with every operator of the representation of H; this set is an algebra) is an abelian algebra. That last condition can be used as definition of multiplicity free for any linear representation of any group.

b) Given $a \in G$, the conjugation class of a by H is the set $A = \{hah^{-1}, \forall h \in H\}$. Given two such classes we define as $A \cdot B$ the set $\{ab, a \in A, b \in B\}$. Condition b) states that for any pair of classes, $A \cdot B = B \cdot A$. It seems easy to extend Wigner's proof to compact groups. Examples of pairs of group and subgroup which satisfy this theorem: $\mathcal{S}(n)$ and $\mathcal{S}(n-1)$, U(n) and U(n - 1).†

From the group law one verifies that the direct product SU(2) × SU(2) and its diagonal subgroup satisfies b); by Wigner's theorem this implies (1.30). It would be interesting to extend, if possible, Wigner's proof to type I locally compact groups.††

---

† After the lecture, Professor G. Mackey gave a proof for compact groups, using his theory of induced representations.

†† Wigner proved in 1941 (paper reproduced in reference A., see 1.6) for finite groups another property equivalent to a) and b). Let $\zeta(g)$ the number of square roots of g in the finite group G, and $v(g)$ the number of elements of G which commute with g. In a finite group $\sum_{g \in G} v(g)^2 - \zeta(g)^3 \geq 0$. The equality occurs if and only if G is simply reducible.

Another property of SU(2) that we have mentioned is that any irrep of SU(2) is equivalent to its contragredient. For any irrep D_j of SU(2) this defines an isomorphism C: $E_j \overset{C}{\to} E'_j$ between the E_j space of D_j and its dual E'_j with the canonical property

$$E_j \overset{C}{\to} E'_j; \; C^T = (-1)^{2j}C \tag{1.31}$$

for the transposed C^T of C. Physicists normalize C by

$$C^T C = 1 \quad . \tag{1.32}$$

We are by now ready to give you a glimpse of the calculus developed independently by Wigner and Racah for the necessity of physics in order to exploit completely rotational invariance in atomic physics (and later on in nuclear physics and many other branches of quantum physics). Many of the numbers observed in atomic spectra (spacing between neighbors in a family of lines, relative intensity of these lines, etc.) turn out to be algebraic functions of the coefficients defined by Wigner and Racah. Since they are so useful, the literature on Wigner and Racah coefficients is abundant. They have been found to possess unexpected symmetries, there are unproven conjectures on them. However, the language of this physics folklore seems to be unknown to the mathematically minded ethnographer.

I hope there will be much discussion on this subject in this Rencontres. The rest of this section should help to start it.

To write Wigner's "three j" coefficients, physicists choose a base in each Hilbert space E_j, carrier of the irrep D_j, for every j. It is formed by the eigen vectors of a chosen U(1) (Cartan subgroup of SU(2)) ordered in terms of decreasing eigen value μ (going from j to $-j$ by integer steps). It is obvious that most of their properties are base independent.

Consider an element of the one-dimensional vector space

$$(E'_{j_1} \otimes E'_{j_2} \otimes E_{j_3})^G = \mathrm{Hom}(E_{j_1} \otimes E_{j_2}, E_{j_3})^G \tag{1.33}$$

and denote it

$$(\overset{\circ}{j}_1 \overset{\circ}{j}_2 \underset{\circ}{j}_3) \quad . \tag{1.34}$$

The isomorphism C and its inverse, defined in (1.31), (1.32), transform the tensor (1.20) into its following images

$$(\overset{\circ}{j}_1 \overset{\circ}{j}_2 \overset{\circ}{j}_3) \in (E'_{j_1} \otimes E'_{j_2} \otimes E'_{j_3})^G = \mathrm{Hom}(E_{j_1} \otimes E_{j_2} \otimes E_{j_3}, \mathbb{C})^G \tag{1.35}$$

$$(\underset{\circ}{j}_1 \underset{\circ}{j}_2 \underset{\circ}{j}_3) \in (E_{j_1} \otimes E_{j_2} \otimes E_{j_3})^G = \mathrm{Hom}(\mathbb{C}, E_{j_1} \otimes E_{j_2} \otimes E_{j_3})^G \tag{1.36}$$

$$(\overset{\circ}{j}_1 \underset{\circ}{j}_2 \underset{\circ}{j}_3) \in (E'_{j_1} \otimes E_{j_2} \otimes E_{j_3})^G = \mathrm{Hom}(E_{j_1}, E_{j_2} \otimes E_{j_3})^G \tag{1.37}$$

and so on.

Equations (1.35), (1.36) show that $(\overset{\circ}{j}_1\overset{\circ}{j}_2\overset{\circ}{j}_3)$ (resp., $(\underset{\circ}{j_1}\underset{\circ}{j_2}\underset{\circ}{j_3})$) belong to a one-dimensional representation of the permutation group of the three factor spaces labeled by j_1, j_2, j_3. Computation shows that the representation is

$$\begin{array}{lll} \square\square\square \text{ (symmetrical)} & \text{if } j_1 + j_2 + j_3 & \text{is even} \\ \begin{smallmatrix}\square\\\square\end{smallmatrix} \text{ (antisymmetrical)} & \text{if } j_1 + j_2 + j_3 & \text{is odd} \quad . \end{array} \tag{1.38}$$

The composition of the two homomorphisms

$$E_{j_1} \otimes E_{j_2} \otimes E_{j_3} \xrightarrow{(\overset{\circ}{j}_1\overset{\circ}{j}_2\overset{\circ}{j}_3)} \mathbb{C} \xrightarrow{(\underset{\circ}{j_1}\underset{\circ}{j_2}\underset{\circ}{j_3})} E_{j_1} \otimes E_{j_2} \otimes E_{j_3} \tag{1.39}$$

is an element of $\text{Hom}(E_{j_1} \otimes E_{j_2} \otimes E_{j_3}, E_{j_1} \otimes E_{j_2} \otimes E_{j_3})^G$ that we denote $(\overset{\circ}{j}_1\overset{\circ}{j}_2\overset{\circ}{j}_3)$ $(\underset{\circ}{j_1}\underset{\circ}{j_2}\underset{\circ}{j_3})$.

Wigner proved (see reference A and Equation (24.18b) of his book quoted in the Introduction)

$$\int_{SU(2)} D_{j_1}(g) \otimes D_{j_2}(g) \otimes D_{j_3}(g) d\mu(g) = (\overset{\circ}{j}_1\overset{\circ}{j}_2\overset{\circ}{j}_3)(\underset{\circ}{j_1}\underset{\circ}{j_2}\underset{\circ}{j_3}) \tag{1.40}$$

where $d\mu(g)$ is the invariant $SU(2)$ measure of mass $\int_{SU(2)} d\mu(g) = 1.$

This also defines for you, up to a sign, which element of the one-dimensional vector space $(E_1 \otimes E_2 \otimes E_3)^G$ has been chosen by physicists for $(\overset{\circ}{j}_1\overset{\circ}{j}_2\overset{\circ}{j}_3)$.

Of course tensors can have their indices contracted (notation $\times$); for instance $(\overset{\circ}{a}\overset{\circ}{b}\overset{c}{\times})(\overset{\times}{c}\overset{\circ}{p}\overset{\circ}{q})$ is the composed homomorphism

$$E_a \otimes E_b \otimes E_p \otimes E_q \xrightarrow{(\overset{\circ}{a}\overset{\circ}{b}\overset{c}{\circ}) \otimes I_p \otimes I_q} E_c \otimes E_p \otimes E_q \xrightarrow{(\overset{\circ}{c}\overset{\circ}{p}\overset{\circ}{q})} \mathbb{C} \quad .$$

Wigner's notation is very handy!

Note that equation (1.40) yields

$$\int_{SU(2)} \chi_{j_1}(g)\chi_{j_2}(g)\chi_{j_3}(g) d\mu(g) = (\overset{\times}{j}_1\overset{\times}{j}_2\overset{\times}{j}_3)(\overset{j_1}{\times}\overset{j_2}{\times}\overset{j_3}{\times}) = \Delta(j_1,j_2,j_3) \tag{1.41}$$

where χ_j is the character of D_j.

Approximate expression, asymptotic expression, for large j's exist for the components of those tensors in the basis described above (see the thorough bibliography of reference A (see 1.6)). Regge (paper reproduced in A (see 1.6)) has found a 72 elements group of symmetry $\sim \text{Aut}(S(3) \times S(3))$ for the set of components

$\binom{\mu_1\mu_2\mu_3}{j_1j_2j_3}$ of $(\overset{\circ}{j_1}\overset{\circ}{j_2}\overset{\circ}{j_3})$.†

In 1941 Racah and Wigner (both papers reproduced in A (see 1.6)) introduced a "six j" symbol (numerical function of six irrep of SU(2)), also known as recoupling coefficient. It is canonical.

Consider the sequence of SU(2)-homomorphisms

$$E_e \xrightarrow{\binom{\circ fa}{e\circ\circ}} E_f \otimes E_a \xrightarrow{I_f \otimes \binom{\circ bc}{a\circ\circ}} E_f \otimes E_b \otimes E_c \xrightarrow{\binom{\circ\circ d}{fb\circ} \otimes I_c}$$

$$E_d \otimes E_c \xrightarrow{\binom{\circ\circ e}{ad\circ}} E_e \quad . \tag{1.42}$$

Since E_e is the carrier of an irrep, this SU(2)-homomorphism must be a multiple of the identity operator on E_e.

Its trace defines (up to a sign that I do not guarantee here) the six j's symbol

$$\begin{Bmatrix} abc \\ def \end{Bmatrix} = (-1)^{b+c-d+e+f} \binom{\overset{\times}{e}fa}{\underset{\times\times}{}}\binom{\overset{\times}{a}bc}{\underset{\times\times}{}}\binom{\overset{\times\times}{fb}d}{\underset{\times}{}}\binom{\overset{\times\times}{ac}e}{\underset{\times}{}} \quad . \tag{1.43}$$

Wigner has shown that for given a, b, d, e, $\left\{\begin{smallmatrix} abc \\ def \end{smallmatrix}\right\}$ is an orthogonal matrix with indices c, f. He also proved the relation (in his book, Chapter 24)

$$\begin{Bmatrix} abc \\ def \end{Bmatrix}^2 = \iiint \chi_a(r)\chi_b(s)\chi_c(t)\chi_d(st^{-1})\chi_e(tr^{-1})\chi_f(rs^{-1})d\mu(r)d\mu(s)d\mu(t) \quad .$$

Asymptotically its value is a rapidly oscillating function of some variables, but an average over some range of one argument yields, when a, b, c, d, e, f form the edges of a tetrahedron, the asymptotic value

$$\begin{Bmatrix} abc \\ def \end{Bmatrix}^2 \to (24\pi V)^{-1}$$

where V is the volume of the tetrahedron.

G. Ponzano and T. Regge (in reference B (see 1.6), first paper) have conjectured precise asymptotic formulae for $\left\{\begin{smallmatrix} abc \\ def \end{smallmatrix}\right\}$ whether or not the value of the arguments can be the length of the edges of a tetrahedron.

Furthermore, Regge has found (paper reproduced in A (see 1.6)) the largest linear group acting on the Z-module generated by the symbols a/2, b/2, c/2, d/2, e/2, f/2 and having $\left\{\begin{smallmatrix} abc \\ def \end{smallmatrix}\right\}$ as invariant. It is the group $S(3) \times S(4)$ which includes the permutation group of the columns.

---

† Most of these symmetries appear naturally (see Bargmann's paper in A (see 1.6)); for the others see G. Flamand, *Ann. Inst. H. Poincaré*, 7, 353 (1967).

Neatest and very symmetrical expressions for the $(j_1 j_2 j_3)$ and $\begin{Bmatrix} j_1 j_2 j_3 \\ j_4 j_5 j_6 \end{Bmatrix}$ symbols can be found in the paper of V. Bargmann (last paper reproduced in book A (see 1.6)) who uses Hilbert spaces of analytic functions as spaces of SU-2 irreps.

2. ATOMIC AND MOLECULAR PHYSICS

2.1. Group Theory and Atomic Physics

The application of group theory to atomic physics is essentially of this type; only the even part $f_+(\vec{r}) = 1/2(f(\vec{r}) + f(-\vec{r})$ (respectively, the symmetric part $f_+(\vec{r}_1,\vec{r}_2) = 1/2(f(\vec{r}_1,\vec{r}_2) + f(\vec{r}_2,\vec{r}_1))$ of $f(\vec{r})$ (resp., $f(\vec{r}_1,\vec{r}_2)$) contributes to the integral over the whole space $\int f(\vec{r})d^3\vec{r}$ (resp., $\int f(\vec{r}_1,\vec{r}_2)d^3\vec{r}_1 d^3\vec{r}_2$). This is the explanation of two empirically known facts (before 1926), the Laporte selection rule for atomic spectra and the partition of the helium spectrum into two independent subsets (attributed to ortho and parahelium). Of course, these examples are the simplest because they are based on invariance under a two element group (Z_2). We will have to use invariance under $SO(3)$, $\mathcal{S}(n)$ and $U(2)$ for atoms and invariance under subgroups of $SO(3)$ for molecules.

2.2. The Correspondence Principle

We had a general description of quantum mechanics, but now we have to know how to study a given physical system. There does not seem to exist an axiomatic formulation of the question, so here again, physics is still an art! However, when the system under consideration, with a finite number of degrees of freedom can be described by classical Hamiltonian mechanics, the "correspondence principle" tells physicists how to treat it quantum-wise.

Let $h(p_k, q_e)$ be the classical Hamiltonian and

$$\frac{dp_k}{dt} = \dot{p}_k = -\frac{\partial h}{\partial q_k}, \quad \dot{q}_\ell = \frac{\partial h}{\partial p_\ell}$$

the Hamiltonian equations. The corresponding observables P_ℓ, Q_ℓ in quantum mechanics form the abstract algebra with unit

$$P_k Q_\ell - Q_\ell P_k = [P_k, Q_\ell] = -i\hbar\delta_{k\ell}\mathbb{1}$$

$$[P_k, P_\ell] = 0 = [Q_k, Q_\ell] \tag{2.1}$$

where $2\pi\hbar$ is Planck's constant.

In the cases we shall study, h is a sum $h = h' + h''$ where h' is a function of the p's and h'' is a function of the q's. Then $H = H' + H''$ where H' and H'' are the same functions respectively of the P's and the Q's. There is yet no synthetic formulation of quantum mechanics as there is for classical mechanics by symplectic manifolds (see, however, work in progress by Kostant, Souriau). We also know that the relations between the classical and quantum treatment of the same problem are not simple (see e.g., Van Hove's work in 1951 comparing the two automorphism groups).

The Hamiltonian operator is the generator of the group of time translations

$$[H,Q_k] = i\hbar\dot{Q}_k, \; [H,P_\ell] = i\hbar\dot{P}_\ell \quad . \tag{2.2}$$

A representation of the algebra defined by (2.1) and (2.2) was obtained, independently of Heisenberg's work by Schrödinger, using the concept of de Broglie's waves. Indeed, the algebra (2.1) is realized by self-adjoint operators of $L(\mathcal{H})$ where $\mathcal{H}$ is the Hilbert space of square integrable functions $\Psi(q_i)$. Then

$$Q_k\Psi = q_k\Psi, \; P_\ell\Psi = \frac{\hbar}{i}\frac{\partial}{\partial q_\ell}\Psi \quad . \tag{2.3}$$

The Ψ are also functions of the time (t) and the Schrödinger equation is

$$H = i\hbar\frac{\partial}{\partial t} \quad . \tag{2.4}$$

This representation raises some analysis problems. On the other hand, von Neumann's theorem (J. von Neumann: "Die Eindeutigkeit der Schrödingerschen Operatoren", *Math. Annalen*, 104, 570 (1937)) tells us that all irreducible representations of the algebra defined by Equation (2.1) are equivalent when e^{iP_k}, e^{iQ_ℓ} are realized by unitary operators.†

Quantum mechanics was also discovered by Dirac who gave the neatest formulation of the "correspondence principle".†† In classical Hamiltonian mechanics one has also a Lie algebra, that of the Poisson brackets (P.B). Let f, g be two functions of the p's and the q's,

$$P.B.(f,g) = \Sigma_\ell \frac{\partial f}{\partial q_\ell}\frac{\partial g}{\partial q_\ell} - \frac{\partial f}{\partial q_\ell}\frac{\partial g}{\partial p_\ell} \quad . \tag{2.5}$$

† For systems with an infinite number of degree of freedom, as they appear in statistical mechanics and field theory, this is far from true. Infinities of irreducible representations of (2.2) have been given first by Friedrichs, Van Hove, Gärding and Wightman, Segal and several other physicists and mathematicians. An excellent thin book on the subject is by A. Guichardet, *Algèbres d'Observables Associées aux Relations de Commutation*, Armand Colin, Paris, (1969). (See also G. Mackey, *Duke Math. J.*, 16, 313 (1949)).

†† Historically, the expression "correspondence principle" had a more restricted meaning.

The Lie algebra of the corresponding quantum observable is

$$[F,G] = i\hbar \quad \text{quantum observable of } \text{P.B.}(f,g) \quad . \tag{2.5'}$$

As you surely know $|\Psi|^2 \prod_k dq_k$, where Ψ is a solution of Schrödinger Equation (2.4), is the density of probability to find the system at the coordinate $\{q_k\}$. This of course appeals very much to physicists. As mathematicians you will like just as well to work with the abstract algebra. As a short, but fundamental illustration of the use of that algebra, let us prove the Heisenberg uncertainty relations.

Let A, B be the self adjoint operators corresponding to the observables a, b. If x> is a given state of the physical system we study, we have seen that <xAx> is the expectation value of "a" for x> and the mean square dispersion of probability is given by

$$(\Delta a)_x = |\langle x,(A - \langle x,Ax\rangle)^2 x\rangle|^{1/2} = |\langle x,\hat{A}^2 x\rangle|^{1/2} = ||\hat{A}x|| \tag{2.6}$$

where

$$\hat{A} = A - \mathbb{1}\langle xAx\rangle \quad . \tag{2.6'}$$

By Schwarz' inequality

$$(\Delta a)_x(\Delta b)_x = ||\hat{A}x|| \cdot ||\hat{B}x|| \geq |\langle \hat{A}x,\hat{B}x\rangle| \geq \frac{1}{2}|\langle x,[A,B]x\rangle| \tag{2.7}$$

If A and B satisfy the same canonical relations as the P's and Q's we do obtain

$$(\Delta a)_x(\Delta b)_x \geq \frac{1}{2}\hbar \quad . \tag{2.8}$$

2.3 Particle of Mass m in a Spherically Symmetric Potential

Let $V(r)$ be a spherical symmetric potential, where r denotes $|\vec{r}|$. The Hamiltonian of the particle is

$$H = \frac{1}{2m}\vec{P}^2 + V(r) \quad , \tag{2.9}$$

which is invariant under the orthogonal group $O(3)$. Using the vocabulary of 1.3, H, $\vec{P}^2$, $V(r)$ are "scalar operators"; $\vec{P}$, $\vec{R}$ <u>and</u> $\vec{R} \times \vec{P} = \vec{L}$ are (polar <u>and</u> axial) vector operators. (So we put an arrow on them!) If, $\vec{a}$, $\vec{b}$, etc., are vectors of the three-dimensional vector space E_3 of the adjoint representation of $O(3)$, we should write the canonical commutation relations (2.1)

$$[\vec{P}(\vec{a}),\vec{Q}(\vec{b})] = i\hbar\mathbb{1}\,\frac{1}{2}\,\beta(\vec{a},\vec{b}) = i\hbar\vec{a},\vec{b}\mathbb{1} \quad , \tag{2.10}$$

where the Cartan-Killing form β has been defined in (1.11).†

---

† See also the Appendix on commutation relations at the end of 2.

From (2.10) and the definition by the correspondence principle of the angular momentum operator (see end of 1.5), $\vec{L} = \vec{R} \times \vec{P}$, we obtain

$$[\vec{L}(\vec{a}),\vec{L}(\vec{b})] = i\hbar\vec{L}(a \wedge b) \quad , \tag{2.11}$$

which confirms that the vector-operator representing the angular momentum is the representation (up to i) of the $O(3)$ Lie algebra on the Hilbert space of our problem.

Some physicists write $\vec{P} \cdot \vec{a}$, $\vec{L} \cdot \vec{n}$ for $\vec{P}(\vec{a})$, $\vec{L}(\vec{n})$. But do not be surprised if in all physics text books an orthonormal basis of vectors notations Q_i, P_j, L_k are used for $\vec{Q}(\vec{e}_i)$, $\vec{P}(\vec{e}_j)$, $\vec{L}(\vec{e}_k)$, etc.

The operators corresponding to the observables which are constants of motion generate the algebra $\{H\}'$, the commutant of H. Hence, the equation that one deduces from (2.10) and the definition of $\vec{L}$

$$\vec{a} \in E_3,\ [L(\vec{a}),H] = 0 \quad \text{or symbolically} \quad [\vec{L},H] = 0 \tag{2.12}$$

means both that the Hamiltonian is invariant under rotations and that the angular momentum is a constant of motion.

The Casimir operator (with the physicists' normalization) of $O(3)$ is $\vec{L}^2 = \Sigma_{i=1}^{3} L_i^2$. As is well known, its values for irreducible representations of $SU(2)$ are $j(j + 1)\hbar^2$ where $2j$ is an integer ≥ 0; and $2j + 1$ is the dimension of the representation. Only integer values of j appear in the $SO(3)$ irrep. When the state vector is an eigenvector of $\vec{L}^2$ with eigenvalue $j(j + 1)\hbar^2$, we say shortly that the corresponding angular momentum is $j\hbar$.

2.4. The Hydrogen Atom

Consider two particles of mass m_1, m_2 electric charge Ze, $-e$ (Z is a positive integer). The total Hamiltonian for this system of two particles is

$$h_{tot.} = \frac{\vec{p}_1^{\,2}}{2m_1} + \frac{\vec{p}_2^{\,2}}{2m_2} - \frac{Ze^2}{r} \quad , \tag{2.13}$$

where $r = |\vec{r}|$ with $\vec{r} = \vec{r}_2 - \vec{r}_1$.

Introduce the center of mass

$$\vec{r}_0 = (m_1\vec{r}_1 + m_2\vec{r}_2)(m_1 + m_2)^{-1} \tag{2.14}$$

and $\vec{r}$ as new variables instead of $\vec{r}_1$ and $\vec{r}_2$; let $\vec{p}_0$ and $\vec{p}$ the conjugate variables. Then

$$h_{tot.} = \frac{\vec{p}_0^{\,2}}{2(m_1 + m_2)} + \left(\frac{\vec{p}^{\,2}}{2m} - \frac{Ze^2}{r}\right) = h_{cm} + h \quad , \tag{2.15}$$

where

$$m = m_1 m_2 (m_1 + m_2)^{-1} \quad . \tag{2.15$'$}$$

The motion of the center of mass is described by h_{cm} while h corresponds to the internal energy of the system. So quantum-wise, we have to study the spectrum of

$$H = \frac{\vec{P}^2}{2m} - \frac{Ze^2}{R} \quad , \tag{2.16}$$

for obtaining the energy of the hydrogen atom levels. The first quantum study of the hydrogen atom was made by Pauli, Z. *Phys.*, 36, 336 (1926) before Schrödinger's equation was published. Pauli did study the abstract algebra generated by $\vec{R}$, $\vec{P}$, H and Equations (2.1), (2.2), and (2.15). The angular momentum $\vec{L} = \vec{R} \times \vec{P}$ is a constant of motion. Another constant of motion is the Runge-Lenz vector

$$\vec{A} = \frac{1}{2}(\vec{L} \times \vec{P} - \vec{P} \times \vec{L}) + \frac{\lambda}{R}\vec{R} \text{ with } \lambda = mZe^2 \quad . \tag{2.17}$$

Note that

$$\frac{1}{2}(\vec{L} \times \vec{P} - \vec{P} \times \vec{L}) = (\vec{R} \cdot \vec{P})\vec{P} - \vec{R}(\vec{P}^2) - i\hbar\vec{P} = \vec{P}(\vec{P} \cdot \vec{R}) - (\vec{P}^2)\vec{R} + i\hbar\vec{P} \tag{2.18}$$

so we can check that

$$[\vec{A},H] = 0, \ [\vec{L},H] = 0 \quad . \tag{2.19}$$

We recall that ε_{ijk} = sign of the permutation $\binom{123}{ijk}$ or 0 if two indices are equal. From now on we will use the Einstein summation convention, i.e., summation of repeated indices is implied, and we find

$$[L_i,L_j] = i\hbar\varepsilon_{ijk}L_k, \ [L_i,A_j] = i\hbar\varepsilon_{ijk}A_k \tag{2.20}$$

$$[A_i,A_k] = -i\hbar 2mH\varepsilon_{ijk}L_k \tag{2.21}$$

$$\vec{L} \cdot \vec{A} = \vec{A} \cdot \vec{L} = 0 \tag{2.22}$$

$$\vec{A}^2 - 2mH(\vec{L}^2 + \hbar^2) = (Ze^2m)^2\mathbb{1} \quad . \tag{2.23}$$

Let us just consider the bound states of the hydrogen atom. They correspond to the spectrum of $H < 0$. Let P_- be the projector on the bound states. For any X write $X^- = XP_-$. From (2.19) when X is $\vec{L}(\vec{a})$ or $\vec{A}(\vec{b})$,

$$P_-XP_- = XP_- = P_-X = X^- \quad .$$

Furthermore, $-2mH^-$ is an inversible positive operator. Let $(-2mH^-)^{-1/2}$ be the positive square root of its inverse and define $K_i^- = A_i^-(2mH^-)^{-1/2}$. Then Equations (2.20′) to (2.23) read

$$[\frac{1}{\hbar} L_i^-,\frac{1}{\hbar} L_j^-] = \frac{i}{\hbar} \varepsilon_{ijk}L_k^-, \ [\frac{1}{\hbar} L_i^-,\frac{1}{\hbar} K_j^-] = \frac{i}{\hbar} \varepsilon_{ijk}K_k^- \tag{2.20′}$$

$$[\frac{1}{\hbar} K_i^-,\frac{1}{\hbar} K_j^-] = \frac{i}{\hbar} \varepsilon_{ijk}L_k^- \tag{2.21′}$$

$$\vec{L}^- \cdot \vec{K}^- = \vec{K}^- \cdot \vec{L}^- = 0 \tag{2.22′}$$

$$\frac{1}{\hbar 2} K^{-2} + \frac{1}{\hbar 2} \vec{L}^{-2} = (\frac{Ze^2m}{\hbar})^2(-2mH)^{-1} \quad . \tag{2.23′}$$

We last define

$$\vec{J}^{(\pm)} = \frac{1}{2\hbar}\vec{L}^{(-)} \pm \frac{1}{2\hbar}\vec{K}^{(-)} \tag{2.24}$$

so the previous equations read

$$[J_i^{(\pm)}, J_j^{(\pm)}] = i\epsilon_{ijk}J_k^{(\pm)}, \quad [J_i^{(+)}, J_j^{(-)}] = 0 \tag{2.25}$$

$$\vec{J}^{(+)^2} = \vec{J}^{(-)^2} = \frac{1}{4}\left(\left(\frac{Ze^2m}{\hbar}\right)^2(-2mH)^{-1} - 1\right) \quad . \tag{2.26}$$

The spectrum of this operator is $j(j + 1) = (n^2 - 1)/4$ with $2j + 1 = n$, positive integer. So the energy spectrum of the bound states of the hydrogen atom is

$$(\text{n positve integer}, \ \epsilon_n = -\frac{(Ze^2)^2}{\hbar}\frac{m}{2n^2} = \frac{-Z^2}{n^2}\left(\frac{e^2}{\hbar c}\right)^2\frac{mc^2}{2} = \frac{-1}{2n^2}(Z\alpha)^2mc^2) \tag{2.27}$$

where

$$\alpha = \frac{e^2}{\hbar c} = \frac{1}{137.03\ldots} \tag{2.28}$$

in rationalized units of charge, α is the fine structure constant, a dimensionless fundamental constant of physics.

Some Physical Comments. The ratio binding energy/electron rest-mass energy is the number

$$\frac{\epsilon_n}{mc^2} = -\frac{(Z\alpha)^2}{2n^2} \quad .$$

The value of every physical observable we can compute will appear as the product of a pure number and the quantity of same physical dimensions built with the constants $e, \hbar, m, c$. Example: length $\hbar/mc = 3.86 \times 10^{-11}$ cm; energy $mc^2 = .51 \times 10^6$ eV; time $\hbar/mc^2 = 1.28 \times 10^{-21}$ sec. The pure number is a function of α only. It is the value of the observable in the unit system $\hbar = m = c = 1$ that we will use, and α is the value of e^2 in this system. For instance

$$\langle\frac{1}{R}\rangle = Z\alpha\frac{mc}{\hbar} \sim \left(\frac{1}{2}\,10^{-8}\ \text{cm}\right)^{-1} = \left(\frac{1}{2}\ \text{Ångström}\right)^{-1} \quad .$$

We have studied not only the bound state of the hydrogen atom $p^+e^-(m_p = 1836\ m_e)$ (the nucleus can also be a deuteron $\sim 2m_p$), but also that of positronium $e^+e^-(m_1 = m_2)$, munomium $\mu^+e^-(m_\mu = 207\ m_e)$, μ-atom, Π-atom, ionized Helium ion $\overset{+}{He}$, etc.

More On The Group Aspect. The states of energy ϵ_n are eigen states of $\vec{J}^{(+)^2}$ and $\vec{J}^{(-)^2}$ and they form the space $\mathcal{H}_n$ of the irrep (j,j) of $SO(4)$; $\mathcal{H}_n$ has dimension

$$(2j + 1)^2 = n^2 \quad . \tag{2.29}$$

The Lie algebra of the physical rotation $(\vec{L})$ is the diagonal of SU(2) $\oplus$ SU(2) = SO(4), so the representation of the rotation group in $\mathcal{H}_n$, space of the irrep $(j,j) = (\frac{n-1}{2}, \frac{n-1}{2})$ of SO(4) reduces to

$$(j,j)|_{SO(3)} = \oplus_{\ell=0}^{2j} D_\ell \tag{2.30}$$

i.e.,

$$\ell = 0, 1, \ldots, n-1 \quad . \tag{2.30$'$}$$

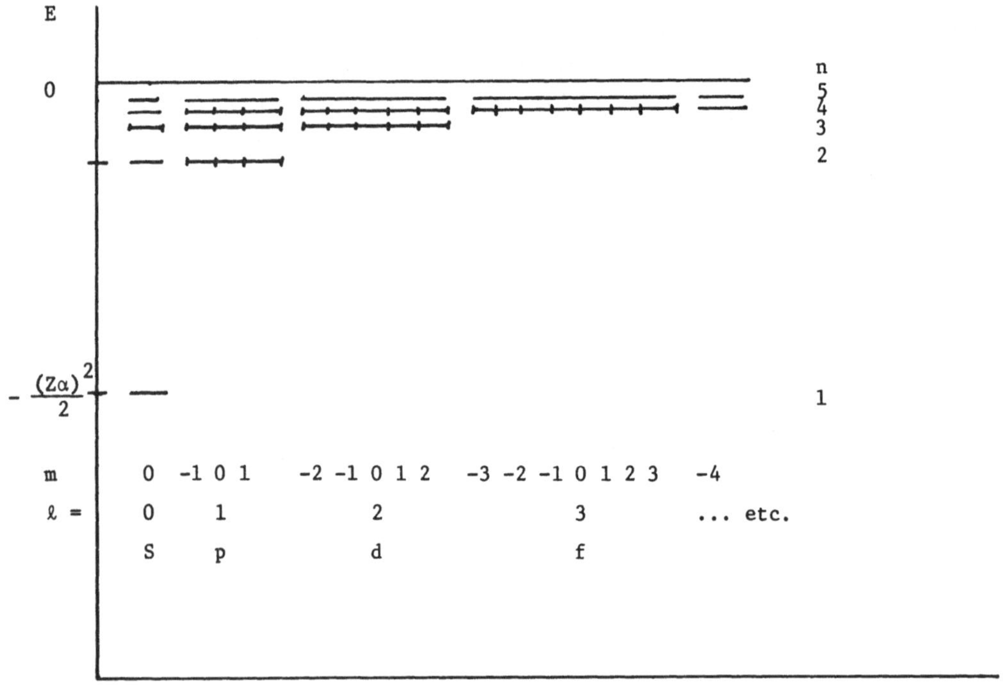

FIGURE 2.1. LOWEST STATES, IN A LINEAR ENERGY SCALE OF THE HYDROGEN ATOM

There is an infinite number of states with energy between $-\varepsilon$ and 0 because of the dependence in r^{-1} of the potential for $r \to \infty$. The eigenvectors of the abelian algebra generated by H, $\vec{L}^2$, $\vec{L}(\vec{e}_3)$ form an orthogonal basis for the Hilbert space of bound states. A complete set of labels for them is the quantum numbers n, ℓ, m; n = positive integer, ℓ and m integers $0 \le \ell \le n-1$, $-\ell \le m \le \ell$, corresponding to the eigenvalues $Z(\alpha)^2/2n^2$, $\ell(\ell+1)$, m of $(mc^2)^{-1}H$, $\hbar^{-2}\vec{L}^2$, $\hbar^{-1}\vec{L}(\vec{e}_3)$

Note that the trivial representation appears once only for each n, and from the Frobenius reciprocity theorem we know that

$$\oplus_{2j=0}^{\infty}(j,j) = U^{D_0} \tag{2.31}$$

the induced representation of SO(4) by the trivial representation of SO(3). In other words, $P_{-}\mathcal{H} = L_2$ (functions on S_3) since the sphere S_3 is the homogeneous space $SO(4)/SO(3)$. (This was exploited by V. Fok, Z. *Physik*, 98, 145 (1935), see also L. Hulthen, Z. *Physik*, 86, 21 (1933).)

From Mackey's theory of induced representations, (2.31) is also the content of the representation of $R^4 \square SO(4) = E_4$ (euclidean group in four dimensions) induced by the trivial representation of $R^4 \square SO(3)$, the stabilizer of any chosen vector $\neq 0$ of R^4. This is an irreducible representation of E_4. We can also consider $P_{-}\mathcal{H}$ as the space of an irrep of $SO(4,1)$ (obtained by deformation of the irrep of E_4 considered above). However, the physical meaning of the generators (representing the elements of Lie algebra) of E_4 or $SO(4,1)$ is not very transparent!

The spectrum of H on $P_{+}\mathcal{H}$, (positive energy) is $(+0,\infty)$. One cannot speak of eigenvectors of H for the positive energy = unbounded states of a proton and an electron and one has to study their scattering. However, since $HP_{+} = P_{+}H$ is a positive operator one can define $\vec{K}^{+} = \vec{A}P_{+}(2mHP_{+})^{-1/2}$ and $1/\hbar\ \vec{L}P_{+}$ and $1/\hbar\ \vec{K}^{+}$, which generate a $SO(3,1)$ Lie algebra as was noted and exploited by V. Bargmann, Z. *Physik*, 99, 576 (1936). Thus, $P_{+}\mathcal{H}$ is a direct integral of (infinite dimensional unitary) irreps of $SO(3,1)$, the Lorentz group. It is also the space of an irrep of the inhomogeneous $SO(3,1)$ that we call the Poincaré group (it is an irrep of spin zero and fixed mass $m > 0$).

What we observe mainly in atoms are their emission or absorption of photons of frequency

$$\nu = \frac{1}{2\pi\hbar}(\varepsilon_{n_1} - \varepsilon_{n_2}) \quad . \tag{2.32}$$

So the wave length is

$$\frac{c}{\nu} = \pi(Z\alpha)^{-2}\left(\frac{1}{n_1^2} - \frac{1}{n_2^2}\right)^{-1}$$

which is $\sim \frac{4}{Z\alpha}10^3$ times the "size" of the atom.

All large enough frequencies of (2.32) were already seen in the spectrum of atomic hydrogen. In fact, there is a "fine structure" which corresponds to a relative splitting in the energy of the states with $\ell \neq 0$ of the order of $\alpha^2 \sim 1/2\ 10^{-4}$.

The theory even predicts the intensity of the spontaneous emission of photons. Since its wave length is large compared to the atomic size, the light wave is a dipole emission† and the probability for spontaneous emission of a photon from

---

† Electromagnetic waves, predicted by Maxwell's equations, were produced by Hertz with an oscillating electric dipole. An example of such a dipole would be a charge $-e$ rotating around a charge $+e$ with a given frequency ν. That light was an electromagnetic wave was a Maxwell hypothesis and Selenyi, by clever experiments, verified in 1913 that light emitted by atoms was a dipole radiation. (Higher multipoles occur for more complicated charge distributions for which the

state $x\rangle$ to state $y\rangle (E_x > E_y)$ is

$$\lambda_{xy} = \frac{4}{3}(E_x - E_y)^3 |\langle x e\vec{R} y\rangle|^2 \quad . \tag{2.33}$$

(Note that $\langle x e\vec{R} x\rangle$ is the expectation value of the electric dipole of a state and, as we shall see, it vanishes.) The intensity of the emitted light by N_x atoms in state $x\rangle$ is

$$i_{xy} = N_x \frac{4}{3}(E_x - E_y)^4 e^2 \sum_k \mathrm{Tr} P_x R_k P_y R_k \quad . \tag{2.34}$$

The Wigner-Eckhart theorem (see beginning of 1.6) predicts that for all vector operators, the matrix elements between two given eigenstates $x\rangle$, $y\rangle$ of $\vec{L}^2$, are colinear.

<u>Example</u>. For $x\rangle$, $y\rangle$ eigenstates of H

$$\frac{-i\hbar}{m} \langle x\vec{P}y\rangle = \langle x[H,\vec{R}]y\rangle = (E_x - E_y)\langle x\vec{R}y\rangle \quad . \tag{2.35}$$

Consider from now on states which are eigenstates of $\vec{L}^2$ (eigenvalue $\ell(\ell + 1)$). Note that

$$\langle x\vec{L}y\rangle = 0 \quad \text{if} \quad \ell_x \neq \ell_y \quad , \tag{2.36}$$

while for a general vector operator

$$\langle x\vec{R}y\rangle = 0 \quad \text{if} \quad \ell_x + \ell_y = 0 \quad \text{or} \quad |\ell_x - \ell_y| > 1 \quad . \tag{2.37}$$

However, we should have taken into account the parity operation

$$\Pi(\vec{r}) = -\vec{r} \quad .$$

The corresponding Π operator satisfies

$$\Pi^2 = I, \ \Pi\vec{R}\Pi = -\vec{R}, \ \Pi\vec{P}\Pi = -\vec{P} \tag{2.38}$$

but, since $\vec{L}$ is an axial vector

$$\Pi\vec{L}\Pi = \vec{L} \quad . \tag{2.39}$$

Eigenvectors of $\vec{L}^2$ have a well defined parity (the eigenvalue of Π). Looking at their realization by spherical harmonics, one finds

$$\Pi\vec{L}^2 = (-1)^{\ell}\vec{L}^2 \quad . \tag{2.40}$$

Thus, when x, y have a well defined angular momentum

$$\langle x\vec{R}y\rangle = -\langle x,\Pi\vec{R}\Pi y\rangle = -(-1)^{\ell_x+\ell_y}\langle x\vec{R}y\rangle$$

so

$$\langle x\vec{R}y\rangle = 0 \quad \text{if} \quad \ell_x + \ell_y = \text{even} \tag{2.41}$$

which is Laporte selection rule we spoke of in 2.1. The set of both equations (2.37) and (2.41) is equivalent to: no electric dipole transition: $\langle x\vec{R}y\rangle = 0 \Leftrightarrow |\ell_x - \ell_y| \neq 1$.

---

dipole moment $\sum_i \vec{r}_i q_i = 0$; see work of Mie, Poincaré, Rayleigh, etc., on multipole expansion - it's applied group theory!) The trouble was: classically atoms should always radiate and use up their energy fast. Quadrupole radiation in atoms can be observed in exceptional cases (rare-earth, atoms in interstellar vacuum). In a radiation field, electromagnetic emission of photons can be induced and become intense: laser!

<u>How Does This Theory Compare With Experiments?</u> It is both very good and very poor. Within an accuracy of 10^{-4} the agreement for the values of the binding energy is perfect. The value predicted by the present theory of quantum electrodynamics gives a correction in $(Z\alpha)^4/n^3$ (i.e., a relative correction of $(Z\alpha)^2/n \sim 10^{-4}$) so that levels with different ℓ and same n have a small difference in energy.†

What is very bad is the counting of the number of levels. This can be seen by putting the hydrogen atom in a constant electromagnetic field ($\vec{F}$, electric and $\vec{B}$, magnetic). Then one must add to H

$$H_{em} = -\frac{3e}{2}\vec{K}\cdot\vec{F} + \frac{e}{2mc}\vec{L}\cdot\vec{B} \quad . \tag{2.42}$$

The effect of $\vec{F}$ (Stark effect) is well reproduced, but not that of $\vec{B}$. Indeed, levels of the same ℓ should split into $2\ell + 1$ levels separated by $eB/2mc$. They do split, but in an even number of levels!! This is due to the electron spin which we have not yet taken into account (see 2.6). One should also take into account the proton spin with effects

$$\sim \langle \frac{e^2}{m_e m_p} \frac{1}{R^3} \rangle \sim \alpha(Z\alpha)^3 m_e m_p^{-1}) \quad .$$

2.5. The Helium Atom

It has a nucleus of charge $Ze = 2e$ (mass $\sim 4m_p$) and 2 electrons. After separation of the center of mass motion, the Hamiltonian for the internal energy is

$$H = H_1 + H_2 + \frac{e^2}{R_{12}} \tag{2.43}$$

where $H_i = P_i^2/2m - Ze^2/R_i$, the hydrogen Hamiltonian and the operator R_{12} corresponds to $r_{12} = |\vec{r}_2 - \vec{r}_1|$ the relative distance of the two electrons. If we neglect the term in e^2/R_{12} (this is a better than 10% approximation) our problem is solved. We will consider only bound states. Let $\mathcal{H}^{(1)}$ be the Hilbert space of the bound states of hydrogen atom. Our simplified helium atom has Hilbert space $\mathcal{H}^{(1)} \otimes \mathcal{H}^{(1)}$ with Hamiltonian $H_0 \otimes I + I \otimes H_0$ where H_0 is that of hydrogen. So the binding energy is $-(Z\alpha)^2(1/n_1^2 + 1/n_2^2)/2$ i.e., the sum of the binding energies for the two electrons.

We assume here that the term e^2/R_{12} is a perturbation in the technical sense (see Kato's book for mathematical rigor). This term breaks the SO(4) invariance, so the electron levels with different ℓ and same n no longer have the

† The difference between the two levels $n = 2$, $\ell = 1$ and 2 predicted by the theory of quantum electrodynamics is essentially $Z^4\alpha^5 2^{-3}$, i.e., $\sim 10^3$ megacycles and the agreement with experiment is of the order of 10^{-1} megacycles $\sim 10^{-15}\, mc^2/\hbar$. Quantum electrodynamics is not yet well defined for the mathematicians! Refined predictions for positronium, muonium, etc., are also very precisely verified.

same energy. (As we shall see later in 2.6, for a given n, E increases with ℓ.) What is left is angular momentum and parity conservation

$$[\vec{L},H] = 0, \ [\Pi,H] = 0 \tag{2.44}$$

and the indistinguishability of the two electrons

$$[S_{12},H] = 0 \tag{2.45}$$

where S_{12} is the operator permuting the two electrons

$$S_{12}^2 = I, \ S_{12}(A \otimes B)S_{12} = B \otimes A \in L(\mathcal{H}^{(1)} \otimes \mathcal{H}^{(1)}) \tag{2.45'}$$

The decomposition of the tensor product $\mathcal{H}^{(1)} \otimes \mathcal{H}^{(1)}$ into the direct sum of symmetrical and antisymmetrical tensor spaces

$$\mathcal{H}^{(1)} \otimes \mathcal{H}^{(1)} = \mathcal{H}^{(1)} \vee \mathcal{H}^{(1)} \oplus \mathcal{H}^{(1)} \wedge \mathcal{H}^{(1)} \tag{2.46}$$

that we also wrote

$$\mathcal{H}^{(2)} = \mathcal{H}_{[2]} \oplus \mathcal{H}_{[1^2]} \tag{2.46'}$$

give the decomposition into eigenspaces of S_{12}. Let x, y be states of the hydrogen atom. Which of the two states $x \vee y = 1/\sqrt{2}(x \otimes y + y \otimes x)$ or $x \wedge y = 1/\sqrt{2}(x \otimes y - y \otimes x)$ yield the smallest expectation value for the positive operator e^2/R_{12}? It is obviously $x \wedge y$ because the two-electron wave function vanishes when e^2/R_{12} is very large (while that of $x \vee y$ has generally a maximum when $R_{12} = 0$). This symmetry character yields a new selection rule for the dipole radiation; the matrix element of the transition operator is proportional to

$$\langle\Psi|\vec{R}_1 + \vec{R}_2|\Psi'\rangle \quad . \tag{2.47}$$

Since $\vec{R}_1 + \vec{R}_2$ is symmetrical, Ψ and Ψ' must have the same symmetry character $\varepsilon = \varepsilon' \ (\varepsilon^2 = 1)$ since

$$\langle\Psi,(\vec{R}_1 + \vec{R}_2)\Psi'\rangle = \langle\Psi,S_{12}(\vec{R}_1 + \vec{R}_2)S_{12}\Psi'\rangle = \varepsilon\varepsilon'\langle\Psi(\vec{R}_1 + \vec{R}_2)\Psi'\rangle \quad . \tag{2.47'}$$

As we announced in 2.1, this shows that the helium levels are to be divided in two sets according to their symmetry characters, and electric dipole transitions occur only within each set. Let me remind you that helium got its name because it was observed in the sun before being observed on earth. Its spectrum appears to be composed of two spectra, one for orthohelium ($\varepsilon = +1$), one for parahelium ($\varepsilon = -1$). This was a complete mystery before quantum mechanics. The explanation was given by Heisenberg in 1926, "(Uber die Spektra von Atomsystem mit zwei Elektronen", *Z. Physik*, 39, 499 (1926)). It also explained that the orthohelium has more levels; those of the type $x \otimes x$, as for instance the lowest level ($n = 1$, $\ell = 0$ for each electron). It is observed that corresponding rays (e.g., transitions $(1,0) \vee (n,\ell) \to (1,0) \vee (n',\ell')$ and $(1,0) \wedge (n,\ell) \to (1,0) \wedge (n',\ell')$ with $n' \neq 1, \ell' \neq 0$) of parahelium are about three times more intense than those of orthohelium. To explain it, the electron spin will have to be taken into account (see also 2.9).

2.6. Pauli Principle. The Electron Spin

We want to pass now to the case of n-electron atoms. The internal energy Hamiltonian is

$$H^{(n)} = \Sigma_{i=1}^{n} H_i + \sum_{1\leq i<j\leq n} \frac{e^2}{R_{ij}} \quad \text{with} \quad H_i = \frac{P_i^2}{2m} - \frac{Ze^2}{R_i} \ . \tag{2.48}$$

Of course $H^{(n)}$ is invariant under the permutation group $S(n)$ of the n electrons. It is also the case for the electric dipole operator $e(\Sigma_i R_i)$ and for all observables. Identical particles cannot be distinguished from each other and every prediction of the theory must be invariant under $S(n)$.

When we consider states of Z (or n) distinguishable particles, we considered (with success for the helium atom) the Hilbert space tensor product of the $\mathcal{H}$ for each particle. Consider again $\mathcal{H}^{(n)} = \overset{n}{\otimes}\mathcal{H}^{(1)}$ for n identical particles. $S(n)$ acts on $\mathcal{H}^{(n)}$ by the representation $s \to S(s)$. Invariance under $S(n)$ of all observables requires that they are in the commutant $\{S(s)\}'$ of the set $\{S(s), s \in S(n)\}$ of operators. As we saw the rank one projectors wich represent physical state are observables of the theory so $\forall s$, $S(s)P_x S(s) = P_x$ for any vector x which represents a state. This requires that the vector x> belongs either to $\mathcal{H}_{[n]} = \overset{n}{\vee}\mathcal{H}^{(1)}$ (completely symmetrical) or to $\mathcal{H}_{[1^n]} = \overset{n}{\wedge}\mathcal{H}^{(1)}$ (completely antisymmetrical). The other spaces $\mathcal{H}_{[\ ]_\lambda}$ of the other factorial representations of $S(n)$ are excluded as space of physical states.

We have used both $\mathcal{H}_{[2]}$ and $\mathcal{H}_{[1^2]}$ for the helium atom. However, the use of $\mathcal{H}_{[n]}(n > 2)$ for atoms does not represent nature. Indeed, the ground state of any atom would have all electrons with the same binding energy (of the order of $(Z\alpha)^2/2$). Experimentally, only two electrons have this binding energy (X-ray spectrum for Z-large enough). The necessary energy (called ionization energy) for removing a first, a second, a kth ..., the Zth electrons of any neutral atom increase irregularly from a fraction of α^2 to $Z(\alpha)^2/2$. Moreover, as we shall see, vectors of some other $\mathcal{H}^{(n)}_{[\ ]_\lambda}$ do appear! The solution to this puzzle is that $\mathcal{H}^{(1)}$ is not the Hilbert space of the bound states of one electron in a constant potential. The electron has another degree of freedom, the spin and the Hilbert space of its states has to be changed into a new

$$\mathcal{H}^{(1)} = L^{(1)} \otimes K^{(1)}$$

where $L^{(1)}$ is the $L_2(R^3,t)$ previously called $\mathcal{H}^{(1)}$ and $K^{(1)}$ is a two-dimensional Hilbert space. Pauli was the first in 1924 to introduce the spin as an intrinsic angular momentum and magnetic moment for the nuclei, but it was Goudsmit and Uhlenbeck who introduced in 1925 the spin as an intrinsic angular momentum $\hbar/2$ for the electron. This explained the number of energy levels which appear in the Zeeman effect, but it did not explain the magnitude of their splitting. Indeed, the magnetic moment produced by an electric charge e moving with an angular momentum

$\vec{j}$ (Ampere's law!) is

$$\vec{\mu}/\frac{e}{2mc} = g\vec{j}/\hbar \tag{2.49}$$

where $e\hbar/2mc$ is the Bohr magneton. For the orbital momentum, $g(|\vec{j}/\hbar|$ integral) $g = 1$, but for the spin $|\vec{j}/\hbar| = 1/2$, g appeared to be 2. This was a mystery solved by Thomas in 1925. It is a relativistic effect.

It is an experimental fact that we have to use Fermi Statistics for electrons, i.e., the Hilbert space of electronic states of an n-electron atom is

$$\mathcal{H}^{(n)}_{[1^n]} = \overset{n}{\wedge}\, \mathcal{H}^{(1)} \subset \oplus_\lambda (L^{(n)}_{[\ ]_\lambda} \otimes K^{(n)}_{[\ ]^c_\lambda}) \tag{2.50}$$

where $\mathcal{H}^{(1)}$ is the (new) one electron Hilbert space defined in (2.48).

Since $\dim K^{(1)} = 2$, the Young diagram of $[\ ]^c_\lambda$ has only two lines of length $\lambda_1 \geq \lambda_2 \geq 0$. Of course $\lambda_1 + \lambda_2 = n$; we will show that $\lambda_1 - \lambda_2$ is the chemical valence.

The diagram of $[\ ]_\lambda$ in $L^{(n)}_{[\ ]_\lambda}$ is the one symmetric through the diagonal. It has two columns $\lambda_1 \geq \lambda_2 \geq 0$. In other words, it has λ_2 lines of length 2 and $\lambda_1 - \lambda_2$ lines of length 1. That means that it cannot be completely symmetrical in more than two electrons, i.e., there can be only two electrons at most in each orbital state; then two electrons must have "different spin states", or more exactly, their spin-state has to be antisymmetrical. This is the Pauli principle, discovered by Pauli (Z. *Physik*, 31, 765 (1925)).

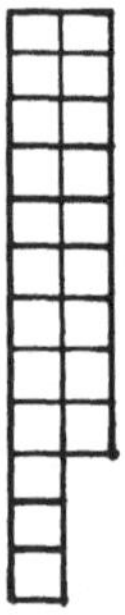

FIGURE 2.2. YOUNG DIAGRAM OF THE ORBITAL PART OF n-ELECTRON STATE: $L^{(n)}_{[\ ]_\lambda}$. HERE $n = 21$.

2.7. Atomic Shell Structure - Periodic Table

We can now clearly describe the lowest state of an n-electron atom. The orbital part is a vector of $\mathcal{H}^{(n)}$ obtained by filling all the lowest energy states, putting only two electrons in each electronic orbital state. Of course, in atoms with $n > 1$ electrons,† the two specific properties of the $1/R$ potential disappear. The number of bounded states is finite, and the SO(4) degeneracy no longer exists, i.e., states with the same n and different ℓ have different energy. The observed order of increasing energy for the states is given in Table 2.1.

TABLE 2.1. THE ELECTRON STATES ARE LISTED IN ORDER OF INCREASING ENERGY

n	1	2	2	3	3	4	3	4	5	4	5	6	5	4	6	7 ...
ℓ	0	0	1	0	1	0	2	1	0	2	1	0	2	3	1	0 ...
Spectroscopist Notation	1s	2s	2p	3s	3p	4s	3d	4p	5s	4d	5p	6s	5d	4f	6p	7s ...
$2(2\ell + 1)$	2	2	6	2	6	2	10	6	2	10	6	2	10	14	6	2 ...
Total	2	4	<u>10</u>	12	<u>18</u>	20	30	<u>36</u>	38	48	<u>54</u>	56	66	80	<u>86</u>	88 ...

Note that for a given n, the energy increases with ℓ. This is of course predicted by computation and can be qualitatively understood. Consider a nucleus of charge Ze and k electrons. At infinity an electron will feel the coulomb potential $(Z - k)e/r$, but when it gets near, the probability of finding some of the k electrons at the distance $r \sim n(Z\alpha)^{-1} \hbar(mc)^{-1}$ is not negligible, the $(k + 1)^{th}$ electron feels a potential above the Coulomb potential. It feels less the difference if it is in a $\ell = 0$ state (more concentrated inside the sphere of radius r) than in a $\ell = 1$ state, $\ell = 2$ state, etc., where the concentration of probability is more and more on the surface of the sphere. Note also from Table 2.1 that "p shells" $(\ell = 1)$ are filled by the 10th, 18th, 36th, 54th, 86th electron and this is just the atomic number of the "rare gas" elements, very inactive chemically, Neon, Argon, Krypton, Xenon, and Radon. We can even give the Mendeleev's periodic table, in terms of electron states, if we know the order of increasing binding energy of the states (n,ℓ). Using the spectroscopic notation

$$\begin{matrix} \ell = 0 & 1 & 2 & 3 & 4 & 5 & 6 \\ s & p & d & f & g & h & i \end{matrix} \quad . \tag{2.51}$$

This order is, from Table 2.1, (n,ℓ): 1s, 2s, 2p, 3s, 3p, 4s $\sim$ 3d, 4p, 5s $\sim$ 4d, 5p, 6s, 5d $\sim$ 4f, etc. The sign $\sim$ indicates roughly the same energy so that the two shells are filled simultaneously. Indeed, for the element of $\ell = 2$ (d-shell) the + (++) sign indicates 1 (or 2) more electrons in the d-shell, taken from the s

† Because of the electrostatic repulsion among the electrons.

shell. We can now construct a periodic Table.

TABLE 2.2. PERIODIC TABLE (UP TO ELEMENT 56)

	$\ell=0$	2	$\ell=1, 2(2\ell+1) = 6$						$\ell=2, 2(2\ell+1) = 10$										$\ell=3, 2(2\ell+1)=14$
electron state	s		p						d										f
	1	2																	
n=1	H 1	He 2	1	2	3	4	5	6											
n=2	Li 3	Be 4	B 5	C 6	N 7	O 8	F 9	Ne 10	1	2	3	4	5	6	7	8	9	10	
n=3	Na 11	Mg 12	Al 13	Si 14	P 15	S 16	Cℓ 17	A 18	Sc 21	Ti 22	V 23	Cr+ 24	Mn 25	Fe 26	Co 27	Ni 28	Cu+ 29	Zn 30	
n=4	K 19	Ca 20	Ga 31	Ge 32	As 33	Se 34	Br 35	Kr 36	Y 39	Zr 40	Nb+ 41	Mo+ 42	Tc+ 43	Ru+ 44	Rh+ 45	Pd++ 46	Ag+ 47	Cd 48	14 rare earths 58 to 71
n=5	Rb 37	Sr 38	In 49	Sn 50	Sb 51	Te 52	I 53	Xe 54	La 57 ↑ Hf 72										
n=6	Cs 55	Ba 56																	

In d-shell, + (or ++) means 1 (or 2) more electrons in n-d (coming from the (n + 1),s) state).

The atomic state of an atom is labeled by the filled states, e.g., Oxygen: $(1s)^2(2s)^2(2p)^4$, i.e., 8 electrons. In general, the electrons fill up all states of lower energy and fill incompletely the last "shell"; for example, in the case of Oxygen, we can add 2 more electrons in the 2p state. The question which arises is which state is the atom ground state for an incomplete shell? We can label this state by a Young diagram (let us do it for the first elements).

TABLE 2.3. YOUNG DIAGRAM OF FUNDAMENTAL STATES

Z =	1	2	3	4	5	6	7
Name	H	He	Li	Be	B	C	N

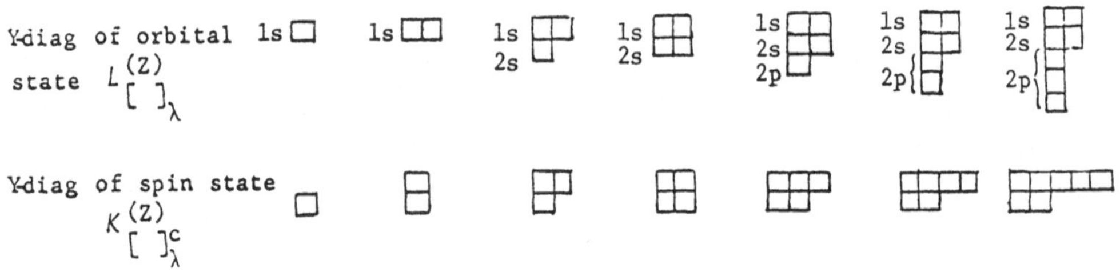

Now we can generalize what we say for helium. Given k electrons in the same energy state (i.e., $k \leq 2(2\ell + 1)$), the lowest energy state of this k electron configuration is the most antisymmetrical in the coordinates ($k = 2, k = 3, \ldots$), so it is the most symmetrical for the spin coordinates ($k = 2, k = 3$, etc.). This is illustrated in Table 2.3 for the 2p-electrons for $f = 1, 2, 3$. When $k = 4$, it is not possible to have a completely antisymmetric tensor on a $2\ell + 1 = 3$ dimensional Hilbert space, so we can give the successive atomic states of p shells.

TABLE 2.4. np-SHELL STATES

nb of electrons k = 1 filled shells	1	2	3	4	5	6
space state				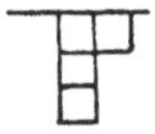	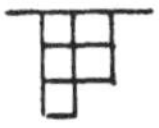	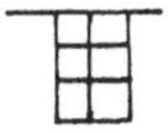
spin state (filled shells)		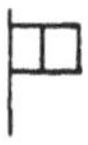	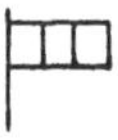	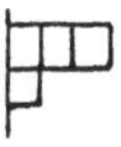	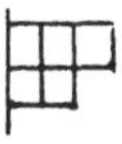	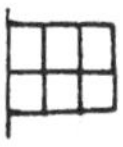
valence	1 (or 3)	2(or 4)	3	2	1	0
n = 2	B	C	N	O	F	Ne
n = 3	Al	Si	P	S	Cl	A
n = 4	Ga	Ge	As	Se	Br	Kr
n = 5	In	Sn	Sb	Te	I	Xe

The ionization energy (energy necessary to extract one electron from the p-shell) is increasing with k, as we expect, along a given p-shell, except for the four electron state, because it is the first one not completely antisymmetric.

Although the energy of ns-states is lower than that of the np-state, a low excited state of atoms with $k = 1$ or 2 is $k = 1$; $(ns)(np)^2$; $k = 2$; $(ns)(np)^3$, that is, an ns-electron goes to an np-state. This increases the valence by two and gives more bounded molecules.

There would be a lot more to say, even from group theory, about the Mendeleev Table. For instance did you notice that the ferromagnetic elements (Ni, Co, Fe and also Mn in alloys are together in an incomplete 3-d shell, etc.? But we shall stop here.

2.8. Atomic States in a Given Shell - Spin Orbit Coupling

The Pauli principle, expression of the Fermi statistics, simplifies the study of atoms. Instead of studying an n-electron system, within a good approximation we can (for not too highly excited states) consider k electrons outside a closed shell which has angular momentum zero, electric charge $(n - k)e$ (with a distribution depending on the electron wave function). This closed shell can be considered as a spherical potential and the Hilbert space of states for k electrons in an $n - \ell$ shell is

$$\mathcal{H}^{(k)} = \overset{k}{\wedge} (\mathcal{H}_{2\ell+1} \otimes K_2) \qquad (2.52)$$

to a good approximation, an electron state is a kth order decomposable antisymmetric tensor, $x \wedge y \wedge z \ldots$ (k factors).

Example. $\ell = 1$, $k = 2$, $\dim \mathcal{H}^{(1)} = \dim \mathcal{H}_3 \otimes K_2 = 6$; and $\dim \mathcal{H}^{(2)} = 15 = \binom{6}{2}$ for $k = 6$, $\dim \mathcal{H}^{(6)} = 1$ (complete shell again). Each decomposable tensor of $\mathcal{H}^{(k)}$ can be given a name or a label. That is what the spectroscopist does using a complete set of observable $\vec{L}^2$, $\vec{S}^2$, $\vec{J}^2 = (\vec{L} + \vec{S})^2$, J_z, that is the orbital angular momentum ℓ, the spin angular momentum s, the total angular momentum j, and its projection on axis j_3.

TABLE 2.5. THE $1S(np)^2$-STATES IN ORDER OF INCREASING ENERGY

Spectrocopist notation	$^3P_0$		$^3P_1$		$^3P_2$		$^1D_2$		$^1S_0$
L and space symmetry	1 		1 		1 		2 		0 
S and spin symmetry	1 		1 		1 		0 		0 
J = total ang. mom.	0		1		2		2		0
2J+1=nb of states	1	+	3	+	5	+	5	+	1 =15
$<\vec{L}.\vec{S}>$	-2		-1		1		0		0

Remark on Table 2.5. Note that the space-antisymmetric  states (P-states) are below the  states as we already emphasized. For the symmetric states, the S-state which feels more the repulsion than the D-states, is above them. Why do the P-states appear in order of increasing J? This is the small spin-orbit effect that we can explain in the following way.

The orbital state of angular momentum $L\hbar$ produces a magnetic moment $(e\hbar/2mc)\vec{L}$, while the spin state of angular momentum $\vec{S}\hbar$ produces a magnetic moment $g(e\hbar/2mc)\vec{S}$ with $g = 2$ (see (2.49)). The interaction between the two magnetic moments has for matrix element†

$$\frac{e^2}{2}\left(\frac{\hbar}{mc}\right)^2 \left\langle \frac{\vec{L} \cdot \vec{S}}{R^3} \right\rangle \quad . \tag{2.53}$$

For a state $|j,\ell,s\rangle$ the expectation value of $\vec{L} \cdot \vec{S}$ is easy to compute from

$$\vec{J}^2 = (\vec{L} + \vec{S})^2 = \vec{L}^2 + 2\vec{L} \cdot \vec{S} + \vec{S}^2 \quad . \tag{2.54}$$

And the expectation value for state $|j,\ell,s,j_z\rangle$ (when j_z is the eigenvalue of J_z) is

$$\langle \vec{L} \cdot \vec{S} \rangle = \frac{1}{2}(j(j + 1) - \ell(\ell + 1) - s(s + 1)) \tag{2.55}$$

where

$$|\ell - s| \leq j \leq \ell + s\ell,\ j + s \ \text{ integers } \geq 0 \quad . \tag{2.56}$$

This explains the value of $\langle \vec{L} \cdot \vec{S} \rangle$ in Table 2.5.

We have seen that for hydrogen the $\langle n\ 1/R^3\ n\rangle \sim (Z\alpha/n)^3$, so the expectation value of the spin orbit term is

$$\sim \langle L \cdot S\rangle \frac{(Z\alpha)^2}{2n^2} \frac{Z\alpha^2}{h} \sim \varepsilon_n \frac{\langle L \cdot S\rangle}{n} Z\alpha^2 \sim 10^{-4}\ \varepsilon_n \quad .$$

This is an order of magnitude. In the sodium atom (alcaline = hydrogen like) spectrum $[(1s)^2(2s)^2(2p)^6](3s)$ (fundamental state) and the $[\ ](3p)$ state has the largest splitting, i.e., nearly 10^{-3}, so the very bright 3p - 3s (yellow) Na-line is a doublet.

2.9. Spin and Euclidean or Galilean Invariance

In Chapter 2, Sections 2.7 and 2.8, we have mainly used spin as a new degree of freedom for the electron. This new degree of freedom has two discrete values (often called "up" and "down" in the physics manual) so the corresponding Hilbert space K_2, of complex valued functions defined in a 2-element set, has dimension 2. The atom Hamiltonian (2.48) is independent of this spin degree of freedom, i.e., it is of the form $H \otimes I$ acting on the space $\oplus_\lambda (L^{(n)}_{[\ ]_\lambda} \otimes K^{(n)}_{[\ ]^c_\lambda})$ of Equation (2.50). The permutations of the n electron spins are represented by operators of the form $I \otimes S(s)$ which commute with H. So eigenstates of H can have well defined $[\ ]^c_\lambda$. The simplest illustration is the helium atom $(n = 2)$.

† This is a short for $\vec{L} \otimes \vec{S}$ applied to $\mathcal{H}^{(k)}_{2\ell+1[\ ]_\lambda} \otimes K^{(k)}_{[\ ]^c_\lambda}$.

The set of states with $[\ ]^c_\lambda = [2] = \square\square$ was called parahelium, with $[\ ]^c_\lambda = [1^2] = \begin{smallmatrix}\square\\\square\end{smallmatrix}$ was called orthohelium. Since $\dim K^{(2)}_{\square\square} = 3$ and $\dim K^{(2)}_{\begin{smallmatrix}\square\\\square\end{smallmatrix}} = 1$, helium states (which are tensor products $x \wedge y$ or $x \vee y$ of different ($x \neq y$ hydrogen states) have the statistical weight (for instance in the Boltzman distribution of thermodynamic equilibrium) of 3 for parahelium and 1 for orthohelium; this explains that spectral lines of the former are three times more intense than the corresponding lines of the latter. Using the considerations of Chapter 1, Section 1.4, on the relations between the unitary groups and the permutation groups, we could also consider the action of the group U(2) on the two-dimensional space $K^{(1)}_2$. Its action on $K^{(n)}_{[\ ]^c_\lambda}$ is through the factorial representation $[\ ]^c_\lambda$, and this action is on $\mathcal{H}^{(n)} \subset L^{(n)}_{[\ ]^c_\lambda} \otimes K^{(n)}_{[\ ]^c_\lambda}$ of the type $I \otimes (\oplus D_{[\ ]^c_\lambda}) = I \otimes (\oplus_s D_s)$; it commutes with the Hamiltonian $H \otimes I$.

These two points of view are formally equivalent for the classification of quantum states, but the $SU(2) \subset U(2)$ has a deeper meaning. It is related to the Euclidean or Galilean invariance. Let G be either group, and $\bar{G}$ its universal covering, i.e., there is a surjective homomorphism $\bar{G} \overset{\pi}{\to} G$. (As we have seen in Chapter 1, Section 1.2, and will see again in Chapter 4, it is an extension of the relativity group which acts through a linear representation on the Hilbert space of states. See also O'Raifeartaigh.) There is also a surjective homomorphism $\bar{G} \overset{\varphi}{\to} SU(2)$. (In the Euclidean case for instance $\bar{G} = R^3 \,\square\, SU(2)$; $\square$ = semi-direct product, where $SU(2)$ is the covering of the rotation group.) This gives us the action of $\bar{G}$ on $\mathcal{H}^{(1)}$; the one particle-state $\Psi(\vec{x},t;\sigma) \in L_2(\vec{x},t) \otimes K_\sigma$ transfoms into

$$(U(\bar{g})\Psi)(\vec{x}t,\sigma) = \Psi(\pi(\bar{g})^{-1} \cdot (x,t);\ \varphi(\bar{g})^{-1} \cdot \sigma) \quad . \tag{2.57}$$

Often physicists prefer to write equivalently $\mathcal{H}^{(1)}$ as the Hilbert space of square integrable function Ψ_σ of $\vec{x},t$ with value in the two-dimensional Hilbert space K_σ. Then (2.56) reads

$$(U(\bar{g})\Psi_\sigma)(\vec{x},t) = \sum_{\sigma'=1,2} D_{1/2}(\varphi(\bar{g}))_{\sigma\sigma'}\Psi_{\sigma'}(\pi(\bar{g})^{-1} \cdot (\vec{x},t)) \quad . \tag{2.58}$$

To summarize, the spin is related to (essentially the rotation part of) Euclidean (and a portion of the larger Galilean) invariance; and it is an intrinsic angular momentum for the electron. We will study it in 4.5. The value $g = 2$ for the corresponding electron intrinsic magnetic moment is, however, a relativistic effect (see Figure 2.5).

Conservation of angular momentum implies only that $\vec{J} = \vec{L} + \vec{S}$ (orbital + spin angular momentum) be a constant of motion. In atoms $\vec{L}$ and $\vec{S}$ are separately conserved to a good approximation only because H is spin-independent (see Equation (2.48).

2.10. Molecules

The interaction which binds N atomic nuclei and n electrons into a neutral molecule (or a charged molecular ion) is essentially the electrostatic (= coulomb) interaction. Instead of treating directly a $N + n$ body problem, one uses the Born-Oppenheimer approximation where the (heavy) nuclei are considered fixed. Take for example the Hamiltonian of the hydrogen molecule (subscript A = I, II for the two protons, $i = 1, 2$ for the two electrons, $r_{A,i} = |\vec{r}_A - \vec{r}_i|$, etc.).

$$H = \frac{1}{2M}(\vec{p}_I^{\,2} + \vec{p}_{II}^{\,2}) + \frac{1}{2M}(\vec{p}_1^{\,2} + \vec{p}_2^{\,2}) - e^2\Big(\sum_{\substack{A=I,II \\ i=1,2}} \frac{1}{r_{A,i}}\Big) + \frac{e^2}{d} + \frac{e^2}{r_{12}} \tag{2.59}$$

where $d = |\vec{r}_I - \vec{r}_{II}|$ the distance between the two hydrogen nuclei is considered as a parameter in the Born-Oppenheimer approximation (and the kinetic energy of the nuclei will be neglected). When d is very large, a state of (2.59) is in $\mathcal{H}^{(2)}$, the tensor product of two hydrogen atom-Hilbert spaces. Consider first the space dependence $L_2^{(2)}(\vec{r},r)$ and the two-dimensional subspace $h = (x \otimes y) \oplus (y \otimes x)$ where x, y are hydrogen states. These two states have the same energy $\varepsilon = \varepsilon_x + \varepsilon_y$. However, in this basis, for h, the Hamiltonian operator $H|_h$ is not exactly diagonal when d is finite because each electron feels also the attraction of the other nucleus, so

$$H|_h = \begin{pmatrix} \varepsilon & \rho \\ \bar{\rho} & \varepsilon \end{pmatrix} \tag{2.60}$$

(since it is Hermitian), and its eigenvalues are $\varepsilon \pm |\rho|$. Hence, the two eigenstates of $H|_h$ are $1/2(x \wedge y)$ and $1/2(x \vee y)$ and they have an energy difference of $2|\rho|$. When $d \to \infty$, $|\rho| \to 0$ and so does $e^2/d - |\rho|$. When $d \to 0$, $e^2/d - |\rho| \to \infty$. But there is a domain for d for which $e^2/d - |\rho| < 0$, and a value of d for which $e^2/d - |\rho|$ is minimum. The ground state is of the type $x \otimes x$, and from Fermi statistics the two-electron spins form an antisymmetrical state. Hydrogen (or alcaline) form a similar type of liaison (covalent bond) with electrons of unfilled shell of atoms. The number of atoms which can be bound to an atomic (spin) state 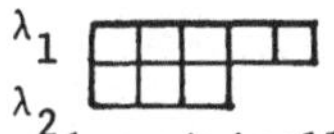 is $\lambda_1 - \lambda_2$ in order to form a closed "spin" shell, as was discovered empirically before 1920, and $\lambda_1 - \lambda_2$ is the "valence" of the atom. Quantum mechanics has explained qualitatively and quantitatively the covalent bond (W. Heitler and F. London, Z. *Physik*, 44, 455 (1927)). It explains, for instance, why the molecules H_2, H_2S, H_2Se are of the form 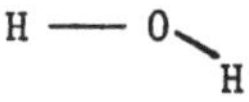 with an angle $\geq 90°$ (the repulsion of the two H atoms makes the angle increasing from 90° for H_2Se, H_2S, H_2O (= 108°). It explains why NH_3 is a trihedron and CH_4 a tetrahedron, why C_2H_4 is flat 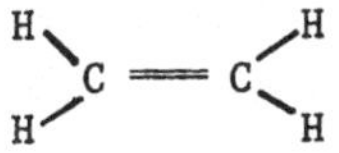 (σ and π electrons). It explains mesomery (e.g., for benzene), etc. Group theory is so useful for explaining molecular spectra! We have to skip this subject for now and simply refer to an elementary but elegant textbook, *Quantum Chemistry*, by Eyring, Walter, and Kimball, Wiley, New York (1944).

The symmetry group of a molecule is a subgroup of $O(3)$, the three-dimensional orthogonal group. When its shape is known experimentally, its symmetry group G is known. Let us refer to Wigner's paper (Göttingen, (March 1930), p. 133) on the characteristic elastic vibration modes of molecule (given by the equivalent classes of G), as examples of the application of group theory. Wigner studied CH_4 (whose group G is $S(4)$) as an illustration. The H. A. Jahn, E. Teller theorem (*Proc. Roy. Soc. Ser. A*, <u>161</u>, 220 (1932)) proves that the electron orbital state of "non-straight" molecules cannot transform as an irrep of G of dimension > 1. (The irrep has dimension 2 for molecules whose atoms are on a straight line.)

We will study here only one very important example.

2.11. Measurement of Spin and Statistics of Nuclei by the Study of Diatomic-Molecule Spectra

The Hamiltonian H of a diatomic molecule can be divided into

$$H = H_{\text{electronic}} + H_{\text{vibration}} + H_{\text{rotation}} + H'$$

where, to a good approximation, H' can be neglected. $H_{\text{electronic}}$ gives the electronic states of the molecule; each such state yields a distance d (between the two nuclei) which minimizes the energy. The invariance group is $O(2)$ or if the two nuclei are identical, $O(2) \times Z_2$. Binding energy for such states are typically a fraction of α^2 (few electron volts). H vibration is essentially the harmonic oscillator Hamiltonian for small oscillations around the equilibrium position fixed by the distance d. The equidistant spacing of the vibration level is small compared to α^2, and the H rotation yields for each d also rotation energies proportional to $\ell(\ell + 1)$, ℓ integer ≥ 0, and small compared to the vibrational energies (rotational bands; in spectrum). If the two nuclei of the molecule are identical, which is the symmetry of the molecular state for the permutation group $S(2)$ of these two nuclei? The symmetry depends only on the <u>spin state</u> of the nuclei, (each of spin j) its $SU(2)$ irrep is

$$D_s,\ 0 \leq s \leq 2j;\quad \square\square\ \ s = 2j,\ 2j-2,\ 2j-4,\ \ldots$$
$$\begin{smallmatrix}\square\\ \square\end{smallmatrix}\ \ s = 2j-1,\ 2j-3,\ \ldots$$

and the rotational state of the system, $\square\square$ for ℓ even, $\begin{smallmatrix}\square\\ \square\end{smallmatrix}$ for ℓ odd.

Since H is independent of the nuclear spin (to a very good approximation) the symmetry character of the nuclear spin state is a constant of motion (with often a lifetime of weeks) and is, as for Helium, called ortho or para. Because of "statistics", the symmetry character of the rotational state is also a constant of motion. So the rotational spectrum of the molecule divides into two independent sets of transitions - those between even ℓ, and those between odd ℓ. The transitions occur in both states as quadrupolar $\ell + 2 \to \ell$, with a (radio-wave) photon energy $\sim (\ell + 2)(\ell + 3) - \ell(\ell + 1) = 4\ell + 6$. The number of $\square\square$ nuclear spin states is $(2j + 1)(2j + 2)/2 = (j + 1)(2j + 1)$. The number of $\begin{smallmatrix}\square\\\square\end{smallmatrix}$ nuclear spin states is $(2j + 1)(2j)/2 = j(2j + 1)$. So if for the molecule the relative intensity of spectral (rotational) lines is ($2j$ integer ≥ 0), $j/j+1$ for ℓ even/ℓ odd, the nuclear spin is j, the statistics of the nuclei is $\left(\begin{smallmatrix}\square\\\square\end{smallmatrix} \times \square\square\right)/\left(\square\square \times \begin{smallmatrix}\square\\\square\end{smallmatrix}\right) = \begin{smallmatrix}\square\\\square\end{smallmatrix}$ Fermi; if it is $j/j + 1$ for ℓ odd/ℓ even, the nuclear spin is j, the statistics of the nuclei is $\left(\begin{smallmatrix}\square\\\square\end{smallmatrix} \times \begin{smallmatrix}\square\\\square\end{smallmatrix}\right)/\left(\square\square \times \square\square\right) = \square\square$ = Bose. Experimentally, only Fermi statistics is found for half odd integral j (as for the electron) and Bose statistics for integral j. We will summarize this important experimental fact by

$$\text{statistics} = (-1)^{2j} \quad . \tag{2.61}$$

For instance when only even ℓ rotational states exist, we conclude that $j = 0$, and the statistics has to be Bose.

Historically, the first nuclear spin measured (F. Rasetti, *Z. Physik*, 61, 598 (1930)) was (in 1929) that of N_{14} (nitrogen molecule N-N). Rasetti found $j = 1$ and Bose statistics. But it was then believed that the universe was made of protons p^+, electrons e^-, and photons γ, (the only particles then known, and that the nucleus N_{14} of charge 7e, contained 14 protons and 7 electrons, thus, half integral spin and Fermi statistics were expected. This measurement started a crisis in physics.

Appendix. On Commutation Relations

Professor Bargmann pointed out to me that I have spoken of the invariance group of the commutation relations only in the context of rotational invariance (see Equation (2.10)). Surely it is worth mentioning the general case: consider the relations

$$[P_i, Q_j] = i\hbar\delta_{ij}\mathbb{1} \tag{2.61}$$

($i,j = 1$ to n). Let $a = (a_1 \ldots a_n)$, $b = (b_1 \ldots b_n) \in R^n$; we can use the tensor operator notations $P(a) = \Sigma_i a_i P_i$, $Q(b) = \Sigma_j b_j Q_j$. Equation (2.61) defines a $2n + 1$ dimensional Lie algebra which is a central non-abelian extension $\underline{g}$ of R^{2n} by R^1 (center of $\underline{g}$). This extension is defined by the antisymmetrical bilinear form on $R^{2n} = R^n \oplus R^n$

$$\sigma(a \oplus b, a' \oplus b') = a \cdot b' - b \cdot a' \tag{2.62}$$

where $a \cdot b = \Sigma_i a_i b_i$. The symplectic group $Sp(n)$ which leaves this form invariant is a group of automorphism of $\underline{g}$.

The corresponding simply connected group G has, up to an equivalence, a unique unitary irrep (von Neumann's theorem). Its Schrödinger realization as operators on the space L^2 of functions of n variables: $x = (x_1 \ldots x_n)$ is $U_a = e^{iP(a)}$ with $(U_a f)(x) = f(x + a)$; $V_b = e^{iQ(b)}$ with $(V_b f)(x) = e^{i\hbar b \cdot x} f(x)$. Here x, $a \in E_n$, $b \in E'_n$ dual of E_n. In the case of Equation (2.10) $n = 3$. Furthermore, the rotation group $SO(3)$ leaves invariant the symmetrical linear form β on E_3 and we used the corresponding identification of E_3 and its dual.

3. NUCLEAR PHYSICS: STRONG AND WEAK INTERACTIONS

3.1. The Set of Known Nuclei

The nuclei are made of protons p and neutrons n. These two particles have similar masses $m_p = 1836.10\ m_e = 938.25$ MeV, $m_n = 939.55$ MeV. The proton has electric charge + e. Both have spin 1/2. We define a nucleus by its number Z of protons and N of neutrons, and denote it by (Z,N); it contains A = Z + N nucleons. Nuclei have bound excited states, which are unstable. The ground state itself may be unstable and the nucleus may transform spontaneously into another nucleus by one of the following types of decay.

a) β^--decay $n \to p^+ + e^- + \bar{\nu}$($\bar{\nu}$ = antineutrino); $(Z,N) \to (Z + 1,N - 1) + e^- + \bar{\nu}$
β^+-decay $(Z,N) \to (Z - 1,N + 1) + e^+ + \nu$ which competes with e^--capture $(Z,N) + e^- \to (Z - 1,N + 1) + \nu$ (which requires less energy).
The mean life τ can vary from 10^{-3} sec to 10^{20} years.

b) α-decay†: $(Z,N) \to (Z - 2,N - 2) + (2,2)$ for $A > 140$ nuclei, τ from seconds to 10^{20} years.
And two much rarer types:

c) neutron emission: $(Z,N) \to (Z,N - 1) + n$ rare, $\tau <$ few seconds,

d) spontaneous fission into two smaller nuclei $(Z,N) \to (Z_1,N_1) + (Z_2,N_2)$.

Let us call nuclei stable if they have a half life of decay $\tau > 10^{20}$ years. 274 stable nuclei are known.

A even	Z even	N even	165		A odd	Z even	N odd	55
	Z odd	N odd	4	(Z = N = 1,3,5,7)		Z odd	N even	50

The much greater abundance of Z even, N even nuclei is strikingly illustrated in Figure 3.1 which gives the number of stable nuclei for given Z (isotopes) and for a given N (isotones).

It is worthwhile to note from Figure 3.1 that nuclei for Z = 20, Z = 50, (N = 82) have definitely more isotopes (or isotones) than their even-neighbors. This is also true, but less strikingly, for N = 20, N = 50 (and also N = 28). The heaviest stable nucleus is Pb_{208}, Z = 82, N = 126. Another striking feature in the distribution of stable nuclei in function of Z and N is that with two exceptions $N - Z \geq 0$ and $N - Z$ is a slowly increasing function of $A = N + Z$:

$N - Z = -1$ for the proton (Z = 1) and He^3(Z = 2)
$N - Z = 0$ for 13 nuclei;
$N - Z = 1$ for 16 nuclei;
$N - Z$ increases with A on the average $(N - Z) \sim 6.10^{-3}\ A^{5/3}$.

† What was first called an α-particle has been identified with a Helium nucleus: (2,2).

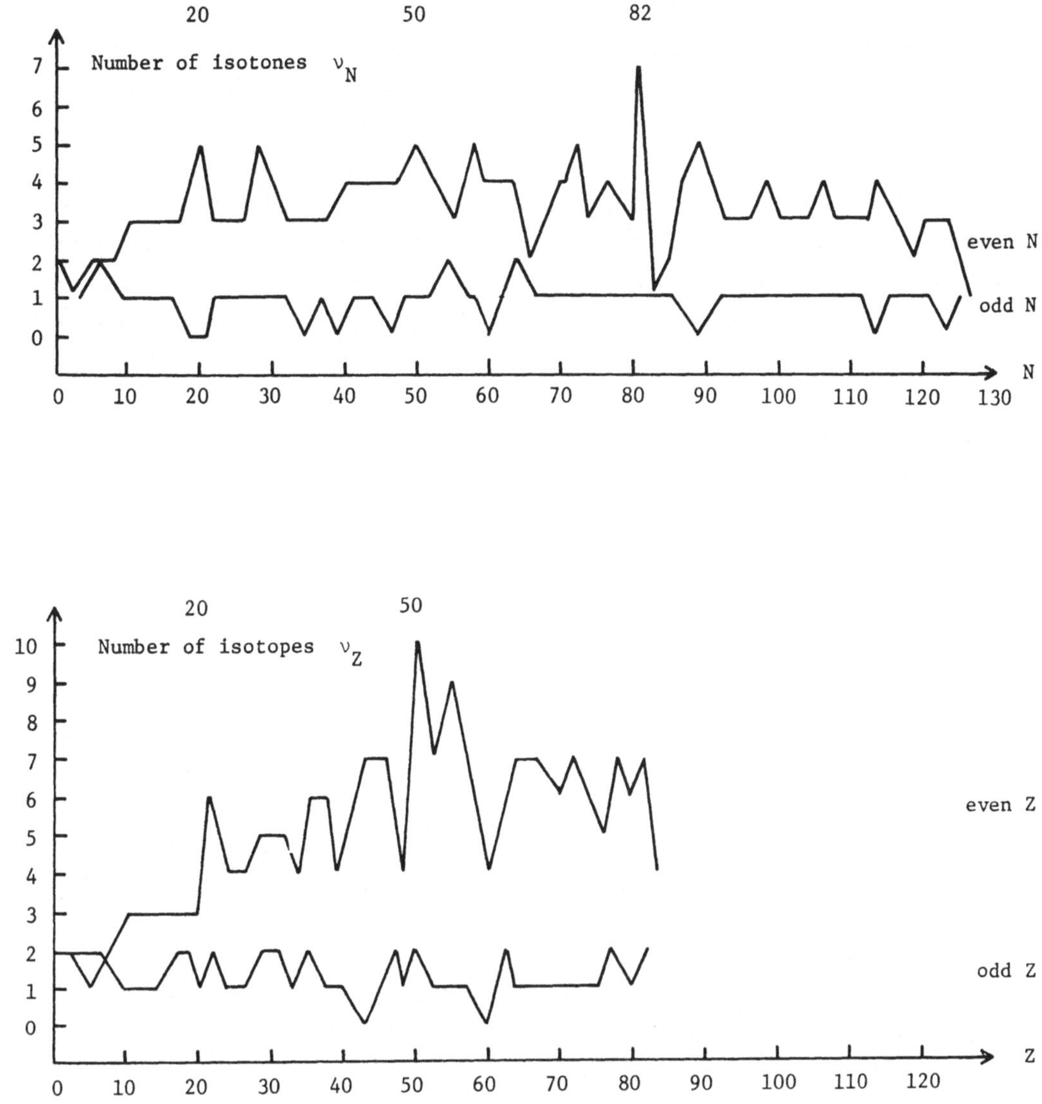

FIGURE 3.1. NUMBER ν_N AND ν_Z OF STABLE NUCLEI (Z,N) AS FUNCTIONS OF Z AND N

Note that there are no stable nuclei for $Z = 43, 61, > 83$, for $N = 19, 21, 35, 39, 45, 61, 89, 115, 123, > 126$ and none for $A = N + Z = 5, 8, 147, \geq 208$. The heaviest stable nucleus is ${}^{126}_{82}\mathrm{Pb}_{208}$ with $Z = 82$, $N = 126$. The most striking feature of Figure 3.1 is that ν_Z and ν_N are mainly 1, sometimes 2 or 0 for odd Z or odd N. Their value is more irregular for even Z or even N, there are relative maxima for $Z = 20 = N$, $Z = 50 = N$, $N = 82$ and also $N = 28$.

Nuclear forces are the most intense in nature, however, they do not bind more than 208 nucleons together†. The reason is that they have short range and also that nucleons obey Fermi statistics which, for condensed states, yield effects similar to repulsion.

More than one thousand different unstable nuclei are known. Those with a lifetime $\tau > .1$ (age of universe) and their decay products are found in nature, some are still produced in nature (C^{14}). All other are man made; more than half of those have Z-even, N-even. There exists a semi-empirical formula (Weizsäcker's) giving the binding energy of the lowest state of (stable or unstable) nuclei in function of Z, N and $A = Z + N$

$$B(Z,N) = Zm_p + Nm_n - m(Z,N) = U_\nu A - U_c Z(Z-1)A^{-1/3}$$

$$- U_s A^{2/3} - U_t \frac{(Z-N)^2}{A} + U_p \frac{(-1)^Z + (-1)^N}{2} A^{-3/4} \qquad (3.1)$$

the values of the constants U are in MeV

$$U_\nu = 14.0 \text{ MeV}, \; U_c = .61 \text{ MeV}, \; U_s = 14.0 \text{ MeV}, \; U_t = 84.2 \text{ MeV}, \; U_p = 34 \text{ MeV}$$

U_ν corresponds to maximal average binding energy by nucleon. The term U_c corresponds to the Coulomb repulsion among Z protons equally distributed in a sphere of radius proportional to $A^{1/3}$. The term U_s corresponds to a surface effect which suggests a short range for nuclear forces; U_t favors a minimum for $|Z - N|$ while U_p corresponds to pairing effects in like nucleons. As we saw, nuclei with even Z and N are more stable and more numerous than those with odd Z and/or odd N. A rule without exception is that all known Z even, N even nuclei have zero spin (= angular momentum at rest).

The distribution of nuclear spin for odd A nuclei is discussed in Section 3.4.

3.2. Isospin

As soon as the neutron was discovered (1932), Heisenberg created a formal language for the study of nuclei. Neutrons and protons are considered as the same particles, the nucleons, which have five degrees of freedom: 3 continuous in space ($\vec{x}$), a two valued one, σ, for the spin and a new one that Heisenberg simply called the fifth degree of freedom, τ, and which distinguishes neutrons and protons

† The existence of neutron stars with a radius of 10 km to 100 km and containing $\sim 10^{57}$ neutrons has been postulated. These stars seem to be observed now as "pulsars". They are indeed gigantic nuclei, but the binding energy is due both to nuclear and gravitational forces.

(*Z. Phys.*, 77, 1 (1932)); since, like the spin, it is two-valued, it is now called isospin.†

This Heisenberg convention has revealed itself more than useful. Indeed it was quickly established that nuclear forces did not distinguish between protons and neutrons: their differences (different electric charge and magnetic moment, small mass difference) are attributed mainly to electromagnetic effects and it is a reasonable approximation to neglect them.

If we denote the Hilbert space of our nucleon states by

$$\mathcal{H}^{(1)} = L_2(\vec{x},t) \otimes K_\sigma \otimes K_\tau \tag{3.2}$$

that of a number A of nucleons is

$$\mathcal{H}^{(A)} = \mathcal{H}^{(1)}_{[1^A]} = P_{[1^A]} \bigoplus_\lambda ((L_2 \otimes K_\sigma)^{(A)}_{[\lambda]} \otimes K^{(A)}_{\tau[\lambda]^c}) \tag{3.3}$$

where $P_{[1^A]}$ is the projector on $\mathcal{H}^{(1)}_{[1^A]}$. A convenient approximation for the study of a nucleus of A nucleons is to replace the sum of 2-particle interactions,†† by an average potential (= sum of 1-particle Hamiltonians) plus a residual 2-particle potential, which is still attractive. Then the analogy with the study of atoms††† allows us to draw qualitative conclusion. Using the same type of argument as in Section 2.6 for atoms, but here with the opposite sign, we know that for the ground state $[\lambda]$ in Equation (3.3) should be as symmetrical as possible, so $[\lambda]^c$ is as antisymmetrical as it can be with the restriction that it has only two lines. This implies that the two lines are as nearly equal as possible

$$\lambda_i \geq 0;\ [\lambda]^c = [\lambda_1,\lambda_2] 0 \leq \lambda_1 - \lambda_2 = \lambda \text{ minimal};\ \lambda_1 + \lambda_2 = A \quad . \tag{3.4}$$

If the nucleus has Z protons and N neutrons (Z + N = A) its states are completely symmetrical in sup{Z,N} particles, so

$$\lambda_1 \geq \sup\{Z,N\} \tag{3.5}$$

and

$$|Z - N| \leq \lambda_1 - \lambda_2 \quad . \tag{3.6}$$

---

† Called isotopic spin since 1936, the name isobaric spin would have been more proper. Anyway it has been shortened into isospin by the natural evolution of language.

†† In fact physicists are more sophisticated: when a sum of 2-particle interaction does not yield a good enough approximation, one adds also the sum of all k-particle $(2 < k \leq A)$ interactions, mainly for $k = A$ (collective effects).

††† There is still a difference. Atoms of n electrons consist of n + 1 particles and as we have seen, the elimination of the center of mass motion is easy: one singles out the nucleus, and the electrons are all treated on the same footing. This elimination is still clumsily carried out in nuclear physics.

So (3.4) can be translated into: the most stable nuclei have as small $|Z - N|$ as possible. As we have seen, this is well verified for light nuclei, where the electromagnetic repulsion of protons in negligible; when this repulsion is taken into account $0 < N - Z$ has to be a slowly increasing function of $A = N + Z$.

In the same approximation in which n, p are considered identical, isobars (nuclei with the same number $A = Z + N$ of nucleons) should be identical. Consider Figure 3.2; it gives the energy spectrum of the known states for $A = 15$,

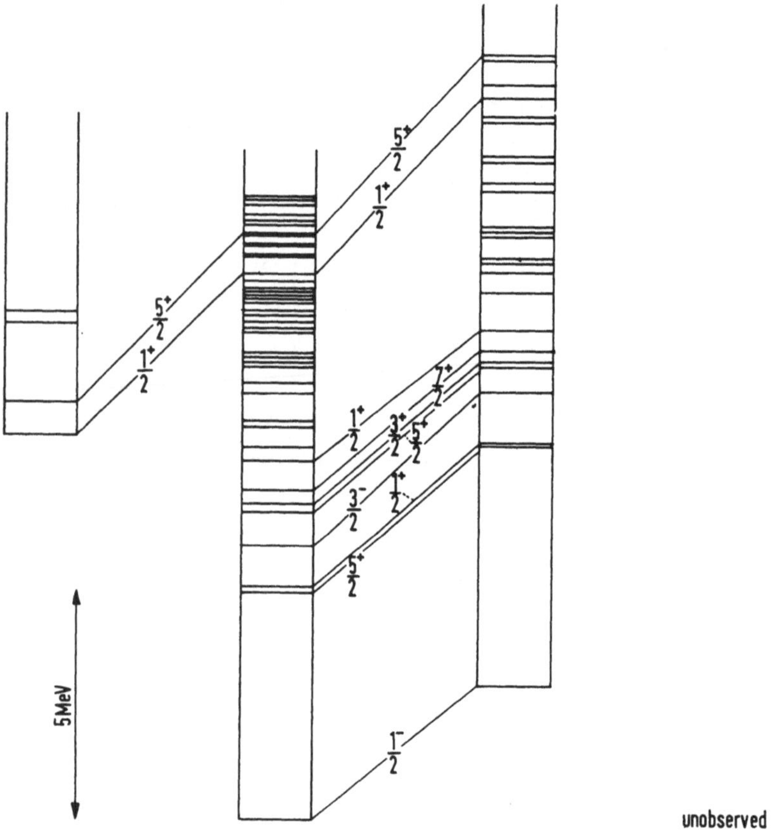

C^{15}	N^{15}	O^{15}	F^{15}
Z = 6	7	8	9
N = 9	8	7	6

j^P is the spin (j, 1/2 integer > 0) and the parity ($\pm$) of the state.

FIGURE 3.2. SPECTRUM OF STATES OF ISOBARS 15

States of C^{15} have isospin $\geq 3/2$. Another conventional notation for nuclei is to use the chemical symbol of the corresponding atom (this gives implicitly the number Z) and write the number of nucleons $A = Z + N$ in superscript.

and the known value of spin and parity of these states. The similarity of the spectra (at least for the low lying part) for $1/2|Z - N| = 1/2$ (i.e. N^{15} and O^{15} nuclei) is striking. The essential difference is a shift upward in energy of ~ 3 MeV for O^{15} which has one more proton than N^{15}. The pairs of corresponding states are called "doublets" of isospin 1/2 states.

Let us explain in detail this point of view, which exploits the relations between the permutation groups $S(n)$ and the unitary groups $U(k)$ that we have recalled in 1.4 and used in 2.9.

Nuclear interaction does not distinguish between protons and neutrons. For a nucleus this implies a property of invariance under the permutations $(\in S(A))$ of its nucleons. We could also have translated this property as follows:

All nuclear physics observables O acting on $\mathcal{H}^{(1)}$ (space of the one particle states for the nucleon) in Equation (3.3) are of the form (3.7), where

$$\mathcal{H}^{(1)} = L_2(\vec{x},t) \otimes K_\sigma \otimes K_\tau \tag{3.2}$$

$$O = \quad N \quad \otimes I \tag{3.7}$$

$$U(2) = \quad I \quad \otimes U(2) \tag{3.8}$$

i.e., they correspond to a trivial action on K_τ, the factor in this tensor product which corresponds to Heisenberg's fifth degree of freedom "proton - neutron".

The action of the group $U(2)$ on $\mathcal{H}^{(1)}$, as defined by (3.8), commutes with every observable: $U(2) \subset \{O\}'$, the commutant of the algebra of "one particle observables". The action of this $U(2)$ can be extended to every $\mathcal{H}^{(A)}$, $(A \geq 0)$, Hilbert space of the A particle states. Therefore, in nuclear physics, when the non-nuclear interactions are neglected, this $U(2)$ is a subgroup of the invariance group. $\mathcal{H}^{(A)}$ has the same decomposition into spaces of factorial representation for $S(A)$ and $U(2)$ and we use the same symbols (Young diagrams) for the corresponding representations.

Since Coulomb repulsion of the protons can be neglected only in light nuclei, it was not expected that isospin conservation could be an interesting concept for heavier nuclei. However, the progress of nuclear physics in the last five years has shown that for nuclei with A up to 100, isospin is indeed a useful concept. For a non-technical review of this question, see W. R. Coker and C. F. Moore, "Isobaric Analog Resonances", *Physics Today*, 22, no. 4, 53 (1969).

3.3. U(4) Invariance

In 1936 Wigner, in his paper "On the consequence of the Symmetry of the nuclear Hamiltonian on the Spectroscopy of Nuclei", *Phys. Rev.*, 51-106 (1937)†

† Reproduced in Dyson's anthology: *Symmetry Groups in Nuclear and Particle Physics*, Benjamin, New York (1966).

studied the approximation where not only isospin dependence of nuclear forces is neglected but also the spin dependence. Then Equation (3.7) and (3.8) can be replaced by

$$\mathcal{H}^{(1)} = L_2(x) \otimes K_\sigma \otimes K_\tau \tag{3.2}$$

$$\mathcal{O} = N \otimes I \otimes I \tag{3.9}$$

$$\mathcal{U}(4) = I \otimes \mathcal{U}(4) \quad . \tag{3.10}$$

In this approximation, nuclear theory is also invariant under the group $\mathcal{U}(4)$ acting on the four dimensional space $K = K_\sigma \otimes K_\tau$ and Equation (3.3), for the Hilbert space of A nucleon states can be replaced by

$$\mathcal{H}^{(A)} = \mathcal{H}^{(1)}_{[1^A]} = P_{[1^A]} \oplus_\lambda (L_{2[\lambda]} \otimes K_{[\lambda]^c}) \tag{3.11}$$

where the $[\lambda]$ are representations of $\mathcal{U}(4)$.

For the most stable states, the property (used in 3.2) of the "residual" two-nucleon force to be attractive implies now that $[\lambda]$ is as symmetrical as possible, so $[\lambda]^c$ is as antisymmetrical as possible, i.e., its Young diagram has its four lines of length $\lambda_1 \geq \lambda_2 \geq \lambda_3 \geq \lambda_4 \geq 0$ (with $\lambda_1 + \lambda_2 + \lambda_3 + \lambda_4 = A$) as nearly equal as possible. For A/4 = integer this implies $\lambda_1 = \lambda_2 = \lambda_3 = \lambda_4 = A/4$. This $\mathcal{U}(4)$ irrep has dim. 1. The restriction of this representation of U(4) (acting on $K_\sigma \otimes K_\tau$) to the subgroup SU(2) × SU(2), yields a spin 0 and isospin 0 for the ground state. As we have seen, the former result is observed for all such nuclei, the latter only for light nuclei $(Z < 17)$ where Coulomb repulsion of protons is not too large. For nuclei with $A = 4n + 2$, the $[\lambda]^c$ representation of lowest lying states is $\lambda_1 - 1 = \lambda_2 - 1 = \lambda_3 = \lambda_4 = n$; it has dimension $\binom{4}{2} = 6$. Its restriction to the subgroup SU(2) × SU(2) decomposes into the direct sum of two three-dimensional representations: one of spin 1, isospin 0, the other of spin 0, isospin 1. In Figure 3.3 (for which $n = 1$) this gives correctly the spin of the lowest state of Li^6 (spin 1) and He^6 and Be^6 (spin = 0). These last two levels form an isospin triplet with the third level (spin 0^+) of Li^6. The other levels whose spin are marked in Figure 3.3 belong to another equivalent representation of $\mathcal{U}(4)$ with an angular orbital momentum (i.e., angular momentum of the space degree of freedom) $\ell = 2$. So the total angular momentum has the possible value $j = \ell = 2$ for the spin 0, isospin 1 states and $\ell - s \leq j \leq \ell + s$ i.e., $= j = 3, 2, 1$ for the spin 1, isospin 0 states i.e., those of Li^6 with no correspondents in He^6 and Be^6.

States belonging to a U(4) irrep are called supermultiplets in physics literature. The study of Galilean invariance of the theory of supermultiplets is very similar to that made in 2.9 for atomic physics.

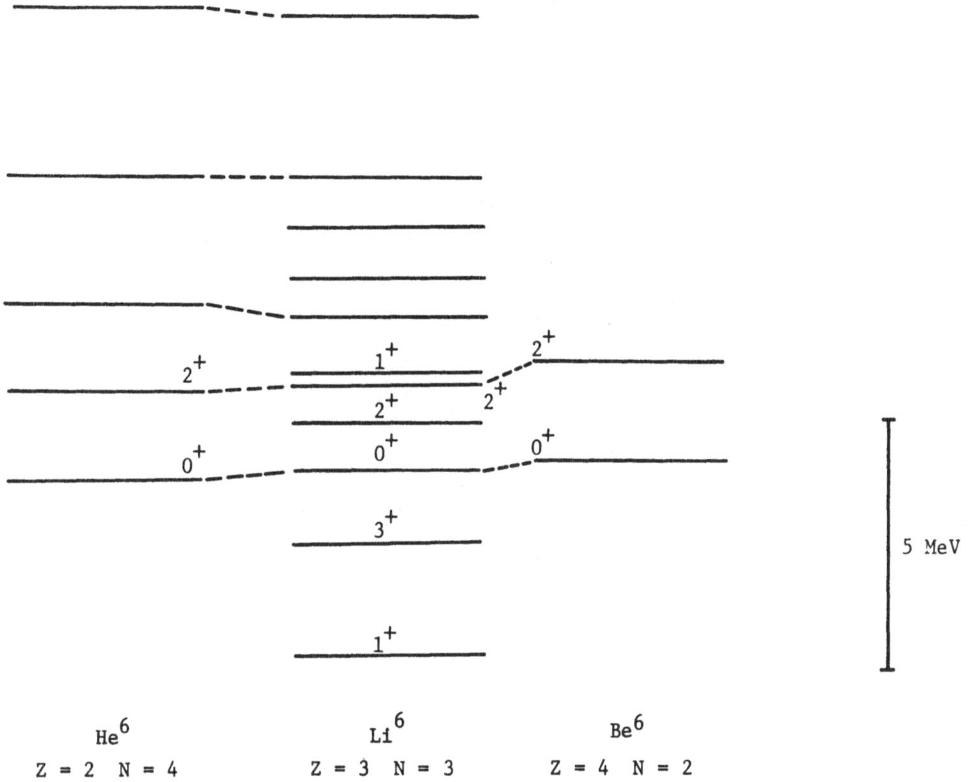

j^p is the spin (j integer ≥ 0) and the parity p(±) of the state.

FIGURE 3.3. SPECTRUM OF LEVELS OF THE NUCLEI WITH 6 NUCLEONS

One has to consider the covering $\overline{G}$ of the Galilean group G, $\overline{G} \xrightarrow{\pi} G$ and also the homomorphism

$$\overline{G} \xrightarrow{\phi} SU(2) \times 1 \subset \mathcal{U}(2) \times \mathcal{U}(2) \subset \mathcal{U}(4) \quad .$$

The invariance group of the theory is the direct product $G \times \mathcal{U}(4)$ and $\overline{G}$ is the subgroup $\overline{G} \xrightarrow{i} G \times \mathcal{U}(4)$ with $i(\overline{g}) = (\pi(\overline{g}), \phi(\overline{g}))$.

The approximation of spin and isospin independence for nuclear forces leading to $\mathcal{U}(4)$ invariance is crude and could not be expected to be very useful for nuclei with a number A of nucleons not very small. However, as for the better approximation of isospin conservation, $\mathcal{U}(4)$ invariance has been usefully applied to nuclei with A up to 100 as shown by the statistical study of the energy of their ground state: P. Franzini and L. A. Radicati, "On the validity of the supermultiplet model", *Phys. Lett.*, 6, 322 (1963). (Reproduced in Dyson's anthology, quoted in 3.3 and 4.)

3.4. Shell Model

We have seen that nuclei with Z or N = 8, 20, 50, 82, and N = 126 seem more stable. Many properties of nuclei (e.g. binding energy per nucleon, magnetic and quadripole moment) single out these numbers. A survey of nuclear tables shows that for A = odd nuclei, the parity and spin of the ground state vary in a regular pattern which suggests very much the filling of shells (as in atomic physics). Ground state, spin j and parity $\pm$ depend only on the value of the odd number Z or N so the order of the levels, with increasing energy, is the same for protons and neutrons. The order obtained can be deduced from the three-dimensional harmonic oscillator spectrum $E_n = n\hbar\omega + E_\circ$ (where ω is a constant) with some modifications.

Consider the set of nine operators $T_{ij} = P_iP_j + Q_iQ_j$ where the P_i and Q_j $(i = 1,2,3)$ satisfy the canonical commutation relation

$$[P_i,Q_j] = i\hbar\delta_{ij} \quad .$$

Its use for the computation of the commutators $[T_{ij},T_{i'j'}]$ shows that the T_{ij} form a representation up to $i\hbar$ of the Lie algebra U(3). The center

$$H = \mathrm{Tr}T_{ij} = T_{11} + T_{22} + T_{33}$$

is the harmonic oscillator Hamiltonian (in convenient units). From $[H,T_{ij}] = 0$ we deduce that there is a U(3) invariance for the three-dimensional harmonic oscillator similar to the SO(4) invariance of the hydrogen atom, 2.4, and we can find the spectrum of H by a similar method:

The ground state $n = 0$ has energy $E_\circ$. The number of states of energy E_n is $1/2(n + 1)(n + 2)$. Their orbital angular momentum ℓ satisfies $(-1)^\ell = (-1)^n$, $0 \le \ell \le n$. This corresponds to the first column of Figure 3.4. The second column gives the spectrum of the Hamiltonian

$$\hbar^{-1}H' = \omega H - \omega'\vec{L}^2 - 2\omega''\vec{L} \cdot \vec{S} \tag{3.12}$$

with $\omega, \omega', \omega''$ positive constants, $\omega > \omega'$ and ω''. Using Equation (2.55) for $s = 1/2$ and when $\ell > 0$, $j = \ell + \varepsilon\, 1/2$, $\varepsilon = \pm 1$ one obtains the energy spectrum

$$\ell > 0 \qquad E_{n,j,\ell} - E_{0,\frac{1}{2},0} = n\omega - \omega'(\ell(\ell + 1) - \varepsilon\omega''(\ell + \tfrac{1}{2}), \ \varepsilon = \mathrm{sign}(j - \ell) \tag{3.13}$$

$$\ell = 0 \qquad\qquad = n\omega \tag{3.13'}$$

This Hamiltonian H' is the one-nucleon Hamiltonian in the average potential produced by the whole nucleus. As in 2.7 we can now "fill the successive shells" for protons and neutrons. Such shell-model for nuclei was proposed in 1949 (see M. Goeppert Mayer and J. H. D. Jensen, *Elementary Theory of Nuclear Shell Structure*, Wiley, New York (1955)). It is very successful in explaining the properties of the

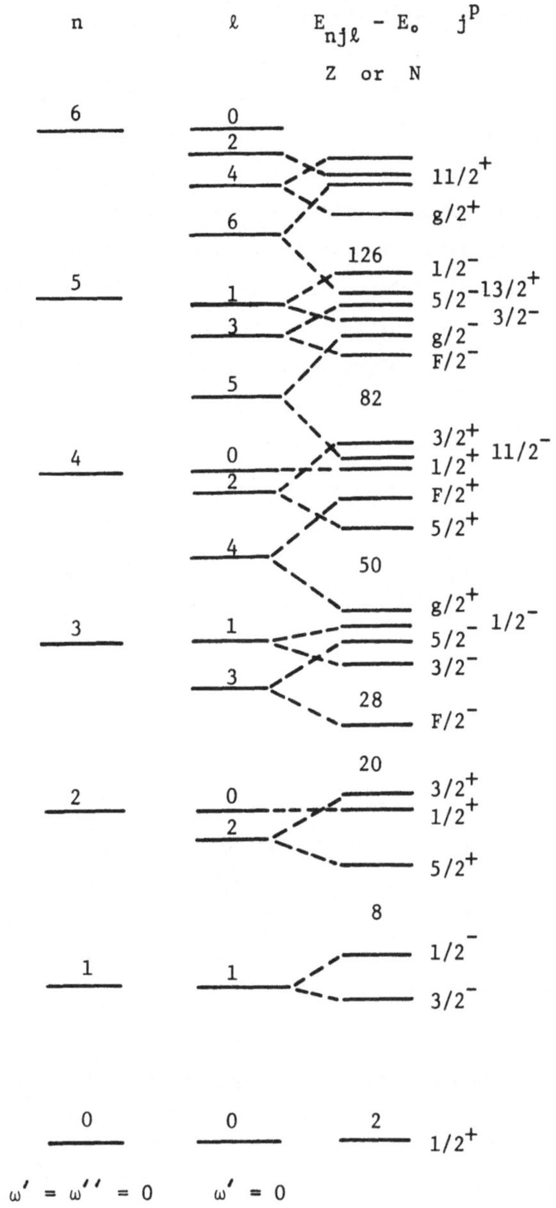

j^P is the spin (j half integer > 0) and the parity (±) of the state.

FIGURE 3.4. ENERGY SPECTRUM OF THE ONE NUCLEON HAMILTONIAN OF EQUATION (3.12)

low lying levels of nuclei, and their decays. A more detailed book on nuclear shell structure is that of A. de Shalit and I. Talmi.

Note that a Z (or N) particle state is to a lesser degree a decomposable antisymmetrical tensor than it is for atoms (see 2.8). There is much more configuration mixing, i.e., the nuclear states are tensors which are linear combination of different decomposable tensors with the same quantum numbers.

Shell-model computations require a classification of states inside the "same shell". That was done between 1942 and 1949 by Racah (*Phys. Rev.*, 76, 1352 (1949)) who introduced the seniority quantum number. (See also work of Flowers and many references in Dyson's anthology quoted in 3.3 and 4.3.)

Part of the problem is to label unambiguously irreps $D_j^{\cdot}$ of SU(2) appearing in the decomposition into a direct sum of irreps of the tensor power of a given irrep $\overset{n}{\otimes} D_j$ (where J is usually integral for atoms, half integral for nuclei). The method is to find a chain of subgroups

$$H_\circ = U(2\bar{j} + 1) \supset H_1 \supset \ldots \supset H_k \supset SU(2) \quad \text{or} \quad SO(3) \tag{3.14}$$

(where SU(2), (2J odd) or SO(3), (2J even) is the subgroup of U(2J + 1) formed by the matrices of D_j) with the following property:

The successive restrictions of the representation of U(2J + 1)

$$\overset{n}{\otimes} \square = \oplus_\lambda \; s_\lambda [\lambda] \qquad \text{(see (1.15}'))$$

(where $\oplus_\lambda$ is over all Young diagrams of n squares and s_λ is the dimension of the corresponding irrep of $S(n)$) to the different H_i, $(0 \le i \le k)$ must finally yield direct sums of SU(2) irreps with multiplicity one. Table 3.1 illustrates a simple example J = 2, n = 3, the U(5) irrep $[3]$ and $[1,1,1]$ restricted to SU(2) yields only multiplicity one. This is not the case for $[2,1]$. One intermediate group is necessary $H_1 = SO(5)$.

TABLE 3.1. DECOMPOSITION OF $\overset{3}{\otimes} D_2$ (3 PARTICLES IN D-SHELL)

representation of U(5)			$(\square)^3 = [3] \oplus 2\,[2,1] + [1,1,1]$
dimension			$5^3 = 35 + (2 \times 40) + 10$
Irrep of U(5)		**Restriction to SO(5)**	**Restriction to SO(3)**
$[3]$	=	one irrep	$[3] = D_6 \oplus D_4 \oplus D_3 \oplus D_2 \oplus D_0$
$[2,1]$	=	$A \oplus B$	$A = D_5 \oplus D_4 \oplus D_3 \oplus D_2 \oplus D_1$
			$B = D_2$
$[1,1,1]$	=	one irrep	$[1,1,1] = D_3 \oplus D_1$

Neglecting the s_λ multiplicity, every inequivalent irrep of SO(3)/or SU(2) has a different genealogy of irreps of H_i. Racah says that they have different seniority quantum number. For distinguishing the different seniorities Racah had to introduce the exceptional Lie group G_2 among the H_i (in the case $J = 3$, $n = 7$)! When J is half integer, one generally takes $H_1 = Sp(2J + 1)$. Of course, nuclear physicists nowadays use more refined models of nuclei (e.g., *Nuclear Structure I, II, III*, by A. Bohr and B. Mottelson, Benjamin). To go into more details is outside the scope of these lectures. We refer again to a non-technical paper by D. R. Inglis: "Nuclear Models", *Physics Today*, 22 no. 6, 29 (1969) for a recent survey.

3.5. The Hadrons

Although SU(2) invariance, through isospin considerations, is more familiar to physicists than $S(n)$ invariance for the study of the property of nuclear interaction not to distinguish between neutrons and protons, is it more fundamental? If one had to deal only with nucleons, the answer is no; both mathematical methods are physically equivalent. However, there are many more particles with strong interaction; they cannot be permuted with the nucleons but they can be attributed an isospin. Let us give as example the π-meson. In 1935, Yukawa predicted the existence of mesons which are to the nuclear interaction what photons are to the electromagnetic interaction. He predicted their electric charge $\pm$, their mass, their lifetime, their decay mode. Soon the particles were discovered but it was a case of mistaken identity with the μ-lepton! The Yukawa particle was discovered in 1947 and is called $\pi^{\pm}$. In 1937, physicists (e.g., Kemmer) showed that 3 states of charge were necessary for the meson, +, 0, -. Indeed, in order that nuclear interaction preserve isospin, they have to be invariant under the corresponding SU(2). In Yukawa's theory the meson field is coupled with the nucleonic current. This current transforms under SU(2) as a tensor operator of the space of the representation $(D_{1/2})^2 = D_1 \oplus D_0$. Then the simplest SU(2) invariant Yukawa coupling which can include electrically charged meson, is of the form

$$\int \vec{j}(x) \cdot \vec{\phi}(x)dx \tag{3.15}$$

where $\vec{j}(x)$ and $\vec{\phi}(x)$ are vector operators for the isospin SU(2) and the interaction is the scalar product of these vectors. The π°-meson so predicted in 1937 was found in 1950.

Already in 1947 two other strongly interacting particles had been found. The generic name "hadron" was given to particles with strong interaction. The rate of discovery of new hadrons has passed from 15 in the fifties to 250 in the sixties. We give their mass spectrum and their spin and parity when known, in Table 3.2;

TABLE 3.2. SPECTROSCOPY OF HADRONS

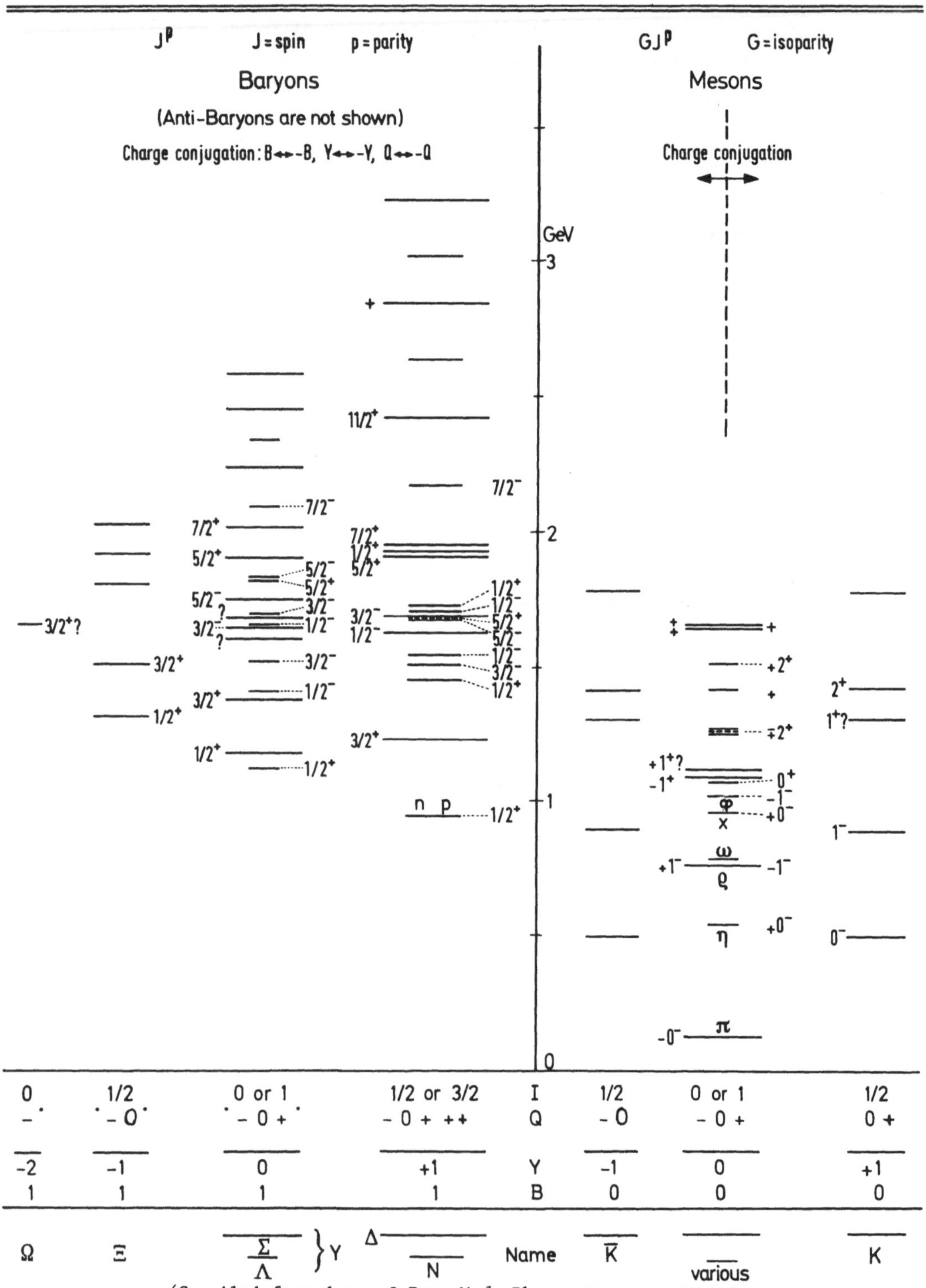

(Compiled from data of *Rev. Mod. Phys.*, January, 1969.)

different columns of this table correspond to different values of the quantum numbers preserved by the nuclear interaction also called strong interaction.

There is a charge <u>b</u> which is for the strong interaction what the electric charge is for the electromagnetic interaction. The "charged" particles have $b = \pm 1$; they are called baryons. The "neutral" particles $b = 0$ are called mesons. It happens that all baryons have half integral spin and all mesons have integral spin. This will be explained in 4.2 (Equation 4.8). There has been no difficulty in attributing an isospin to every baryon. Particles inside the same isospin multiplet have the same spin and parity, their masses are equal within 1% (exceptionally 3% for the π-mesons) and they have different electric charges. Isospin conservation allows us to predict some missing members of a multiplet which are then looked for and have always been found. Instead of using the value of their electric charge q the $2t + 1$ states of the same isospin multiplet can be labeled by the value t_3 of the isospin $SU(2)$ generator in the direction "3". These two labels are equivalent but different; the difference $q - t_3$ depends on the multiplet; since both q and t_3 are conserved by strong interaction, their difference

$$y = 2(q - t_3) \tag{3.16}$$

is a new strong interaction quantum number which can be attributed to each isospin multiplet†; it has integral values, and it is called the hypercharge. To include it, one had enlarged the $SU(2)$ invariance group to a $U(2)$ group††. In 5.1, we will see how $U(2)$ was enlarged further.

In Table 3.2 we have left out the antibaryons, each one is to the corresponding baryon what antiproton is to proton (see Dirac quotation in introduction). Antibaryons are obtained from baryons by the involution called C which changes the sign of the charges b, q, y and leaves invariant the mass and the spin (for more detail on C and its relation with parity, see 4.6). To each isospin multiplet with values t, y, b correspond the C-conjugated multiplet $t, -y, -b$. A new quantum number is then necessary for the self-conjugated multiplets $(b = 0, y = 0)$. I introduced it in 1953††† and called it "isoparity". It is denoted G in Table 3.2.

Taking into account all quantum numbers introduced for hadrons, the invariance group should be written $(U_1 \times U_2) \,\square\, Z_2(C)$ where U_1 corresponds to the baryonic charge, $Z_2(C)$ is the two element group generated by C and $\square$ means the semi-direct product. The action of C is equivalent to the complex conjugation of the matrices of $U(1)$ and $U(2)$. Irreps of this group when $b = 0 = y$, and by

---

† Relation (3.16) was guessed by M. Gell-Mann, *Phys. Rev.*, <u>92</u>, 833 (1953); see also T. Nakano and K. Nishijima, *Prog. Theo. Phys.*, <u>10</u>, 587 (1953).

†† And not to $U_1 \times SU(2)$ because of the relation $(-1)^y = (-1)^{2t}$ implied by (3.16). See L. Michel reference LM I of 4.9.

††† L. Michel, *N. Cim.*, <u>10</u>, 319 (1953).

(3.16), t is integral, are faithful representations of $SO(3) \square Z_2(C)$ which is isomorphic to the direct product $SO(3) \times Z_2(C)$.

Finally, let us emphasize that all particles in Table 3.2 with the exception of the proton (and the antiproton) are unstable. Neither their lifetime nor their decay modes are indicated. Most of them are even unstable by the strong interaction with a lifetime of 10^{-23} to 10^{-22} sec. They are often called resonances instead of particles because they do not completely fit this latter concept. (See 4.4.) The particles stable for strong interaction are the lowest states of the columns in Table 3.2 and the first excited states of the column $y = 0$, $b = 1$ (Σ) and $b = 0$ (η). However, to be stable, or unstable does not seem so fundamental! Indeed if the mass difference between the lowest states of two neighboring columns of baryons in Table 3.2 ($\Delta y = 1$) were $> m_k$ the highest of these lowest states would be unstable for strong interaction. The Σ is stable only because $m_\Sigma - m_\Lambda < m_\pi$; if for instance $m_\Delta - m_N < m_\pi$ were satisfied, the Δ would also be stable. The stability of η is due to the fact that both η and π have spin 0, parity - (invariance under P, see 4.7a) and that η-isoparity = + while π-isoparity = -.

3.6. The Other Particles and the Other Interactions

There are only nine known "elementary" particles which are not hadrons, i.e., have no strong interaction.

The photon, γ, with mass zero, spin 1 (see 4.4).

The 8 leptons μ^+, μ^-; e^+, e^- (electrons) and their associated zero mass neutrinos $\nu_{\mu^+}\nu_{\mu^-}\nu_{e^+}\nu_{e^-}$; they all have spin 1/2. Only the μ are unstable.

$$\mu^\pm \to e^\pm + \nu_{e^\pm} + \nu_{\mu^+}$$

because $m_\mu = 207\, m_e$.

All particles have electromagnetic interaction even when they have no electric charge (e.g. $q = 0$; baryons have magnetic moments) but it seems that Ampère's hypothesis† that the whole electromagnetic interaction is through the electromagnetic current $j^\mu(x)$ is well verified; the interaction Hamiltonian is

$$H_{em} = e \int A^\mu(x) j_\mu(x) d^3\vec{x} \tag{3.17}$$

where $A^\mu(x)$ is the photon field (electromagnetic potential vector). In units for which $\hbar = c = 1$, the universal constant e is given by $e^2 = \alpha = 1/137.039$ (see 2.4). Electromagnetic interaction is about 100 times weaker than strong interaction. This is also the order of magnitude of mass difference in an isospin multiplet.

---

† Called nowadays "minimal coupling" in the jargon.

There is another universal interaction, shared by all particles (except the photon), characterized by a universal constant G, introduced by Fermi†, whose value is

$$\frac{G}{\sqrt{2}} = 1.01 \times 10^{-5} \times m_p^2 \tag{3.18}$$

Since it is much weaker than the electromagnetic interaction, this interaction is simply called the "weak" interaction. Fermi postulated as early as 1934 that this interaction has some analogy with the electromagnetic interaction. For instance there are four electrically charged ($\pm$) "weak" hadronic currents, respectively vectors and axial vectors for the Lorentz group, $v_{\mu}^{(\pm)}(x)$, $a_{\mu}^{(\pm)}(x)$ which interacts with the leptons through a leptonic current $\ell^{\mu}_{(\pm)}(x)$, and the interaction Hamiltonian being

$$H_w = \frac{G}{\sqrt{2}} \sum_{\varepsilon=\pm 1} \int \ell^{\mu}(\varepsilon)(x) h_{\mu}(\varepsilon)(x) d^3\vec{x} \tag{3.19}$$

with

$$h_{\mu}(\varepsilon)(x) = v_{\mu}(\varepsilon)(x) - a_{\mu}(\varepsilon)(x), \quad (\varepsilon = \pm 1) \tag{3.20}$$

Equation (3.19) has indeed some similarity with (3.17). The fact that h_{μ} is a linear combination of a vector and an axial vector will explain the parity violation of the weak interaction (see 4.7.b).

R. P. Feynman and M. Gell-Mann (*Phys. Rev.*, 109, 193 (1958)) have found a very deep relation among the three interactions. From the unitary representation of $\mathcal{U}(2)$ on $\mathcal{H}$, the Hilbert space of hadrons, one obtains the representation F of the $\mathcal{U}(2)$-Lie algebra on $\mathcal{H}$. The operators corresponding to the observables y and t_3 are the self-adjoint operators

$$Y = F(y) \quad \text{and} \quad T_3 = F(t_3) \tag{3.21}$$

Since $q = t_3 + 1/2\, y$ (Equation 3.16) for all hadronic states, this relation has also to be true for the self adjoint operators representing these observables, so

$$F(q) = Q = \int j^{\circ}(x) d^3\vec{x} = T_3 + \frac{1}{2} Y \tag{3.22}$$

Note that $\partial_{\mu} j^{\mu}(x) = 0 \Rightarrow Q$ is time independent $\Leftrightarrow [H,Q] = 0$. However, Q here is the total electric charge of the hadronic part of the world, it is not conserved since weak interaction can transfer it to the leptonic part of the world. It is conserved only in the approximation which neglects weak interaction.

The beauty of the discovery by Feynmann and Gell-Mann is that, when electromagnetic and weak interactions are neglected, the vector part of the weak hadronic currents $v'_{\mu}(\varepsilon)(x)$ (Equation 3.20) and the electric current $j_{\mu}(x)$ of the

† E. Fermi, "Versuch einer Theorie der β-strahlen", *Z. Physik*, 88, 61 (1934).

hadrons are images of the same tensor operator for the U(2) group of invariance of strong interaction for, respectively, the vectors $t_\pm$ and q of the vector space of the complexified U(2)-Lie algebras

$$t_\pm \wedge y = 0 = y \wedge t_3, \; t_\pm \wedge t_3 = \pm t_\pm \quad . \tag{3.23}$$

This implies that

$$T_\pm = F(t_\pm) = \int v'^0(\pm)(x) d^3\vec{x} \tag{3.24}$$

The isospin group, which was introduced in 3.2 in such a formal and abstract fashion, becomes a physical reality since it is generated by the space integral of the weakly interacting hadronic currents! The addition of the electric charge generates the full U(2) group. When electromagnetic and weak interactions are not neglected, $\partial_\mu v^\mu(x)$ as well as $\partial_\mu j^\mu(x)$ do not vanish and the representation of U(2) on $\mathcal{H}$ becomes: 1) time-dependent for the physicists (just as $[P,Q] = i\hbar\mathbb{1}$ is true at any time with time-dependent P, Q); 2) undefined for the mathematicians (as Coleman and other physicists have shown). Have you noticed the v' instead of v in Equation (3.24)? I have shortened a long story. The Feynman-Gell-Mann hypothesis really needs the enlargement of the U(2) group to SU(3) as we will explain in 5.1 and 5.3.

To stay inside U(2), one has to decompose h_μ of Equation (3.19)

$$h_\mu(\varepsilon)(x) = h'_\mu(\varepsilon)(x) \cos\theta + h''_\mu(\varepsilon)(x) \sin\theta \tag{3.25}$$

where $h'(\varepsilon)$ has hypercharge $y = 0$ and $h''(\varepsilon)$ has $y = \varepsilon$ and θ is the Cabibbo angle†. The same decomposition appears separately for the $v_\mu(\varepsilon)(x)$ and the $a_\mu(\varepsilon)(x)$ part of h (Equation (3.20)). The angle θ has a value 15° so the $|\Delta y| = 1$ weak transitions are slower than those with $|\Delta y| = 0$ by a factor $tg^2\theta$.

They have also a different "selection rule" for isospin. As we just said, v' is a vector-operator for the SU(2) isospin group. This is also true for a' and h'. Hence

weak transitions $|\Delta Y| = 0$ satisfy $|\Delta T| = 0$ or 1
while weak transitions $|\Delta Y| = 1$ satisfy $|\Delta T| = 1/2$

i.e., h'', v'', a'' are SU(2)-spinor operators.

We have also to mention two other charges conserved by <u>all</u> known interactions (as the baryonic and electric charges). They are the two leptonic charges which seem separately conserved: that of the e-type with value $\varepsilon = \pm 1$ for e^ε, ν_{e^ε} and zero for μ, ν_μ; that of the μ-type: with value $\varepsilon = \pm 1$ for μ^ε, ν_{μ^ε} and zero for e, ν_e.

---

† N. Cabibbo, *Phys. Rev. Lett.*, <u>10</u>, 531 (1963).

4. RELATIVISTIC INVARIANCE. THE DISCRETE SYMMETRIES C. P. T.

4.1. The Poincaré Group and its Automorphisms; Zeeman Theorem

Physicists call Poincaré group the inhomogeneous Lorentz group†. We will denote its connected component by $\mathcal{P}_0$. It is the semi-direct product $\mathcal{T} \square \mathcal{L}$ of the connected Lorentz group $\mathcal{L}_0$ by the translation group $\mathcal{T}$. It has a trivial center. It is a 10 parameter real Lie group. Its universal covering $\overline{\mathcal{P}}_0$ is the semi-direct product $\mathcal{T}_\square SL(2,C)$, whose center is a two element group generated by ω = "the rotation by 2π". The group law of $\overline{\mathcal{P}}_0$ is given explicitly in Equation (4.10).

We call P the space reflection $P(\vec{r},t) = (-\vec{r},t)$ and T the time reflection $T(\vec{r},t) = (-\vec{r},t)$, $\mathcal{D}$ the group of dilations $\{\alpha > 0, \alpha(\vec{r},t) = (\alpha\vec{r},\alpha t)\}$. We call $Z_2(P)$, $Z_2(T)$, $Z_2(P) \times Z_2(T)$ the group generated by respectively P, T, P and T. We denote by $\mathcal{P}$, $\mathcal{P}^\uparrow$, $\mathcal{F}^\uparrow$, $\mathcal{F}$ the groups generated by $\mathcal{P}_0$ and respectively $Z_2(P) \times Z_2(T)$, $Z_2(P)$, $Z_2(P)$ and $\mathcal{D}$, $Z_2(P) \times Z_2(T)$ and $\mathcal{D}$. We call $\mathcal{P}$ the full Poincaré group.

It can be proven†† that all automorphisms of these groups are continuous and, if Aut G is the automorphism group of G,

$$\text{Aut } \mathcal{P}_0 = \text{Aut } \mathcal{P}^\uparrow = \text{Aut } \mathcal{P} = \text{Aut } \mathcal{F}^\uparrow = \text{Aut } \mathcal{F} = \mathcal{F} \quad . \tag{4.1}$$

Given any group G, we denote by In.Aut G the group of inner automorphisms and by Out G the quotient Out G = Aut G/In.Aut G. Note that here $\mathcal{P}_0$ = In.Aut $\mathcal{P}_0$ and that $\mathcal{F}$ is the semi-direct product

$$\mathcal{F} = \mathcal{P}_0 \square (Z_2 \times Z_2 \times \mathcal{D}) = \mathcal{P}_0 \square \text{Out } \mathcal{P}_0 \quad . \tag{4.2}$$

Binary Relation on Space Time $\mathcal{E}$

Given $x \neq y$ two distinct points of $\mathcal{E}$, we define the notations:

x T y = (y is inside the light cone of x)

s L y = (y is on the light cone of x)

---

† Lorentz transformations were introduced by Vogt in 1882 and applied by Lorentz to electromagnetism. H. Poincaré (*C. R. Acad. Sci., Paris*, 140, 1504 (1905)) required that they form a group with the rotation group and, from it deduced physical consequences. In *Rend. Circ. Mat. Palermo*, 21, 129 (1906) he included the translations and studied physical implications of invariance under the group we call here Poincaré group.

†† L. Michel, "Relations entre symétries internes et invariance relativiste", lectures published in *Application of Mathematics to Problems in Theoretical Physics, Cargèse 1965*, Lurçat editor, Gordon and Breach (1967) referred to as LM III. We will also refer to my lectures in Istanbul (1962) and Brandeis (1965) as LM I and LM II. They are both published by Gordon and Breach (Book of the lectures, for each school).

$x\ S\ y$ = (y is outside the light cone of x)

$x < y$ = (y is inside the future light cone of x)

$x <\cdot y$ = (y is on the future light cone of x).

Given a permutation f of the points of E, it is said to preserve the binary relations $x\ R\ y$ if $x\ R\ y \Rightarrow f(x)\ R\ f(y)$. E. C. Zeeman (*J. Math. Phys.*, 5, 490 (1964)) proved the following theorem:

Theorem 1.

The necessary and sufficient condition that f and f^{-1}, permutations of E, preserve the relation $x < y$ or the relation $x <\cdot y$, is $f \in F^{\uparrow}$.

Zeeman also established the corollary (proof published in LM II, p. 297):

Corollary 1.

The necessary and sufficient condition that f, permutation of E, preserves the three relations $x\ T\ y$, $x\ L\ y$, $x\ S\ y$ is $f \in F$.

4.2. Relativistic Invariance and Internal Symmetries†

A physical theory is relativistic if its automorphism group G contains P_0. We are also interested in other symmetry groups, subgroups of G, and called internal symmetry groups. Note that if we consider "passive" invariance, the dilations $D \subset G$.

If P_0 is a subgroup of G, one can consider $C = C_G(P_0)$, the centralizer of P_0 in $G = \{g \in G,\ p \in P_0,\ gpg^{-1} = p\}$, $N = N_G(P_0)$, the normalizer of P_0 in $G = \{g \in G,\ p \in P_0,\ gpg^{-1} \in P_0\}$.

That $\mathrm{Aut}\ P_0 = F$ is the semi-direct product (2) and that P has no center imply that

$$N = P_0 \,\square\, (N/P_0) \qquad , \tag{4.3}$$

and there is a canonical homomorphism

$$N/P_0 \xrightarrow{f} \mathrm{Out}\ P_0 = Z_2(P) \times Z_2(T) \times D \quad . \tag{4.4}$$

And for instance $Z_2(P) \subset \mathrm{Im}\ f$ means that parity is preserved in the theory. We also see that $D \cap \mathrm{Im}\ f$ will give information on the mass spectrum. Indeed, a theory of mass zero particle has D in its automorphism group. If $D \subset \mathrm{Im}\ f$ and if there is a particle of mass $\neq 0$ then there are particles with the same properties and any $m > 0$ for the mass value.

---

† We also refer the reader to the paper with the same title: L. Michel, *Phys. Rev.*, 137B, 405 (1965).

O'Raifeartaigh (*Phys. Rev. Lett.*, 14, 519 (1965)) has proven the following theorem when G is a connected Lie group:

Theorem 2.

If the restriction of an irrep (= unitary irreducible representation) of G to P_0 has an isolated point in the mass spectrum, it is the whole mass spectrum.

There have been too many papers written by physicists proving "theorems" much weaker than the following trivial lemma (LM III, p. 450).

Lemma.

Let P_0 be a subgroup of G. If there exists $p \in P_0$, $p \notin T \subset P_0$, such that $\forall g \in G$, $gpg^{-1} \in P_0$, then P_0 is an invariant subgroup of G. Indeed consider the homomorphism f, $G \xrightarrow{f}$ permutations of (G/P_0) giving the action $\forall x, g \in G$, $gP_0 \overset{x}{\rightsquigarrow} xgP_0$ of G on its homogenous space G/P_0. Then $p \in \mathrm{Ker}\, f$ so $P_0 \cap \mathrm{Ker}\, f$ is an invariant subgroup of P_0 containing p; it is P_0 and $P_0 \subset \mathrm{Ker}\, f$; that implies $\forall q \in P_0$, $\forall g \in G$, $qg = gP_0$.

In my opinion, the preceding considerations are physically very poor, indeed P_0 acts on space time so if $G \supset P_0$ is an automorphism group of the theory, Zeeman's theorem implies that in order to preserve causality, G can act on space time only through a quotient subgroup either of F or P if we forget dilations. This led us to consider G as an extension of P.

We are interested in quantum mechanics. So we must use the existence of the *-algebra A of observables.

We refer the reader to the remarkable paper of Haag and Kastler, "An Algebraic Approach to Quantum Field Theory", *J. Math. Phys.*, 5, 848 (1964); there are physical arguments for A to be a C*-algebra.† Let A be its representation (obtained by a Gelfand-Segal construction) by operators on $\mathcal{H}$, the Hilbert space of states, A' its commutant, A'' the enveloping W*-algebra, $Z = A' \cap A''$ its center. The spectral resolution of Z yields superselection rules††. For instance, if the spectrum is discrete, $\mathcal{H} = \oplus_\lambda \mathcal{H}_\lambda$ and the only vectors of $\mathcal{H}$ which represent states are those belonging to one of the $\mathcal{H}_\lambda$. The $\mathcal{H}_\lambda$ are called superselection sectors.

Assume that P_0 is a subgroup of Aut A, which is implementable (i.e., its elements can be realized by operators of $L(\mathcal{H})$).

† This proposition was made by I. E. Segal, more than ten years earlier.

†† Concept introduced by G. C. Wick, A. S. Wightman, E. P. Wigner, *Phys. Rev.*, 88, 101 (1952). See the preprint of Doplicher, Haag and Roberts for the most recent study of this question.

Let U(p) be a realization of the automorphism $p \in P_0$ by an operator $U(p) \in L(\mathcal{H})$. It has to be unitary in order to be an automorphism: $\forall A \in A$; $(UAU^{-1})^* = (UA^*U^{-1})$. If V is any element of the group $U(A')$ of the unitary operators of A', U(p)V is just as good for representing the Poincaré transformation p. So the set:

$$E = \{U(p)V,\ p \in P_0,\ V \in U(A')\} \quad , \tag{4.5}$$

forms a group of unitary operators which is a "central extension" of P_0 by $U(A')$ i.e.,

$$E/U(A') = P_0, \text{ quotient group} \quad . \tag{4.6}$$

and

$$\forall V \in U(A'),\ \forall U \in E,\ V \rightsquigarrow UVU^{-1} \text{ is an inner automorphism of } U(A') \ . \tag{4.6'}$$

One can prove (see Moore's lectures), that any Polish topological group E satisfying (4.6) and (4.6′) is either the direct product $U(A') \times P_0$ or are of the form (see also LM II):

$$E_\alpha = (U(A') \times \overline{P}_0)/Z_2(\alpha,\omega) \quad , \tag{4.7}$$

where the two element group is generated by the element (α,ω) with $\alpha \in {}_2U(Z)$, the group of square roots ($\neq$ e) of the unit, in the group $U(Z)$ which is the center of $U(A')$.† Which is the extension E_α chosen by nature?†† The answer is the extension defined by (4.7) with

$$\alpha = e^{i\pi(B+\Sigma_i L_i)} , \ \alpha^2 = I \tag{4.8}$$

---

† Equation (4.7) implies some topology as explained in Moore's lectures. In "Sur les extensions centrales du groupe de Lorentz inhomogène connexe", *Nucl. Phys.*, 57, 356 (1964), I have studied the same problem for abstract groups: any abelian group A is the direct sum $A = D \oplus K$ where D is the maximal divisible subgroup and K is a reduced subgroup (no infinitely divisible elements $\neq 1$). One has the relations: $H^2(\overline{P}_0,A) = H^2(SL(2,\mathbb{C}),A)$, $H^2(P_0,A) = H^2(L_0,A) = {}_2K + H^2(L_0,D)$, $H^2(SL(2,\mathbb{C}),D)^{\text{Aut }\mathbb{C}} = 0$ and of course $H^2(L_0,D)^{\text{Aut }C} = {}_2D$. Indeed Aut $\mathbb{C}$ the group of automorphisms of the complex field act on SL(2,C) and on L_0 (exactly Aut L_0/L_0 = Aut $\mathbb{C}$). So it acts on $H^2(SL(2,\mathbb{C}),A)$ (through a trivial action on A) since the group of inner automorphisms of L_0 acts trivially. Following the usual convention, also used in Chapter 1, $H^2(SL(2,\mathbb{C}),A)^{\text{Aut }\mathbb{C}}$ is the subgroup of fixed elements. So if $H^2(SL(2,\mathbb{C}),A) \neq 0$, the automorphisms of $\mathbb{C}$ do not pass the non-trivial extensions and the corresponding extensions are very pathological. I found this a sufficient argument for considering in physics only the extensions of Equation (4.7).

†† This was the question that Lurçat and myself asked and answered in *N. Cim.*, 21, 57 (1965) and *Comptes Rendus of the Conference of Aix-en-Provence*, p. 183, C.E.A. Saclay editor, (1962).

where B is the baryonic charge operator and L_i the (different) leptonic charges (see 3.6). Indeed this choice of extension implies the observed relation between spin and charges:

$$(-1)^{2j} = (-1)^{b+\Sigma_i \ell_i} \quad , \tag{4.9}$$

where j is the angular momentum of any state and b, ℓ_i are its baryonic and different leptonic charges. Note that Equation (4.9) shows that the integer or half integer nature of spin form a superselection rule.

4.3. Irrep of $\overline{P}_0$

All irreps (= unitary linear irreducible representations) of $\overline{P}_0$ are known. In 1937, Wigner† showed, by extending Frobenius' methods for finite groups to $\overline{P}_0$, that irrep of $\overline{P}_0$ are characterized by an orbit of $\overline{L}_0$ on T' the dual of T and an irrep of the corresponding little group (= stabilizer). The non-degenerate $\overline{L}_0$ invariant symmetric bilinear form on T (= Minkowski pseudo-Euclidean scalar products) yields an isomorphism of $\overline{L}_0$ space between T and T'.

To be explicit, we denote by $\underline{a}$, $\underline{b}$, ... and A, B, ... respectively the elements of T and $SL(2,\mathbb{C}) = \overline{L}_0$. Let $(a^0,\vec{a}) = (a^0,a^1,a^2,a^3)$ the coordinates of $\underline{a}$ in a basis of T. Consider the isomorphism between $T = R^4$ and the additive group of 2×2 hermitian matrices

$$\underline{a} \longleftrightarrow \tilde{a} = \begin{pmatrix} a^0 + a^3 & a^1 - ia^2 \\ a^1 + ia^2 & a^0 - a^3 \end{pmatrix} .$$

As we saw, the group $\overline{P}_0$ is the semi-direct product $T_\square SL(2,C)$ with the $SL(2,C)$ action on T

$$A \in SL(2,C) \quad , \quad \underline{a} \longleftrightarrow \tilde{a} \overset{A}{\rightsquigarrow} A\tilde{a}A^* \longleftrightarrow A\underline{a} \quad .$$

The Minkowski pseudo-Euclidean scalar product is

$$(\underline{a},\underline{b}) = a^0b^0 - a^1b^1 - a^2b^2 - a^3b^3$$

and the Minkowski "length" of $\underline{a}$ is

$$\underline{a}^2 = (\underline{a},\underline{a}) = \text{determinant } \tilde{a} \quad .$$

We denote by $(\underline{a},A)$ the elements of $\overline{P}_0$ with $\underline{a} \rightsquigarrow (\underline{a},1)$ the canonical injection

† E. P. Wigner, *Ann. of Math.*, 40, 149 (1939) reproduced in F. J. Dyson, *Symmetry Groups in Nuclear and Particle Physics*, Benjamin, New York (1966). Wigner's paper was the first one giving a complete family of irreps of a non-compact non-semi-simple Lie group.

$T \to \overline{P}_0$ and $A \rightsquigarrow (0,A)$ an injection† of $\overline{L}_0 = SL(2,C) \rightsquigarrow \overline{P}_0$. The $\overline{P}_0$ group law is

$$(\underline{a},A)(\underline{b},B) = (\underline{a} + A\underline{b},AB) \tag{4.10}$$

We will use the same notation for elements of T and T'.

It is useful to introduce the notion of stratum. When a group G acts on a set M, all the points with conjugate stabilizers form a stratum: in other words, a stratum is the union of all orbits of the same type (i.e., isomorphic as G-homogeneous spaces). The action of $\overline{L}_0$, decomposes T or T' in four strata. See Figure 4.1.

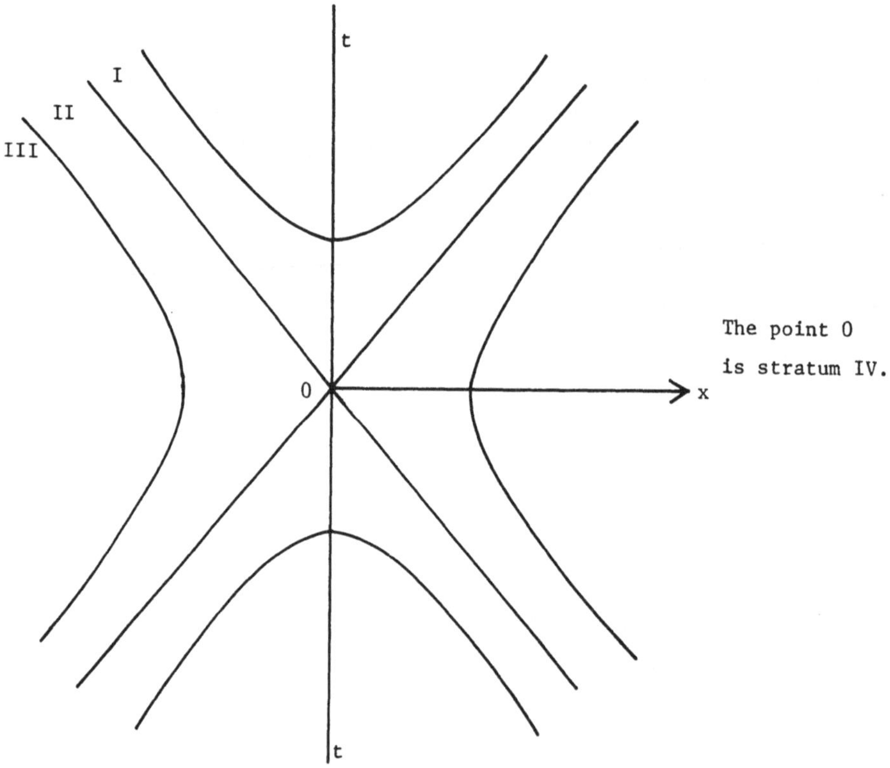

FIGURE 4.1. STRATA ON T BY THE ACTION OF $\overline{L}_0$

---

† This injection is unique up to a conjugation in $\overline{P}_0$. Indeed Wigner, in his paper of 1939, showed that for the abstract groups (with the action of $\overline{L}_0$ on T just defined), $H^1(\overline{L}_0,T) = 0$.

Table 4.1 lists these strata, the corresponding little groups (defined up to a conjugation in $\overline{L}_0$) and the corresponding irreps of $\overline{P}_0$.

TABLE 4.1. STRATUM OF T' DUE TO THE ACTION OF $\overline{L}_0$; IRREPS OF $\overline{P}_0$

Stratum	Little Group	Irrep of $\overline{P}_0$
I $\underline{p}^2 = m^2$	SU(2)	I_a $m > 0$ (i.e., $p^0 > 0$), 2j integer ≥ 0 I_b $m < 0$ (i.e., $p^0 < 0$), 2j integer ≥ 0
II $\underline{p}^2 = 0$	E(2)	II_a sign of p^0, 2λ integer
$\underline{p} \neq 0$	(2-dimensional Euclidean group)	II_b sign of p^0; Ξ positive number, $\omega = 1$ II_c sign of p^0; Ξ positive number, $\omega = -1$
III $\underline{p}^2 < 0$	$\widetilde{SL}(2,\mathbb{R}) = \widetilde{SO}(2,1)$ $\sim$ = double covering	III_a $m^2 < 0$, ascendant and descendant discrete series, $\pm j$ III_b $m^2 < 0$, principal series $i\sigma, \rho \in R$ III_c $m^2 < 0$, supplementary series, $0 \leq \sigma \leq \frac{1}{4}$
IV $\underline{p} = 0$	$SL(2,\mathbb{C}) = \overline{P}_0$	IV Irreps of $SL(2,\mathbb{C})$ (see Stein lectures) are irreps of $\overline{P}_0$ with T trivially represented.

ω is the non-trivial element of the center of $\overline{P}_0$; in I, ω is represented by $(-1)^{2j}$ and in II by $(-1)^{2\lambda}$. Wigner constructed the irreps of type I and II. Those of $\widetilde{SL}(2,\mathbb{R})$ needed for type III were given by Bargmann, *Ann. of Math.*, 48, 568 (1947) and those of $SL(2,\mathbb{C})$ (type IV for $\overline{P}_0$) were first given by I. M. Gelfand, M. A. Naimark, *Acad. Sci. USSR J. Phys.*, 10, 93 (1946) and *Isv. Akad. Nauk SSSR Ser. Mat.*, 11, 91 (1947).

Wigner's method is a particular case of Mackey's theory of induced representations. Since the (measurable) axis $t'Ot$ and Ox are a set of representatives of the orbits, a theorem by Mackey† insures that this method yields all irreps of $\overline{P}_0$. As we will see, the only irreps of $\overline{P}_0$ which correspond to known particles are those of mass $m \geq 0$ (I_a and II_a in Table 4.1). Wigner, in his paper, has given the following realization:

An $\overline{L}_0$ invariant measure on the orbit Ω: $\underline{p}^2 = m^2 \geq 0$, $p^0 > 0$, is $d^3\vec{p}/p^0 = d\Omega$. Consider the functions f defined on Ω with value in a $2j + 1$

† G. W. Mackey has described his theory in a book for physicists: *Induced Representations of Groups and Quantum Mechanics*, Benjamin, New York (1968). The needed theorem is Theorem B, p. 43.

dimensional Hilbert K_j, such that

$$\|f\|^2 = \int_\Omega \langle f(\underline{p}),f(\underline{p})\rangle \frac{d\vec{p}}{p^0} < \infty \quad , \tag{4.11}$$

where $\langle f(\underline{p}),f(\underline{p})\rangle = \Sigma_\alpha \bar{f}_\alpha(\underline{p}) f_\alpha(\underline{p})$ is the hermitian scalar product in K_j. Then

$$(U(\underline{a},A)f_\alpha)(\underline{p}) = \Sigma_\beta e^{i\underline{a}\cdot p} Q_{\alpha\beta}(\underline{p},A) f_\beta(A^{-1}\underline{p}) \quad , \tag{4.12}$$

where the unitary $(2j + 1) \times (2j + 1)$ matrices Q satisfy

$$Q(\underline{p},A)Q(A^{-1}\underline{p},B) = Q(\underline{p},AB) \quad . \tag{4.13}$$

When $j = 0$ or when $m = 0$, dim $K_0 = 1$, and the Q are complex numbers of unit module. In all cases, when $A, B \in L_p$, the little group of p, Equation (4.13) show that the Q form an irrep of L_p.

4.4. Particle States and Irrep of $\overline{P}_0$

What is a particle? This word is used very much by physicists. This word is attributed to the electron, the photon, and the 270 hadrons of Table 3.2 in 3.5, and also to nuclei (an "α-particle" for instance means a Helium nucleus) and even to atoms or ions. The meaning of this word is in full evolution; it was clear up to ten years ago. Let us try to define this word.

"A particle is a physical system which can be isolated and cannot be decomposed into subsystems without destroying it."

This concept is very clear for stable particles: electrons and positrons $e^\pm$, protons and antiprotons $p^\pm$, photons γ, neutrinos and antineutrinos ν, $\bar{\nu}$ and also stable nuclei (deuteron, α-particle, ${}_6C_{12}$) and the fundamental states of atoms or molecules formed with these nuclei.

Invariants which can be attributed to these particles are the $\overline{P}_0$ invariants; mass and spin, and the Hilbert space of the states of a particle is the carrier of the irrep of $\overline{P}_0$ of mass m, spin j (or for $m = 0$, of helicity λ).† For example, proton or antiproton $(m_p,1/2)$, electron or positron $(m_e,1/2)$, neutrinos $(0,-1/2)$, antineutrinos $(0,1/2)$.

To good approximation this concept of particle can be extended to unstable particles whose lifetime is long enough to study them isolated $(\tau > 10^{-21}$ sec). Strictly speaking, because of Heisenberg's uncertainty relations between energy and time, the Hilbert space of states carries the rep $\int^{\oplus}_{\Gamma(\Delta m)} (m,j)dm$ where Γ is a curve

† The $\overline{P}_0$ invariant λ is called helicity by elementary particle physicists but was called circular polarization by its discoverer, Fresnel, in the 1820's. It just happens that the photon is the only known particle whose space of states is the carrier of a reducible representation $(m = 0, \lambda = 1) \oplus (m = 0, \lambda = -1)$ of P_0.

with a mean spread of Δm. For weak decays, $\Delta m/m < 10^{-14}$ and for electromagnetic decays, $\Delta m/m < 10^{-5}$.

It is not clear that this concept of particle can be usefully extended to the strongly unstable resonances (most of the "hadrons" of Table 3.2 in 3.5). Indeed their lifetime τ might be as short as 10^{-23} sec (and $\Delta m/m$ reach 0.15 for the ρ-meson) so they do not exist isolated since the range of strong interaction is 10^{-13} cm (of the same order as $3 \times 10^{10} \times 10^{-23} = \tau c$). However, we shall here call them particles.

We have used also the word particle as a collective name for different particles with the same spin and not very different masses and similar properties, so they can be considered approximately as identical. This was the case of "the nucleon" with the isospin degree of freedom corresponding to the two states p and n; the π-meson with the three states π^+, π^0, π^-.

There is another degeneracy for most particles. It has been found (or it is expected) that they come in pairs with the same representation (m,j) of $\overline{\mathcal{P}}_0$, but all charges are opposite within the pair. Such pairs are called charge-conjugate pairs, the two values of the corresponding degree of freedom are labeled "particle" and "antiparticle". Particles with all charges zero are called self-charge-conjugated, indeed there is no degeneracy under charge conjugation C for them (examples: $\gamma, \pi^0, \eta^0, \rho^0, \omega^0, \phi^0, x^0$, etc.).

Elements of the enveloping algebra $E(\mathcal{P}_0)$ of the Lie algebra of $\mathcal{P}_0$ are observables which we will call the kinematical observables of the particle.

The $\mathcal{P}_0$ Lie algebra is (with $M_{\mu\nu} = -M_{\nu\mu}$)

$$[iP^\mu, iP^\lambda] = 0, \quad [iP^\lambda, iM^{\mu\nu}] = iP^\mu g^{\lambda\nu} - iP^\nu g^{\lambda\mu} \quad . \tag{4.14}$$

$$[iM^{\mu\nu}, iM^{\rho\sigma}] = iM^{\mu\rho} g^{\nu\sigma} + iM^{\nu\sigma} g^{\mu\rho} - iM^{\mu\sigma} g^{\nu\rho} - iM^{\nu\rho} g^{\mu\sigma} \quad . \tag{4.14'}$$

The P^λ, $M^{\mu\nu}$ are the self-adjoint operators on $\mathcal{H}$ representing energy momentum and relativistic angular momentum. Pauli (unpublished) and Bargmann and Wigner (*Proc. Nat. Acad. Sci.*, (1967)) considered in $E(\mathcal{P}_0)$:

$$W_\lambda = -\frac{1}{2}\varepsilon_{\mu\lambda\nu\rho} P^\mu M^{\nu\rho} = -\frac{1}{2}\varepsilon_{\lambda\mu\nu\rho} M^{\mu\nu} P^\rho = (*M \cdot P)_\lambda \tag{4.15}$$

which satisfies

$$[W^\lambda, P^\mu] = 0, \quad [W^\lambda, W^\mu] = i\varepsilon^{\lambda\mu\nu\rho} P_\nu W_\rho \quad , \tag{4.16}$$

$$[W^\lambda, M^{\mu\nu}] = iW^\mu g^{\lambda\nu} - iW^\nu g^{\lambda\mu} \quad . \tag{4.16'}$$

Note that

$$\underline{P} \cdot \underline{W} = P^\lambda W_\lambda = 0 \quad . \tag{4.17}$$

The center of $E(\mathcal{P}_0)$ is generated by $\underline{P}^2 = P^\lambda W_\lambda$ and $\underline{W}^2 = W^\lambda W_\lambda$. For irrep of $\overline{\mathcal{P}}_0$ these operators are the following multiples of the unit:

$$m > 0, \qquad \underline{P}^2 = m^2 \mathbb{1},\ \underline{W}^2 = -m^2 j(j+\underline{1})\ \mathbb{1}$$

$$m = 0 \quad \text{IIa} \qquad \underline{P}^2 = 0, \quad \underline{W}^2 = 0 \quad \underline{W} = \lambda \underline{P},\ \lambda \text{ helicity}$$

$$m = 0 \quad \text{IIb or IIc} \qquad \underline{P}^2 = 0, \quad W^2 = -\Xi < 0 \ .$$

4.5. Particle Polarization

In most experiments, the energy and momenta of the particles are measured (i.e.: monokinetic beam, target at rest, bubble chamber track curved in a magnetic field). The additional observables to be measured in order to have a complete knowledge of the particle state are called "the polarization". Since $\underline{W}$ commutes with $\underline{P}$, it is the polarization operator. However, the $\underline{W}$ components do not commute with each other. Equations (4.14), (4.14′), (4.15), (4.16), (4.16′) show that

$$P_0,\ P_1,\ P_2,\ P_3,\ W_3,\ W^2 \tag{4.18}$$

generate a maximal abelian subalgebra of $E(\overline{\mathcal{P}}_0)$. It is easy to interpret† the $\underline{W}$ observables of a particle (m,j) when one remarks that $\mathcal{H}(m,j)$, the one-particle space of states, is a direct integral

$$\mathcal{H}(m,j) = \int_\Omega^\oplus K_j(\underline{p})\,\frac{d^3\vec{p}}{p^0} \ , \tag{4.19}$$

of $2j+1$ dimensional Hilbert spaces K_j. Given $\underline{p}$, introduce an orthonormal tetrad of vectors in the Minkowski space (i.e., in T')

$$\underline{n}^{(0)} = \underline{p}m^{-1}, \underline{n}^{(i)} (i = 1,2,3, \alpha,\beta = 0,1,2,3) \ , \tag{4.20}$$

Minkowski scalar product

$$\underline{n}^{(\alpha)}, \underline{n}^{(\beta)} = g^{\alpha\beta} \ , \tag{4.20′}$$

right hand orientation

$$\varepsilon^{\lambda\mu\nu\rho} n_\lambda^{(\alpha)} n_\mu^{(\beta)} n_\nu^{(\nu)} n_\rho^{(\delta)} = -\varepsilon^{\alpha\beta\gamma\delta} \ . \tag{4.20″}$$

Introduce then the self-adjoint operators on $K_j(\underline{p})$

$$S^i = -\frac{1}{m}\,\underline{n}^{(i)} \cdot \underline{W}(\underline{p}) \tag{4.21}$$

---

† See for instance L. Michel, *N. Cim. Suppl.*, 14, 99 (1959) for more details and the treatment of the case $m = 0$.

where $\underline{W}(\underline{p})$ is the integrand of

$$\underline{W} = \int_{\Omega}^{\oplus} \underline{W}(\underline{p}) \frac{d^3\vec{p}}{p^0} \ . \tag{4.21'}$$

Of course, Equation (4.17) implies

$$S^0 = -\frac{1}{m}\underline{n}^{(0)} \cdot \underline{W}(\underline{p}) = 0 \ . \tag{4.21''}$$

The $S^{(i)}$ satisfies the commutation relations

$$[S^{(i)},S^{(j)}] = i\varepsilon_{ijk}S^{(k)} \ , \tag{4.22}$$

so they generate a SU(2) Lie algebra, that of the little group of $\underline{p}$. For $\underline{p} = (m,0)$ (particle at rest), $S^{(i)} = M^{0i}(\underline{p})$. This shows the relation between angular momentum and polarization.

For a particle of energy momentum $\underline{p}$, the polarization density matrix $R(\underline{p})$ is a $\mathrm{Tr}R(\underline{p}) = 1$, hermitian $R^*(\underline{p}) = R(\underline{p}) \geq 0$, $(2j+1) \times (2j+1)$ matrix which is transformed by $L \in SU(2)$ (defined by Equation (4.22)) as

$$L\underline{p} = \underline{p},\ R(\underline{p}) \rightsquigarrow Q(\underline{p},L)R(\underline{p})Q^*(\underline{p},L) \ , \tag{4.23}$$

Let

$$R(\underline{p}) = (2j+1)^{-1}\,\mathbb{1} + \sum_{\ell=1}^{2j} R^{(\ell)}(\underline{p}) \ , \tag{4.24}$$

be the decomposition of $R(\underline{p})$ into a sum of irreducible SU(2)-tensor operators. The $R^{(\ell)}(\underline{p})$ are called the multipoles of the polarization matrix. Note that $R(\underline{p})$ and the $R^{(\ell)}(\underline{p})$ can be given a completely covariant form.

$$R^{(1)}(\underline{p}) = -\frac{1}{2}s_\alpha\frac{W^\alpha}{m}, R^{(\ell)}(\underline{p}) = \frac{(-1)^\ell}{m^\ell}s_{\alpha_1\alpha_2\ldots\alpha_\ell}W^{\alpha_1}W^{\alpha_2}\ldots W^{\alpha_\ell} \ . \tag{4.25}$$

Where $s_{\alpha_1\ldots\alpha_\ell}$ is a completely symmetrical ℓ^{th} order tensor satisfying

$$(\text{partial trace}) = s^\alpha_{\alpha\beta\gamma\ldots} = 0,\ \underline{p}_\alpha s^\alpha_{\beta\gamma\ldots} = \underline{p}\cdot s = 0 \ . \tag{4.26}$$

This is obtained† from the equivalent form of relation (4.21)

$$W(\underline{p}) = m\Sigma_i S^{(i)}\underline{n}^{(i)} \ . \tag{4.27}$$

From now on, we shall consider only the dipole polarization (which is the only one existing for a spin 1/2 particle). Its evolution is given in a macroscopic

---

† For more details see C. Henry and E. De Rafael, *Ann. Inst. H. Poincaré*, 2A, 87 (1965).

($\sim$ slowly variable in space time) electromagnetic field F (with $F^i = E^i, F^{ij} = \varepsilon_{ijk}B^k$ and $*F$ the polar tensor of F) by the equation†

$$\dot{\underline{s}} = \frac{e}{m} M \cdot \underline{s}, \text{ with } M = F + P(\frac{(g-2)}{2} F + g'{*F})P \quad , \tag{4.28}$$

and $P = \mathbb{1} - \underline{u} \otimes \underline{u} = P_{\underline{u}^\perp}$ where $\underline{u} = \underline{p}/m$, the quadri-velocity, which satisfies the Lorentz equation:

$$\dot{\underline{u}} \sim \frac{e}{m} F \cdot \underline{u} \quad . \tag{4.28'}$$

The $\cdot$ means the proper time derivative; M and F are skew symmetric tensors so (4.28) and (4.29) correspond to infinitesimal Lorentz transformations which of course preserve the Minkowski products:

$$\underline{u}^2 = 1, \ \underline{u} \cdot \underline{s} = 0, \ 0 \leq \delta = (-\underline{s}^2)^{1/2} \leq 1 \quad , \tag{4.29}$$

where δ is called the degree of (dipole = vector) polarization. The magnetic moment of the particle is $\mu = (g/2)e/m(\hbar = c = 1)$ and $\mu' = g'\ e/m$ is its electric dipole moment. (For neutral particles, write directly μ and μ'.) Note that $g = 2$ is a remarkable value of g which simplifies Equation (4.28). This, as first noted by Thomas in 1926, is characteristic of the Poincaré group and does not happen for Galilean invariance (as we commented in 2.6 and 2.9).

Consider a reaction between particles $A + B \to C + D + \ldots$ where all energy momenta are exactly known. Then the transition probability between pure polarization states is $\lambda = |\langle C;D,\ldots|S(p_A,p_B,p_C,p_D\cdots)|A,B\rangle|^2$ where the isometry

$$S(p_A,p_A,p_C,p_D\cdots) \in \mathrm{Hom}\,(\mathcal{H}_A \otimes \mathcal{H}_B, \mathcal{H}_C \otimes \mathcal{H}_D \otimes \ldots) \quad . \tag{4.30}$$

More precisely, S is the restriction of a unitary operator, defined on $\mathcal{H}$ the Hilbert space of physics and called S-matrix in the physics literature. In the general case of given polarization density matrices

$$\lambda(p_A,p_B;p_C,p_D,p\ldots) = \mathrm{Tr}R_{C,D\ldots}(p_C,p_{D\ldots})SR_{A,B}(p_A,p_B)S^* \quad , \tag{4.30'}$$

where $R_{A,B}(p_A,p_B)$ (resp., $R_{C,D}(p_C,p_{D\ldots})$ are hermitian operators†† on $K_{j_A}(p_A) \otimes K_{j_B}(p_B)$ (resp., $K_{j_C}(p_C) \otimes K_{j_D}(p_D) \otimes \ldots$) which reduce to $\mathbb{1}(2j_A + 1)^{-1}(2j_B + 1)^{-1}$ (etc.) when no polarization is observed. If one observes the polarization of only one of the particles, Equations (4.24), (4.25) and (4.30′) show that $\lambda(p_A,p_B;p_C,p_D\cdots)$

---

† V. Bargmann, L. Michel, V. Telegdi, *Phys. Rev. Lett.*, 2, 435 (1959).

†† Practically, for all experiments, there is no correlation between the states of initial particles so $R_{A,B}(p_A,p_B) = R_A(p_A) \otimes R_B(p_B)$.

depends linearly on the different polarization tensors of this particle $(S_\alpha, S_{\alpha\beta}, S_{\alpha\beta\gamma}, \ldots)$.†

4.6. Invariance Under $P \times Z_2(C)$; PCT Theorem

If a physical theory is invariant under a group, say P_0, one can transform the theory by an automorphism $\alpha \in \mathrm{Aut}\, P_0$, (replace everywhere $g \in P_0$ by $\alpha(g)$). If α is an inner automorphism, by definition of P_0 invariance, the transformed theory is equivalent. If α is not an inner automorphism, the transformed theory might not be physically equivalent. If it is, then one can enlarge the invariance group, in order to include this automorphism. It is obvious that dilations are not an active invariance of physical theories (except when only zero masses occur). What can be said about P, T (and their product PT)? We will assume invariance under P and T and also under C, the charge conjugation, and in the next section, see if these invariances are respected in nature.

It seems a reasonable assumption that P does not act on A', the commutant of A, the representation on $\mathcal{H}$ of the algebra of observables (see Equation (4.5)). We do know the action of T on A', because T has to be represented by an antiunitary operator (see 1.2), i.e., by $U(T) = V(T)K$ where $V(T)$ is a unitary operator and K is a complex conjugation (whose choice cannot be canonical). K (as well as U(T)) induces an anti-linear automorphism on the algebra $L(\mathcal{H})$ i.e.,

$$K\lambda AK = \bar{\lambda}KAK,\ KABK = KAK\ KBK,\ K(A + B)K = KAK + KBK \quad , \tag{4.31}$$

since $K^2 = \mathbb{1}$. Note that if $U = (U^{-1})^*$ is unitary, so is KUK. We assume that T leaves $U(A)$ and $U(Z)$ globally invariant, but acts as an anti-linear automorphism. Finally, we can introduce U(C), the charge conjugation operator on $\mathcal{H}$. By definition C acts trivally on P_0 and anti-commute with all charges. More generally, physical properties of C tell us how it must act on A' which corresponds essentially to internal symmetry. Let

$$D = Z_2(P) \times Z_2(T) \times Z_2(C) \quad , \tag{4.32}$$

$$P_c = P \times Z_2(C) = P_0 \,\square\, D \tag{4.33}$$

In LM I, I gave the proof kindly tailor made by J. P. Serre for us physicists, (Theorem 1, p. 183).

$$H^2(P_c, U(Z)) = H^2(D, U(Z)) \oplus {}_2U(Z)^D \quad , \tag{4.34}$$

(see a similar theorem in Moore's lecture), where ${}_2U(Z)^D$ is the group of the

---

† If the polarization of more than a final particle is observed one has also to introduce polarization correlations.

square roots of the unit of $\mathcal{U}(Z)$ invariant under every element of D. We check that $e^{(i\pi(B+\Sigma_i L_i))} = e^{(-i\pi(B+\Sigma_i L_i))}$ is such an element so relation (4.9) is preserved.

What is the extension in (4.34) chosen by nature? Probably none, as we will see in the next section because P, C, PC (and probably T) are not automorphisms of the physical laws of nature. However, we can consider for D in (4.34), a subgroup of that of (4.32).

Let us first consider parity. Irreps of $\mathcal{P}^{\dagger} = \mathcal{P}_0 \,\square\, Z_2(P)$ are easily deduced from those of $\mathcal{P}_0$. For $m > 0$, and $m = 0$, $\lambda = 0$, there are two irreps of $\mathcal{P}^{\dagger}$, $(m,j,\pm)$ or $(0,0,\pm)$ with opposite parity (eigenvalue of $U(P)$) whose restriction to $\mathcal{P}_0$ is irreducible. For mass zero, $\lambda \neq 0$ irrep of $\mathcal{P}^{\dagger}$ are denoted by $(0,|\lambda|)$ because their restriction to $\mathcal{P}_0$ reduces to

$$(0,|\lambda|)_{\mathcal{P}_0} = (0,|\lambda|) \oplus (0,-|\lambda|) \quad . \tag{4.35}$$

Note that, as projective representations of $\mathcal{P}_0$, $(m,s,+)$ and $(m,s,-)$ are equivalent. More generally, since $g \to g^2$ is a surjective homomorphism of $\mathcal{U}(Z)$, $H^2(Z_2(P),\mathcal{U}(Z)) = 0$. So to speak of the parity of a state is not a canonical statement; only relative parity can be defined for states in the same superselection sector. By convention, the parity of the vacuum is taken +1, as well as that of the proton, the neutron, the electron, the Λ^0.

Wigner in his Istanbul lectures in 1962 (same reference as LM I) has studied the projective irreps of $\mathcal{P}$ (and even $\mathcal{P}_c$). This study can be easily transferred to the study of the extension of $\mathcal{P}$ by $\mathcal{U}(Z)$ (and then by $\mathcal{U}(A')$, from general results of group extension by a non-abelian kernel, as explained in LM I). This is not the case for $\mathcal{P}_c$ because $U(C)$, as unitary operator, does not act on the phase of the projective representation, but C as charge conjugation acts non-trivally on Z. We just give here the following results: $U(T)^2$, $U(CPT)^2$, $U(PT)^2$ are canonical (since $\mathcal{U}(Z)$ is divisible and $\mathcal{U}(Z) \ni g \to g^2$ is surjective) and are $\in {}_2\mathcal{U}(Z)$. For non-zero mass states, a choice different from

$$U(T)^2 = U(PT)^2 = U(CPT)^2 = (-1)^{2j} \quad , \tag{4.36}$$

will require that irrep of $\mathcal{P}_c$ restricted to $\mathcal{P}^{\dagger}_{\square} Z_2(C)$ are not irreducible. This would correspond to a new degree of freedom for particles which is not observed in nature.†

In usual quantum field theories, relations (4.36) are always satisfied. This is related also to the two following theorems:

---

† See Wigner discussion in his Istanbul notes and for a recent review see H. Goldberg, *N. Cim.*, 60, 509 (1969).

Theorem 3.

The good connection between spin and statistics† is a consequence of the Wightman axioms:†† covariance under $\mathcal{P}_0$ of finite component quantum fields, existence of vacuum, positivity of energy and "locality".

Theorem 4.

These axioms also imply invariance under CPT.†††

4.7. How to Observe Violation

4.7.a. Action of P, T, C on Observables

Let us summarize in Table 4.2 the action of the automorphisms P, T, PT of $\mathcal{P}_0$, on invariants of this group. C acts trivally on them, but exchanges particles and antiparticles. The self conjugated particles are eigenstates of C. For instance, consider quantum electrodynamics; C is an automorphism of this theory. The electromagnetic interaction Hamiltonian is:

$$H_{em} = \int j^{\mu}(x)A_{\mu}(x)d^3\vec{x} \quad . \tag{4.37}$$

By definition of C,

$$U(C)j^{\mu}(x)U(C)^{-1} = -j^{\mu}(x) \quad , \tag{4.38}$$

i.e., the electromagnetic current changes sign. So H_{em} is invariant under C if also

$$U(C)A_{\mu}(x)U(C)^{-1} = -A_{\mu}(x) \quad . \tag{4.39}$$

---

† i.e., integral (resp. half integral) spin fields describe particles which satisfy Bose = Completely symmetrical (resp. Fermi = antisymmetrical) statistics. This was proven by Pauli; his last publication in the subject is "Exclusion principle, Lorentz group and Reflection of space time and charge", p. 30 in *Niels Bohr and the Development of Physics*, Pauli editor, Pergamon, New York (1955). There he also proves the CPT theorem, first proven by Lüders and Schwinger.

†† See R. F. Streater and A. S. Wightman, *PCT, Spin and Statistics and All That*, Benjamin, New York (1964); R. Jost, *General Theory of Quantized Fields*, American Mathematical Society, Providence (1965).

††† From weaker axioms (Haag-Araki theory of local observables), H. Epstein, *J. Math. Phys.*, 8, 750 (1967), has proven the CPT invariance of the S matrix. For infinite component fields, neither the connection between spin and statistics, nor the CPT invariance are implied by $\mathcal{P}_0$ invariance. For a counter example, see e.g., I. Todorov, *8th Nobel Symposium*, Wiley (1968).

By definition, U(C)0> = 0> where 0> is the vacuum. So

$$U(C)A_\mu(x)0> = -A_\mu(x)0> \quad , \tag{4.40}$$

i.e., a photon has charged conjugation -1. We have added in Table 4.2 the transformation of the electromagnetic field

$$F_{\mu\nu}(x) = (\partial_\wedge A(x))_{\mu\nu} = \partial_\mu A_\nu(x) - \partial_\nu A_\mu(x) \quad . \tag{4.41}$$

For T, time reversal, the space part $\vec{j}$ of the e.m. current $j^\mu(x)$ changes sign (as a velocity) while the time component (whose space integral is the electric charge) does not. Hence the time reversal property of A_μ, of H_{em} (invariant), of $F^{0i} = E^i$ (electric field) and $F^{ij} = \varepsilon_{ijk}B^k$ (magnetic field).

Consider Equation (4.28). The quadrivector $mF \cdot u = (-\vec{B}\cdot\vec{p}, -p^0\vec{E} - \vec{p}\times\vec{B})$ transforms under t as d/dt mu. Hence, except for the term in g' (electric dipole) Equation (4.28) is invariant under P, T, PT. The term in g' is incompatible with both P and T.

TABLE 4.2. COVARIANCE UNDER P,T,C, OF THE INVARIANTS OF $\mathcal{P}_0$ AND THE ELECTROMAGNETIC FIELD

Physical Observable	P	T	PT	C	CPT
$\underline{p}_i \cdot \underline{p}_j$, $\underline{s}_i \cdot \underline{s}_j$	+	+	+	+	+
$\underline{p}_i \cdot \underline{s}_j$	-	+	-	+	-
$(\underline{p}_i,\underline{p}_j,\underline{p}_k,\underline{p}_\ell)$, $(\underline{p}_i,\underline{p}_j,\underline{s}_k,\underline{s}_\ell)$	-	-	+	+	+
$(\underline{p}_i,\underline{p}_j,\underline{p}_k,\underline{s}_\ell)$, $(\underline{p}_i,\underline{s}_j,\underline{s}_k,\underline{s}_\ell)$	+	-	-	+	-
helicity λ	-	+	-	+	-
$\vec{E}$ (electric field)	-	+	-	-	+
$\vec{B}$ (magnetic field)	+	-	-	-	+

(a,b,c,d) means determinant of the four components of four vectors.

4.7.b. Parity Violation

The consequence of invariance under P is called parity conservation. Consider two states S_1, S_1' of a physical system corresponding to each other through an "active" plane symmetry Σ, and S_2, S_2' two other states of the same system also symmetric to each other through Σ. Let $\lambda_{12} = \mathrm{tr} R_1 R_2$ and $\lambda'_{12} = \mathrm{tr} R'_1 R'_2$ be the respective probabilities of transitions $1 \to 2$.

$$\text{Parity conservation} \Rightarrow \lambda_{12} = \lambda'_{12} \ . \tag{4.42}$$

If an experiment yields $\lambda_{12} \neq \lambda'_{12}$, it proves parity violation. Since $\Sigma\lambda_{12} = \lambda'_{12}$, $\Sigma\lambda'_{12} = \lambda_{12}$, it means that $\lambda_{12} = a + b$, $\lambda'_{12} = a - b$, where

$$a = \frac{1}{2}(\lambda_{12} + \lambda'_{12}) \text{ is a scalar, } b = \frac{1}{2}(\lambda_{12} - \lambda'_{12}) \text{ is a pseudoscalar.} \tag{4.43}$$

So in a two particle decay of a <u>polarized</u> particle $\underline{p} \to \underline{p}_1 + \underline{p}_2$, (or more generally in a decay where only two energy momenta are observed) P conservation $\Rightarrow$ the angular distribution of decay products depends only on the even polarization multipoles $s_{\alpha\beta}$, $s_{\alpha\beta\gamma\sigma}$, ...†

In 1957, the following experiment was performed. Co^{60} nuclei at rest ($\underline{p} = m,\vec{0}$) were polarized in a magnetic field $\vec{B}$; this gives them a dipole polarization only: $\underline{s} = (0,\lambda\vec{B})$. So P is a symmetry of Co^{60} state $P(m,\vec{0}) = (m,\vec{0})$, $P(0,\lambda\vec{B}) = (0,\lambda\vec{B})$. Those nuclei decay spontaneously (β^- radioactivity) emitting electrons of energy momentum $\underline{q}(q^0,\vec{q})$ with an angular dependence proportional to $\underline{s} \cdot \underline{q} = -\lambda\vec{B} \cdot \vec{q} = -\lambda Bq \cos\theta$. This decay proved parity violation.

Similarly, in the spontaneous decay of zero spin π mesons ($\underline{p}_\pi^2 = m_\pi^2$):

$$\pi^\pm \to \mu^\pm + \nu_{\mp} \ ,$$

into a spin 1/2 μ-lepton and a massless $\nu_\pm$ (- for neutrinos, + for antineutrinos). The μ-lepton has a polarization s_μ (which can depend only on the observed quantitatives $\underline{p}_\pi = \underline{p}_\mu + \underline{p}_\nu$, $p_\nu^2 = 0$; remember $s_\mu \cdot p_\mu = 0$; see LM II).

$$\underline{s}_\mu = \mp\left(\frac{m_\pi^2 + m_\mu^2}{m_\pi^2 - m_\mu^2}\frac{\underline{p}_\mu}{m_\mu} - \frac{2m_\mu}{m_\pi^2 - m_\mu^2}\underline{p}_\pi\right) \ , \tag{4.44}$$

where $\mp$ depends on the sign of the (electric charge of) $\mu^\pm$. This proves C and also P violation (by observation of a pseudoscalar $p_\pi \cdot s_\mu$ in the decay).

By the same type of argumentation we verify that those experiments are compatible with CP invariance. Note that in π-decay, the μ-polarization $\underline{s}_\mu$ (given by (4.44′)) satisfies $\underline{s}_\mu^2 = -1$ (complete polarization). Then P_0 invariance (through angular momentum conservation) requires that the accompanying $\nu_\mp$ is emitted in a pure helicity state $\lambda = \mp 1$. All observation on neutrinos helicity suggest that $\nu_\mp$ has helicity $\mp$ for both ν_μ and ν_e.

---

† See Equation (4.25) and, at the end of 4.5, the property for λ to be <u>linear</u> in s_α, $s_{\alpha\beta}$, $s_{\alpha\beta\gamma}$... .

This shows that the set of neutrino states in not invariant under P or C, and it implies that all reactions with neutrinos violate P and C. But neutrino-less (in fact, non-leptonic) decays of hyperons also violate parity. Example: $\Lambda^0 \to p^+ + \pi^-$, the angular distribution depends on $\underline{s}_\Lambda \cdot \underline{p}_p = -\underline{s}_\Lambda \cdot \underline{p}_\pi$ (since $\underline{s}_\Lambda \cdot \underline{p}_\Lambda = 0$).

4.7.c. Time Reversal Invariance

It would be better to call it "velocity reversal" since $T(\vec{p}/p^0 = \vec{v}) = -\vec{v}$.

Let S_1 and S_2 be two states of a physical system and S_1^T, S_2^T the corresponding states obtained by a T active transformation ($p^0 \leadsto p^0$, $\vec{p} \leadsto -\vec{p}$, $s^0 \leadsto s^0$, $\vec{s} \leadsto -\vec{s}$, $\lambda \leadsto \lambda$, etc.). Then

$$T \Rightarrow \lambda_{12} = \lambda_{21}^T \quad . \tag{4.45}$$

Note the reversal of time ordering for the two transitions. A precise experiment comparing the cross section of the two inverse reactions $\to$ and $\leftarrow$

$$\gamma + d^+ \rightleftarrows p^+ + n \quad , \tag{4.46}$$

is in progress. (The rates are equal for pure states; since polarization is not observed one has to divide the rate by the dimension of the polarization space K_j for the particles $\to (2\ 1/2 + 1)^2 = 4$, $\leftarrow 2 \times 3 = 6$.)

Consider an elastic process (same initial and final particles) such as $\pi^- + p^+ \to \pi^- + p^+$ and compare the final polarization $\underline{s}'_p$ of the proton with the initial polarization of the proton target in another experiment. We must have

$$\lambda(\vec{p}_\pi, \vec{p}_p \to \vec{p}'_\pi, \vec{p}'_p, \vec{s}'_p) = \lambda(-\vec{p}'_\pi, -\vec{p}'_p, -\vec{s}'_p \to -\vec{p}_\pi, -\vec{p}_p) \quad . \tag{4.47}$$

There is an approximate condition of T invariance, in perturbation theory which is based on the following expansion of the "S-matrix".

$$S = I + iH + O(H^2) \quad , \tag{4.48}$$

where H has to be a self-adjoint operator (write $SS^* = S^*S = 1$, in first order in H). In this form, we have for orthogonal states (i.e., $R_i R_j = 0$)

$$\mathrm{Tr} R_j S R_i S^* \sim \mathrm{Tr} R_j H R_i H = \mathrm{Tr} R_i H R_j H \quad , \tag{4.49}$$

i.e., <u>in this approximation</u>

$$\lambda_{ij} = \lambda_{ji} \quad . \tag{4.50}$$

Then in this approximation, (4.45) reads $\lambda_{12} = \lambda_{12}^T$. Even in this approximation there is no positive evidence of violation of time reversal in physics, with perhaps the exception of K^0-decay (next section).

Note that in Equation (4.28), using Table 4.2, the term in g' (electric dipole) is not compatible with time reversal invariance (or with P invariance). So the existence of an electric dipole for an elementary particle would prove violation of both P and T. Experimentally g'_{neutron} is known to be $\leq 10^{-9}$ and $g'_{\text{electron}} \leq 10^{-12}$.

Note that PT invariance has a simple formulation. For example

$$\lambda(p_1 + p_2 \to p_1' + p_2') = \lambda(p_1' + p_2' \to p_1 + p_2) \quad ,$$

for spinless particles or for pure states (then change $\underline{s} \to -\underline{s}$, $s_{\alpha_1 \ldots -\alpha_k} \to (-1)^k s_{\alpha_1 \ldots \alpha_k}$)). This is known in physical literature as the "principle of detailed balancing".

4.8. CP Violation

CP violation was first observed by I. H. Christenson, C. W. Cronin, V. L. Fitch and R. Turley, *Phys. Rev. Lett.*, 13, 138 (1964) in K^0-decay. Many experiments have confirmed it.

The state of a K^0 or $\bar{K}^0$ (= anti-K^0, Y = -1) can be described by the Hilbert space $\mathcal{H}^{(1)} = L_2(R^3,t) \otimes K_2$ where K_2 is the vector space of functions defined on the two element set (Y = 1, Y = -1). Then C is of the form $I \otimes C$ while P is of the form $P \otimes I$ so $PC = P \otimes C$. We assume that $P^2 = 1$, $C^2 = 1$, PC = CP so $(PC)^2 = 1$ (as we have seen in 4.6, for spin 0, another assumption will increase the degree of freedom of K's). So we can write

$$\mathcal{H}^{(1)} = \mathcal{H}_+^{(1)} \oplus \mathcal{H}_-^{(1)} \quad , \tag{4.51}$$

$$CP\mathcal{H}_\pm^{(1)} = \pm\mathcal{H}_\pm^{(1)} \quad . \tag{4.51'}$$

Now it is easy to deduce the action of CP on states of two π^0. These are two identical self-conjugated particles hence any state of $2\pi^0$ is eigenstate with value +1 for C. The tensor product of the representation (m,0) of $P^\dagger$, by itself yields

$$\overset{2}{\otimes}(m,0) = \overset{\infty}{\underset{\ell=0}{\otimes}} \overset{\oplus}{\int_{2m}^{\infty}} (m,\ell)dm \quad , \tag{5.52}$$

with symmetry

$$\square\!\square \text{ for even } \ell, \quad \begin{smallmatrix}\square\\\square\end{smallmatrix} \text{ for odd } \ell \quad . \tag{4.52'}$$

Only the $\square\!\square$ = symmetric states are allowed by Bose statistics. And (by an argument essentially similar to that yielding Equation (2.40), P acts in the space of Equation (4.51) by multiplication by $(-1)^\ell$ in each direct summand. So states of $2\pi^0$ are eigenstates of C, P, CP with eigenvalue +1.

For states of $\pi^+ + \pi^-$, one has to consider these two particles as identical in order to apply Bose statistics, but in the two different possible states of charge (+ and -). So states of $\pi^+\pi^-$ of total spin ℓ, are eigenstates of C, P, CP with eigenvalue $(-1)^\ell$, $(-1)^\ell$, 1. When CP was believed to be preserved, it was predicted that states of $\mathcal{H}_+^{(1)}$ in Equation (4.51) would decay into 2π while states of $\mathcal{H}_-^{(1)}$ would decay into 3π states which are eigenstates of CP with eigenvalue -1 (as e.g. all $3\pi^0$ states). This was exactly observed and the states of the two spaces $\mathcal{H}_+^{(1)}$ and $\mathcal{H}_-^{(1)}$ were also called "short" and "long" because the 2π-decay is faster.

In 1964, the above quoted experiment proved that the long lived meson also decays into 2π (with a rate $\sim 10^6$ slower than the short lived).

We do know that the universe around our galaxy is not CP invariant, but the influence of this asymmetry (which could depend on the relative velocity of the K-meson with respect to the galaxy, or the earth) seems to be ruled out by more precise experiments.

Must we conclude that there is a small violation $((10^{-6})^{1/2} = 10^{-3}$ in amplitude) of CP in the transition $K \to 2\pi$? Another possibility could be that CP is conserved in this transition but that the two observed meson with exponential decay: short-lived K_S and long-lived K_L are non-orthogonal states with respectively a large c_S and a small c_L component in $\mathcal{H}_+^{(1)}$. Then the branching ratio

$$b_S = \frac{K_S \to 2\pi 0}{(K_S \to 2\pi^+ + \pi^-)} \quad \text{and} \quad b_\ell = \frac{K_L \to 2\pi 2}{K_L \to \pi^+ + \pi^-} ,$$

should be equal, since they would be the branching ratio of all the states in $\mathcal{H}_+^{(1)}$. The value of b_S is $\sim 1/2$ (as predicted by the selection rule $\Delta\vec{T} = 1/2$, see 3.6). The first measured values of b_ℓ were around 10 to 12, but a value zero appears in another experiment. The present experimental evidence is still an incompatible set but "optimists" say it is compatible with $b_L \sim b_S \sim 1/2$.

So it is possible that CP violation is due to a still undetected interaction, to which no particle transition or spontaneous decay can be attributed, and which has to be superweak.

CP violation has also been observed in $K_L \to \pi^\pm + \ell^\mp + \nu$ (where $\nu\ell = \mu$ or e) decay; there is a relative difference of 3.10^{-3} in the two C or CP conjugated rates. But CP violation has not yet been observed anywhere else.

Of course physicists have proposed many theories (about thirty not yet ruled out by the meager experimental data) to explain CP violation. There is no possibility to give more details here.

To conclude, let us just remark that there is no evidence against CPT violation and there is one fact which suggests that CPT is a "much better" invariance than CP: a small upper limit of the $K^0 - \bar{K}^0$ mass difference is well known. It is $m_{K^0} - m_{\overline{K}^0} < 10^{-14} m_K$. Such a perfect equality cannot be due to chance and suggests an invariance in nature which contains C. However, we have seen that C, CT, CP are ruled out, so CPT is the likely candidate in agreement with the CPT theorem 4.6.

Remark on Galilean Invariance

We dealt in Section 4 with relativistic Poincaré invariance only. Although we sometimes spoke in Chapters 2 and 3 of Galilean invariance, such invariance was not thoroughly used in atomic and nuclear physics. E. Inönü and E. P. Wigner characterized the irreps of $\overline{G}$, the covering of the Galilee group, in 1952 (*N. Cim.*, 9, 705).

Their results did not fit with physics. V. Bargmann (*Ann. of Math.*, 59, 1 (1954)) showed that for central extensions of the G Lie algebra $\mathcal{g}$, $H^2(\mathcal{g},R) = R$. For each irrep of $\overline{G}$, this yields a family of projective irreps depending on one parameter m which corresponds to the mass of the particle.

See also O'Raifeartaigh's lectures where it is shown that projective irreps of an invariance group also appear in classical mechanics.

5. THE INTERNAL SYMMETRIES OF HADRONS

5.1. SU(3) Symmetry

5.1.a. The Octets

Table 3.2 of "elementary particles" in 3.5 is reminiscent of similar tables of atomic and nuclear spectra.

So, before a dozen of baryons and as many mesons were known, physicists were searching for a larger symmetry than that of U_2 (isospin and hypercharge) which we have described in 3.5. There is no point and no time to tell here about the ill-fated choices except to mention that of Sakata, with a U(3) group whose fundamental representation was spanned by p, n, Λ, the first three known baryons. (S. Sakata, *Prog. Theor. Phys.*, 16, 686 (1956).)

Just as Heisenberg proposed to consider neutron and proton as two states of the same spin 1/2 particle, the nucleon, by neglecting their very small mass difference (or more precisely attributing it to an electromagnetic self-mass effect), the eight known spin $\frac{1}{2}^+$ baryons p, n, Λ^0, Σ^-, Σ^0, Σ^+, Ξ^-, Ξ^0 could be considered as eight states of the "same" particle although the mass difference is of the order of 15 percent instead of 0.15 percent.

By 1961, seven pseudoscalar mesons (0^-) were known, with the same group-
ing in isospin and hypercharge $y = 1$, $t = 1/2$, K^+K^0; $y = -1$, $t = 1/2$, $K^-\bar{K}^0$; $y = 0$,
$t = 1$, $\pi^+\pi^0\pi^-$ but the spread in mass was much larger.

M. Gell-Mann and Y. Ne'emann independently proposed to use SU(3) as a
classifying group; the eight $\frac{1}{2}^+$ baryons and, predicting a $y = 0$, $t = 0$ pseudoscalar meson which was discovered a few months later and called η^0, the eight 0^-
mesons form two octets = eight dimension space E_8 of the adjoint representation of
SU(3) (⊞ in Young diagram notation). For instance the Hilbert space of states of
one baryon is the tensor product $L(m, \frac{1}{2}^+) \otimes K(⊞)$ where $L(m, \frac{1}{2}^+)$ is the space
of the irrep $(m, \frac{1}{2}^+)$ of P, the Poincaré group and $K(⊞)$ the octet space E_8.
SU(3) is an exact symmetry when the baryon mass differences are neglected. We can
say that strong interactions will be decomposed into two parts: a strong SU-3 invariant part and a semi-strong part invariant under the subgroup $U_2(T,Y)$ only.
This fits the reduction[†]

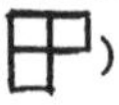

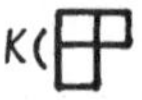

$$⊞ \text{ of } SU(3)\big|_{U(2)} = \substack{⊞ \\ (1,\frac{1}{2})} \oplus \substack{\cdot \\ (0,0)} \oplus \substack{\square\square \\ (0,1)} \oplus \substack{⧨ \\ (-1,\frac{1}{2})} \quad (y,t) \qquad (5.1)$$

$$\dim \quad 8 \quad = \quad 2 \quad + \quad 1 \quad + \quad 3 \quad + \quad 2 \quad .$$

But, would it be possible to consider the SU(3) breaking semi-strong interaction as
a perturbation of the very strong interaction? Surely, if you are an optimist.
After all 15 percent (effect in baryon mass) is small compared to 1.

Let us now study the mass splitting within the SU(3) multiplet.

5.1.b. The Mass Operator

The simple hypothesis for the mass operator M is that it can be decomposed into

$$M = M_0 + M'(y) \quad , \qquad (5.2)$$

where M_0 is a "scalar" tensor operator and $M'(y)$ is the image of y (of the Lie
algebra of SU(3)) by an octet = E_8-tensor operator. Let E be the space of an
irreducible representation of SU(3). Because SU(3) is of rank two, or equivalently has two zero roots (which are zero weights for ⊞)[††]

$$\dim \text{Hom}(E \otimes E_8, E)^{SU(3)} \leq 2 \quad . \qquad (5.3)$$

[†] For $u \in U(2)$, the black column means $(\det u)^{-1}$ while ⊟ means $(\det u)$.

[††] If $\lambda_1 \geq \lambda_2 \geq 0$ are the number of squares in the first and second line of the
Young diagram of an irrep of SU(3), one also uses the notation $(\lambda_1 - \lambda_2, \lambda_2)$
for the irrep of SU(3). The contragredient of (p,q) is (q,p), so (p,p) is
self-contragredient, as in ⊞ = (1,1), while $\square\square\square$ = (3,0) of dimension 10,
has for contragredient ⊞⊞ denoted $\overline{10}$ by the physicists.

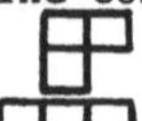

More precisely, it is 2, except for the trivial irrep, for which it is zero, and it is 1 for the irreps whose Young diagram has only 1 line (i.e., $(\lambda_1,0)$), or two equal lines $\lambda_1 = \lambda_2$, (i.e., $(0,\lambda_2)$): for example,  and its contragredient  which are also denoted 10 and $\overline{10}$ because they are of dimension 10. This is also true for  = (1,0) and  = (0,1) denoted 3 and $\overline{3}$; and  = (2,0),  = (0,2) denoted 6 and $\overline{6}$.

Another way to interpret (5.3) is to say that on the Hilbert space K of an irrep of SU(3) there are at most two linearly independent octet-tensor operators. Thus, in the approximation where U(2) is an exact symmetry (i.e., neglect of electromagnetic and weak interactions) the particle masses in a multiplet depend on three parameters (one, the expectation value of M_0, and two at most for $M_1(y)$). From 1.5, we know that we can take for each E, as linearly independent octet-tensor operators F and $D = F \vee F$, where $x \rightsquigarrow F(x)$ is the representation (up to the factor i) of the SU(3) Lie algebra on $\mathcal{H}$, the Hilbert space of hadronic states; it satisfies $F \wedge F = iF$. Explicitly, for any $\underset{\sim}{p}_1, \underset{\sim}{p}_2 \in K$ the space of an irrep of SU(3) in $\mathcal{H}$ and for any octet-tensor operator T

$$<\underset{\sim}{p}_1, T(x)\underset{\sim}{p}_2> = <\underset{\sim}{p}_1, (\alpha F(x) + \beta D(x))\underset{\sim}{p}_2> \quad . \tag{5.4}$$

In the physics literature α/β is called the F/D ratio. If the octet part $M'(y)$ (see Equation (5.2)) of the mass operator has no matrix elements between two subspaces of $\mathcal{H}$ carriers of inequivalent SU(3) irreps,† this implies that

$$M = M_0 + M_1F(y) + M_2D(y) \quad , \tag{5.5}$$

where M_0, M_1, M_2 are SU(3) scalar operators. The operators $F(y)$ and $D(y)$ commute and their common eigenspaces are U(2) multiplets, so they are functions of Y and $T(T+1)$, the generators of the center of the enveloping algebra of $U_y(2)$. By definition $F(y)$ is proportional to Y, the hypercharge operator, and by computation one finds

$$D(y) = \vec{T}^2 - \frac{1}{4}Y^2 - \frac{1}{3}K \quad , \tag{5.6}$$

where K is the (quadratic) Casimir operator of SU(3). So with a convenient change in the definition of the scalar operators, on a SU(3) multiplet the mass of a state of hypercharge y, isospin t

$$m = m_0' + m_1' + m_2(t(t+1) - \frac{1}{4}y^2 \quad . \tag{5.7}$$

Applied to the octet of Baryons N, Λ, Σ, Ξ this yields a relation between their four masses

$$\frac{1}{2}(m_N + m_\Xi) = \frac{1}{4}(3m_\Lambda + m_\Sigma) \quad , \tag{5.8}$$

(Gell-Mann, Okubo mass relation) which is well verified within few MeV (for mass $> 10^3$ MeV!).

---

† There are exceptions to this rule: see 5.1.d, the vector mesons. To convey the main idea, we simplify here too much.

For mesons (zero baryonic charge) because of the charge conjugation between particles and antiparticles, M_1' must be zero. The Gell-Mann Okubo mass relation for pseudo-scalar-mesons

$$m_K = \frac{1}{4}(m_\pi + 3m_\eta) \quad , \tag{5.9}$$

is verified only within 50 MeV, about 1/10 of the K and η mass. Optimistic physicists have found good reasons why this relation should be better verified by m^2 (instead of m).

5.1.c. The First Baryon Decuplet

When SU(3) was proposed as symmetry group in 1961 only the first N and Σ excited states, $\Delta(j^P = \frac{3}{2}^+, t = 3/2)$, $\Sigma^*(j^P = \frac{3}{2}^+, t = 1)$† were known. Gell-Mann putting them in a 10 representation, predicted a Ξ^*, $(j^P = \frac{3}{2}^+, t = 1/2, y = -1,$ excited state of Ξ) and finally a particle $\Omega(j^P = \frac{3}{2}^+, t = 0, y = -2$. As we have seen in the 10 (i.e., $\square\square\square$) irrep, the mass must depend linearly on two parameters (one for M_0 and only one for $M'(y)$), so in this decuplet the Gell-Mann and Okubo relation predicts for the mass m_y of the states of hypercharge y,

$$m_y = m_{\Sigma^*} - (m_{\Sigma^*} - m_\Delta)y \quad . \tag{5.10}$$

A few months later (in 1962) the predicted Ξ^* was found with a mass of 1530 MeV (to be compared to the predicted value (1385 + (1385 - 1236) = 1534 MeV!). It was later established that its spin is 3/2 and it has the same relative parity as Ξ. But the Ω^-, which should be stable against strong and electromagnetic decay, since it would be the lowest hadronic state with b = 1, y = -2, was frantically looked for and not found ... immediately. Many physicists had given up hope and given explanation why the Ω did not exist, before it was found in 1964 after two and a half years of feverish impatience. The Ω mass is 1672 MeV, (to be compared to the predicted 1677 MeV). Its spin has not yet been measured since less than a score of Ω particles have been observed up to now. If it had not been looked for where it was predicted, when would the Ω have been observed by chance?

5.1.d. Other SU(3) Multiplets

The known experimental data at a given date give a deformed view of the SU(3)-multiplets. In the baryon case for instance, no excited states of the Ω are known yet, although $\frac{5}{2}^+$ and other decuplets probably exist. Some octets have been tentatively identified, although too few excited Ξ states are yet known and their quantum numbers are not measured.

---

† Is often called also Y*. We denote by j^P the spin j and parity p.

The mesons seem to prefer to occur in nonets. Indeed a $q = 0$, $y = 0$, $t = 0$, 0^- meson is known in addition to the octet of 0^-. A nonet of 1^-: ρ, ω, ϕ, K^*, $\bar{K}^*$, is very well known. The mass formula could not apply to the known "octet" and the ϕ was predicted. The ω and ϕ are orthogonal states of "mixed configuration" $q = y = t = 0$, $\omega = 1\rangle \cos\alpha + 8\rangle \sin\alpha$, $\phi = 8\rangle \cos\alpha - 1\rangle \sin\alpha$, where $1\rangle$ is a SU(3) singlet and $8\rangle$ is the octet vector $q = y = t = 0$. A nonet of 2^+ is also well established and an octet of 1^+ is likely. There is some possibility of a 27-plet (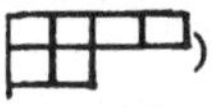) of baryons (not indicated in the Table 2.3, for the experimental data are still preliminary). It is to be noted that only irreps of the adjoint group $SU(3)/Z_3$ do appear.

5.1.e. Cross-Sections and Decays of Resonances

SU(3) invariance implies ratios of resonances decay rates (measured by the natural width and the different branching ratios) into lighter hadrons. This yields remarkably good predictions and explains strange facts such as the small branching ratio for the decay of ϕ into 2π.

For two octet-particle reactions $A + B \to C + D$, one can deduce that the scattering amplitude belongs to the representation

$$8 \otimes 8 = 27 \oplus 8 \oplus 1 \oplus 8 + 10 + \overline{10} \qquad (5.11)$$

symmetric (27, 8, 1) antisymmetric (8, 10, $\overline{10}$)

which yields seven arbitrary parameters. There are less in $8 \otimes 8 \to 8 \otimes 10$. The way to correct for the mass difference is not obvious and the predictions are not spectacular.

An anthology of original papers in SU(3) has been published by Gell-Mann and Ne'emann, *The Eightfold Way*, Benjamin, New York (1964). There is also a book on this subject by M. Gourdin, *Unitary Symmetry*, North-Holland, Amsterdam (1967).

5.2. Geometry on the SU(3)-Octet†

We give here some geometrical properties of the adjoint representation of SU(3).

We have defined in (1.18), (1.19), and (1.19′) the SU(3) invariant scalar product (x,y), the Lie algebra product $x \wedge y$, and the symmetric algebra product

† Full proofs and more results are given in a preprint of L. Michel and L. Radicati, with this title. It also contains some generalizations to SU(n).

$x \vee y$ for any pair of elements $x, y \in E_{n^2-1}$, the real vector space of the adjoint irrep of SU(n). We restrict ourselves here to $n = 3$ and call E_8 the octet space. Its elements can be realized as 3×3 traceless hermitian matrices. They satisfy the equation

$$x^3 - (x,x)x - \mathbb{1} \det x = 0 \quad , \tag{5.13}$$

whose coefficients obey the relation

$$4(x,x)^3 \geq 27(\det x)^2 \quad . \tag{5.14}$$

We find that

$$\det x = \frac{2}{3}(x,x \vee x) \quad , \tag{5.15}$$

so (5.14) can also be written

$$(x,x)^3 \geq 3(x,x \vee x)^2 \quad . \tag{5.16}$$

Orbits of SU(3) on E_8 are in a bijective correspondence with the pairs of real numbers (x,x), $(x,x \vee x)$ satisfying (5.16). When $(x,x)^3 > 3(x,x \vee x)^2$, x is called a regular element of E_8 and its isotropy group G_x is $U(1) \times U(1)$. Its Lie algebra is a Cartan subalgebra and it is generated by x, and $x \vee x$. When $(x,x)^3 = 3(x,x \vee x)^2$, x is called an exceptional element and its isotropy group is U(2). We will also call such x a q-vector or a pseudo-root. We will use from now on only normalized vectors : $(x,x) = 1$. Those vectors r satisfying $(r \vee r,r) = 0$ are the root-vectors. Every pseudo-root vector is of the form

$$q = \pm\sqrt{3}\, r \vee r \quad , \tag{5.17}$$

and also satisfies

$$\sqrt{3}\, q \vee q = \mp q \quad . \tag{5.18}$$

We call it positive or negative (normalized) q-vector. We denote by f_x, d_x the linear mappings $a \overset{f_x}{\leadsto} x \wedge a$, $a \overset{d_x}{\leadsto} x \vee a$. Then

$$[f_a,f_b] = f_{a \wedge b}, \; [f_a,d_b] = d_{a \wedge b} \quad , \tag{5.18}$$

so for $\forall a, b$ of a Cartan subalgebra C_x, the $f(a)$ can be diagonalized simultaneously on a basis z_k of the complexified E_8. Since C_x is left stable by f_a and d_a, we decompose $f_a = f_a'' \oplus f_a^{\perp}$, $d_a = d_a'' \oplus d_a^{\perp}$ on $C_x \oplus C_x^{\perp}$. Then

$$f_a'' = 0, \; f_a^{\perp} z_k = i\,(r_k,a) z_k, \; k = 1,\dots,6 \tag{5.20}$$

$$d_a^{\perp} z_k = (r_k \vee r_k,a) z_k = \frac{1}{\sqrt{3}}(q_k,a) z_k, \; k = 1,\dots,6 \tag{5.20'}$$

where r_k are the six unit roots of C_x and $q_k = \sqrt{3}\, r_k \vee r_k$ are the three positive unit pseudo roots of C_x.

The two eigenvalues of d_a'' are $\pm 1/\sqrt{3}$.

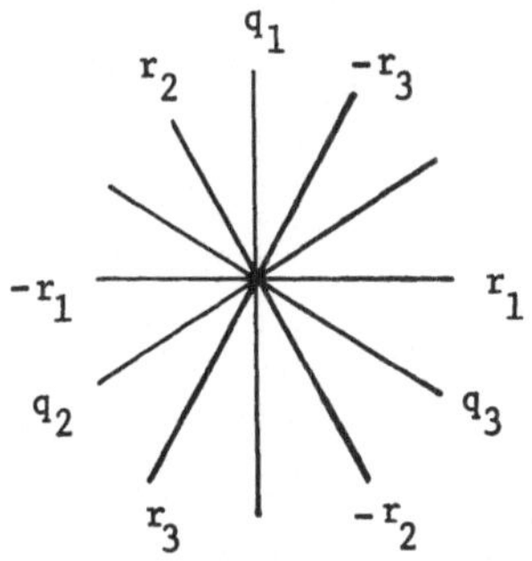

FIGURE 5.1. Roots $\pm r_i$ and pseudo-roots $q_i = \sqrt{3}\, r_i \vee r_i$ of a Cartan subalgebra. The SU(3)-Weyl group $S(3)$ permutes the three q_i.

Lemma

Every two-plane of E_8 contains at least a root. Indeed, the continuous odd function $(x, x \vee x)$ of x on the unit circle $(x,x) = 1$ of the two-plane has at least a zero. There are linear manifolds of root vectors.

For example: given a pseudo-root q, and using the same notation for a Lie subalgebra of SU(3) and its vector space (subspace of E_8)

$$E_8 = U_q(1) \oplus SU_q(2) \oplus U_2(q)^{\perp} \quad ,$$

where the three- and four-dimensional $SU_q(2)$ and $U_2(q)^{\perp}$ spaces contain only root-vectors. An octet of particles form an orthonormal basis of the complexified E_8, which diagonalizes the f_a for all $a \in C_{(y,q)}$ the Cartan algebra generated by the hypercharge and the electric charge directions since Y, Q are generators of $U_y(2) \subset SU(3)$. The Gell-Mann-Nishijima relation

$$Q = T_3 + \frac{1}{2} Y \quad , \tag{5.21}$$

among generators of $U(2) \subset SU(3)$ is translated in the octet geometry; y, $-q$ are unit positive pseudo-roots, $Q = -\, 2/\sqrt{3}\ F(q)$, $Y = 2/\sqrt{3}\ F(y)$,† t_3 is a root, $T_3 = F(t_3)$. We give in Figure 5.2 the corresponding roots of the two lowest octets of particles and also the weight of the lowest decuplet of baryons.

† The factors $2/\sqrt{3}$ are found from the condition that the spectra of Q and Y are the set of integers. Equation (5.21) implies that q and y are normalized pseudo-roots of opposite sign. The choice of sign here $+y$, $-q$ is conventional and corresponds to Figure 5.2.

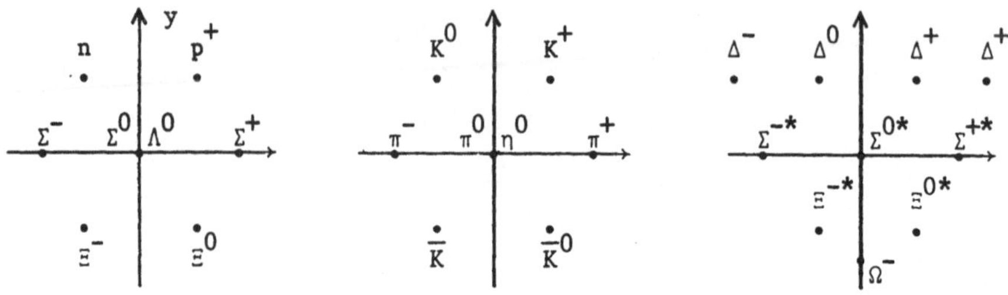

FIGURE 5.2. ROOTS OF OCTETS OF PARTICLES AND WEIGHT OF THE DECUPLET $\Sigma_0 = t_3 = \pi_0 \Lambda_0 = y = \eta_0$, corresponds to the two zero roots.

5.3. Electromagnetic and Weak Interactions in SU(3)

5.3.a. Electromagnetic Interaction

As shown by Equation (5.21), the electric charge operator Q is a generator of $U(2) \subset SU(3)$, so it is also a generator of $SU(3)$ and as we have seen

$$Q = \frac{2}{\sqrt{3}} F(-q) \quad , \tag{5.26}$$

where, as we have seen, q is a pseudo-root. The $SU_q(2)$ is called the U-spin group in the literature, and we can speak of U-spin multiplets, which have the same electric charge $u = 1/2$, p^+, Σ^+ and also Ξ^-, Σ^-, $u = 1$; n, Ξ^0, $1/2\ \Sigma^0 + \sqrt{3}/2\ \Lambda^0$; $u = 0$, $\sqrt{3}/2\ \Sigma^0 - 1/2\ \Lambda^0$. The electric charge is the integral of the time component of the electromagnetic current

$$Q = e \int j^0(x) d\vec{x} \quad , \tag{5.27}$$

and $\partial/\partial^\mu \ j^\mu(x) = 0 \Rightarrow Q$ is a constant (more generally P invariant) operator. Of course $j^0(x)$ could have any $SU(3)$ covariance, with the condition that the integral of the non-octet part vanish. The simplest hypothesis is to assume that the electromagnetic current $j^\mu(x)$ is the image in the direction $-q$ of an octet-tensor operator,

$$e \frac{2}{\sqrt{3}} j^\mu(x;-q) \tag{5.28}$$

(compare with Equation (5.26)). This allows us to draw many conclusions. The magnetic moment of the particle of a multiplet is given by the expectation value of an octet-tensor operator in the direction $-q$. It thus depends on two constants only for an octet (one for a decuplet) and the particles of the same u-spin multiplet have the same magnetic moment. For example one predicts $\mu_{\Sigma^+} = \mu_{p^+}$ which is well-verified. Measurements of μ_{Λ^0}, μ_{ε^+} and μ_{Ξ^-} are in progress, as well as the rate of $\Sigma^0 \to \Lambda^0 + \gamma$ which is related (as a "magnetic dipole" transition) to the values of μ in this octet. The ratio of rates of electromagnetic decay can be predicted. For example:

$$\frac{\text{rate } \pi^0 \to 2\gamma}{\text{rate } \eta^0 \to 2\gamma} = \frac{(t_3,q)^2}{(y,q)^2} \times \text{ ratio of phase-space} = 3 \times \text{ ratio of phase-space} , \quad (5.29)$$

(using (5.26) and $(y,t_3) = 0$). The observed ratio $\phi \to \mu^+ + \mu^-$, $\omega \to \mu^+ + \mu^-$ is a good confirmation of the mixing angle. Finally ratios of photo production cross-sections can also be predicted successfully.

The mass differences inside a $U_y(2)$ multiplet are thought to be of electromagnetic origin. They are quadratic in $j^\mu(x;-q)$ but to a good approximation it seems that only the scalar and octet part are important, so to a good approximation the mass operator (5.5) can be written, when one adds electromagnetic effects,

$$M = M_0 + M_1 \frac{2}{\sqrt{3}} F(y) + M_2 D(y) + M_3 \frac{2}{\sqrt{3}} F(-q) + M_4 D(-q)$$

and inside an SU(3)-multiplet the values of the masses are given by

$$m = m_0' + m_1' y + m_2'(t(t+1) - \frac{1}{4} y^2) + m_3' q + m_4'(u(u+1) - \frac{1}{4} q^2) \quad (5.30)$$

which is well verified for baryons.

5.3.b. Weak Interaction

Cabibbo generalized to SU(3) the Gell-Mann Feynmann hypothesis on the vector part of the weak current $v_\mu^\pm(x)$ coupled to the leptonic current $\ell^{\mp\mu}(x)$ (see 2) by the assumption that $v_\mu^\pm(x)$ and $j_\mu(x)$ are images by the <u>same</u> octet-tensor operator (that we shall denote $v_\mu(x)$) of three different directions: $-q, c_\pm$. Explicitly

$$\text{em current} = \frac{2}{\sqrt{3}} e v_\mu(x;-q) \quad ,$$

$$\text{weak current} = \frac{G}{\sqrt{2}} v_\mu(x;c_\pm) \quad , \quad (5.31)$$

(where G is the Fermi constant). The second Cabibbo assumption is that the axial-vector parts of the weak current $a_\mu^\pm(x)$ are images of another octet-tensor operator, in the same direction $c_\pm$. The total weak current

$$h_\mu^\pm(x;c_\pm) = v_\mu^\pm(x;c_\pm) - a_\mu^\pm(x;c_\pm) \quad , \quad (5.32)$$

is thus also image by an octet-tensor operator. See Cabibbo's original paper (*Phys. Rev. Lett.*, <u>10</u>, 531 (1963)) in *The Eightfold Way* anthology (p.207) for the predictions.

The ± subscript corresponds to the electric charge of the current, i.e.,

$$[Q, h_\mu^\pm(x)] = \pm h_\mu^\pm(x) \quad , \quad (5.33)$$

and using the fact that Q is an SU(3) generator, $Q = 2/\sqrt{3}\, F(-q)$, we can write this equation in the form (1.9)

$$-\frac{2}{\sqrt{3}}\,[F(q),h_\mu(x,c^\pm)] = -\frac{2}{\sqrt{3}}\,h_\mu(x,q\wedge c^\pm) = \pm h_\mu(x,c_\pm) \quad , \tag{5.34}$$

from (5.34) we get

$$q \wedge c_\pm = \mp\frac{\sqrt{3}}{2}\,c_\pm \tag{5.34'}$$

which means that $c_\pm$ are eigenvectors of $F(q)$. Writing $c_\pm = 1/\sqrt{2}(c_1 \pm ic_2)$ Equations (5.31, 5.34′) imply that c_1 and c_2 are unit vectors $\in U_q(2)$, so they are root-vectors, as we have seen in (5.21). Equation (5.34′) is equivalent to $q \wedge c_1 = c_2$, $q \wedge c_2 = -c_1$ which in turn implies

$$\sqrt{3}\,c_1 \vee c_1 = \sqrt{3}\,c_2 \vee c_2 = \sqrt{3}\,c_3 \vee c_3 = c \quad , \tag{5.35}$$

where

$$c_3 = c_1 \wedge c_2 \quad . \tag{5.35'}$$

This means that c, c_1, c_2, c_3 span $U_c(2)$; note also that c, $c_3 \in U_q(2)$. The pseudo root c is called the "weak hypercharge" or "Cabibbo hypercharge". It is a conserved quantity for weak interactions. It commutes with q, $c \wedge q = 0 \Rightarrow (c,q) = -1/2$. However, it does not commute with y; indeed, there are weak transitions violating hypercharge conservation. This lack of commutation is expressed by the $\neq -1/2$ value of

$$(y,c) = 1 - \frac{3}{2}\sin^2\theta \quad , \tag{5.36}$$

where θ is the Cabibbo angle. As we have seen in 3.6, its experimental value is 15 degrees and it is rather well verified that $v_\mu^\pm$ and $a_\mu^\pm$ define the same direction c of weak hypercharge.† The value of this angle is empirically given by

$$\mathrm{tg}\theta = m_\pi/m_k \quad ,$$

Cabibbo's theory not only explained the relative slower rate†† (by $\mathrm{tg}^2\theta$) of the weak transition violating the hypercharge y, but also explained that the super allowed $\Delta T = 0$ nuclear β-decay were slower than the $\mu \to \varepsilon + \nu + \bar{\nu}$ decay by a ratio $\cos^2\theta$.

The "computation" of this angle θ is one of the challenging present problems of physics. It is worth while to point out a purely algebraical relation, giving q as function of y and c.

Given two non-commuting (normalized positive) pseudo-roots y and c, there is always a unique pseudo-root which commutes with both of them

$$\lambda q = \sqrt{3}\,y \vee c + \frac{1}{2}(y + c) \quad , \tag{5.37}$$

---

† To be more precise, the angle of c_v and c_a with y is the same but c_v and c_a could be at a small angle between each other and this has been exploited as a possible explanation of CP violation.

†† To be accurate, it is not the rate but the probability transition = rate/phase space volume, since the phase space volumes, which should be equal in an exact SU(3)-symmetry, are in fact unequal.

where

$$\lambda = -(1 - (y,c)) \quad . \tag{5.37'}$$

The most commonly proposed form of non-leptonic weak interaction is

$$H_{N.L.} = \frac{G}{\sqrt{2}} \sum_{\varepsilon=\pm 1} \int h^{\mu}(x,c_{\varepsilon}) h_{\mu}(x,c_{-\varepsilon}) d^3\vec{x} \quad , \tag{5.38}$$

with the drawback that $H_{N.L.}$ is the image of a reducible tensor operator with some component in the "27" irrep of SU(3). The $\vec{\Delta}T = 1/2$ rule when $|\Delta Y| = 1$ for those weak transitions suggests that this 27 component is negligible compared to the octet component. The proposal of Radicati†

$$H_{N.L.} = \frac{G}{\sqrt{2}} \int (h^{\mu}(x) \vee h_{\mu}(x))(c) d^3\vec{x} \quad , \tag{5.39}$$

makes $H_{N.L.}$ the component along the weak hypercharge c of an irreducible octet-tensor operator. It is compatible with the known experimental data.

5.4. Critical Orbits of a G-Invariant Function on a Manifold M††

Given a group G acting on a set M, the set of all points of M which have conjugated little groups is called a stratum. So a stratum is the union of all orbits of the same type. Inclusion gives a partial ordering of all subgroups, modulo a conjugation, of a group. It corresponds to an (inverse) ordering on the strata. The set of fixed points form the minimal stratum (maximal isotropy group = G). If in the action of G on M there are no fixed points, there might be several minimal strata.

For example, in 5.2 we have seen that in the action of SU(3) on the unit sphere S_j of the octet space, there is the open dense general stratum $|(x \vee x,x)| < 1/\sqrt{3}$, containing a one parameter family of six-dimensional orbits (little group $U(1) \times U(1)$ and a minimal stratum made of two four-dimensional orbit $(x \vee x,x) = \pm 1/\sqrt{3}$. In this paragraph we want to consider

a) the smooth††† action of a <u>compact</u> Lie group G on a smooth manifold M. This action is given by the smooth mapping (= manifold morphism)

$$G \times M \xrightarrow{\phi} M \quad \text{with} \quad \phi(g_1,\phi(g_2,m)) = \phi(g_1 g_2,m) \quad ,$$

b) a real smooth function $M \xrightarrow{f} R$ which is G invariant, that is, the function is constant on every G orbit of M

---

† L. Radicati in *Old and New Problems in Elementary Particle Physics*, Academic Press, New York (1968).

†† This part is entirely a common work with Radicati, partly published in *Coral Gables Conferences 1968*, partly circulated in a preprint.

††† We use the word smooth for infinitely differentiable.

$$g \in G,\ m \in M,\ f(\phi(g,m)) = f(m) \quad .$$

The differential at $m_1 \in M_1$ of a smooth mapping $M_1 \overset{\Psi}{\to} M_2$ is denoted $d\Psi_{m_1}$; it is a linear mapping (with $m_2 = \Psi(m_1)$).

$$T_{m_1}(M_1) \xrightarrow{d\Psi_{m_1}} T_{m_2}(M_2) \tag{5.40}$$

where $T_{m_i}(M_i)$ is the tangent plane of M_i at m_i. So $df_p \in T'_p(m)$ the dual vector space of $T_p(m)$. We call critical point, the $p \in M$ such that $df_p = 0$.

The stabilizer (= little group = isotropy group) G_m in $m \in M$ is a closed and therefore compact subgroup of the compact group G. As is well known,† one can choose local coordinates in a neighborhood V_p of p such that the action of G_p is linear. Let $E_p(M)$ be the vector space corresponding to this linear representation of G_p; so $V_p \subset E_p(M)$. Since G_p is compact and M real, this linear action can be made orthogonal so $E_p(M)$ is a euclidean space. We can then identify df_p with a vector of $E_p(M)$ that we shall call $(\text{grad } f)_p$. The G-orbit of $p, G(p)$, is the image of $g \overset{\phi^{(p)}}{\rightsquigarrow} \phi(g,p)$; it is a submanifold of M; its tangent plane in p, denoted $T_p(G(p))$, is the image of $d\phi_e^{(p)}$ where e is the unit of G. The isotropy group G_p transforms $G(p)$ into itself. Similarly $T_p(G(p))$ is an invariant subspace of $E_p(M)$. The orthogonal subspace $N_p(G(p)) = T_p(G(p))^{\perp} \subset E_p(M)$ is also invariant and it is called the "slice" at p. <u>Note that</u> $(\text{grad } f)_p \in N_p$. Indeed, by definition, for $x \in T_p(M)$, $((\text{grad } f)_p, x) = \lim_{\alpha\to 0}[(f(p+\alpha x) - f(p))]\alpha^{-1}$. The bracket is 0 when $p + \alpha x \in G(p)$, the orbit of p, so it stays zero at the limit, when $x \in T_p(G(p))$.

<u>Note also that</u> $(\text{grad } f)_p$ <u>is invariant by</u> G_p. Let $g \in G_p$; $(g \cdot (\text{grad } f)_p, x) = ((\text{grad } f)_p, g^{-1} \cdot x) = \lim_{\alpha\to 0} \alpha^{-1}(f(p + \alpha g^{-1} \cdot x) - f(x))$, and since $g^{-1} \cdot p = p$, $f(p + \alpha g^{-1} \cdot x) = f(g^{-1} \cdot (p + \alpha x)) = f(p + \alpha x)$, so $\forall x \in E_p(M)$, $(g \cdot (\text{grad } f)_p, x) = ((\text{grad } f)_p; x)$. If the slice $N_p(G(p))$ has no vectors invariant by G_p, then $(\text{grad } f)_p = 0$. We can summarize this by the:

<u>Theorem 1</u>

Let G be a compact Lie group acting smoothly on the smooth real manifold M. If for $p \in M$, the canonical linear representation of G_p on the slice N_p does not contain the trivial representation of G_p, then $G(p)$ is a critical orbit for

† Consider a Riemann metric on M; it is transformed by the action of G_p. By averaging with a G_p-invariant measure, one obtains a G_p-invariant Riemann metric and G_p transforms into each other the geodesics from the fixed point p. In the neighborhood V_p of p, take geodesic coordinates.

any real valued G-invariant smooth function on M (where here again we denote by the same symbol, e.g., SU(2), the vector space of the Lie subalgebra, and also the group!).

Example 1. We have studied the action of SU(3) on $S_7 \subset E_8$. Let q be a unit q-vector, $G_q = U_2(q)$, $T_q(M) = \{q\}^{\perp} \subset E_8$, $T_q(G(q)) = U_2(q)^{\perp}$, $N_q(G(q)) = SU_2(q)$ and $U_2(q)$ acts linearly on it, without fixed vectors.

Example 2. p is an isolated fixed point in M. So there is a neighborhood V_p of p with no other fixed points and $N_p = E_p(M)$ has no invariant $G = G_p$ vector.

This proves that p is a critical point for every G invariant function on M.

We shall now assume moreover, that M is compact. Then there is one stratum (called generic stratum) which is open dense in M; the minimal strata are closed and compact. Let C be a connected component of a minimal stratum; let $p \in C$, $F_p \subset E_p$ be the linear subspace of G_p fixed points. Because G_p is maximal, the points of $V_p \cap F_p$ have G_p as stabilizer so they belong to C. Given a G-invariant real valued smooth function f, let $n = (\text{grad } f)_p$. As we have seen $n \in F_p$ so for small enough $|\varepsilon|$, $p + \varepsilon n \in C$. We can write

$$(n,n) = \lim_{\varepsilon \to 0} \varepsilon^{-1}(f(p + \varepsilon n) - f(p)) \tag{5.41}$$

so if f is constant on C, every $p \in C$ is a critical point of f. If f is not constant on C it has at least an orbit of maxima and an orbit of a minima. Let p a point of such an orbit, and $n = (\text{grad } f)_p$. Then, in Equation (5.41),

$$f(p + \varepsilon n) - f(p) \begin{array}{l} \geq 0 \text{ if } f \text{ is minimum} \\ \leq 0 \text{ if } f \text{ is maximum} \end{array} \text{ at } p \quad ,$$

which means that (n,n) either has the sign of $\pm\varepsilon$ (+ at minimum, - at maximum) which is impossible, or must be zero.

Theorem 2.†

Let G be a compact Lie group, acting smoothly on the real compact manifold M, and let f be a real valued G-invariant, smooth function on M. Then f has at least a critical point for each connected component C of each minimal stratum.

† To prove this theorem, that Radicati and I conjectured, we received great help from A. Borel, C. Moore, and R. Thom.

We will now be interested in a particular function on a sphere: let G be a compact Lie group, E the real vector space carrier of a linear representation $g \rightsquigarrow R(g)$, irreducible over the reals. So R (up to an equivalence) is an orthogonal representation and it is self-contragredient. We denote by $(\underline{x},\underline{y})$ the invariant Euclidean scalar product in E. Let us assume that (with $\vee$ the symmetrical tensor product) $\dim(\mathrm{Hom}\, E \vee E, E)^G = 1$. As we have seen in 1.5, there is a unique (up to a constant factor) symmetrical algebra

$$x \otimes y \xrightarrow{\Psi} x_T y \quad \text{where} \quad \Psi \in \mathrm{Hom}(E \vee E, E)^G \tag{5.42}$$

with $x_T y = y_T x$.

Since the representation is self-contragredient and the tensor product is associative

$$(x_T y, z) = (x, y_T z) = \{x,y,z\} \quad . \tag{5.43}$$

Hence, the invariant $\{x,y,z\}$ is a completely symmetrical G-invariant trilinear form on E. Let $f(\{x,y,z\})$ be a function on the unit sphere $S = \{x \in E, (x,x) = 1\}$.

Using λ as a Lagrange multiplier, critical points of f are given by the equation

$$\mathrm{grad}(f(\{x,x,x\}) + \lambda(1 + (x,x)) = 3f' x_T x - 2\lambda x = 0 \quad , \tag{5.44}$$

where f' is the derivative of the one variable function f; e.g., if $f = \{x,x,x\}$, $f' = 1$. In other words, critical points of f are given by solutions of

$$x_T x = \lambda x \quad ,$$

i.e., the idempotents (or nilpotents for $\lambda = 0$) of the symmetrical algebra.

5.5. SU(3) × SU(3) Symmetry

Physicists have considered symmetries higher than SU(3) for the hadronic world. Of course they are coarser, but still useful as we shall see. The SU(3) × SU(3) symmetry becomes an exact symmetry of the hadronic world when the masses of 0^- mesons are neglected. Note that it is not much more drastic to say that those masses are equal to zero than to say that they are equal as is already implied by SU(3). As a matter of fact, a much milder approximation than SU(3) is to neglect only the π-meson mass (only 140 MeV, and this is smaller than the 0^--meson mass differences). This corresponds to a SU(2) × SU(2) × U(1) subgroup of SU(3) × SU(3). We give in Figure 5.3, a scheme of the lattice of symmetry groups which have been considered for hadronic physics, but in this section we limit ourselves to SU(3) × SU(3) and its subgroup. (See also O'Raifeartaigh lectures for the higher symmetries.) At the level of the middle line of Figure 5.3, a new feature appears; a mixing of internal symmetry and relativity invariance. It is very mild for SU(3) × SU(3) since it concerns only the parity operator. The total symmetry group to consider is the semi-direct product by $Z_2(P)$

$$(P_0 \times SU_3 \times SU_3) \square Z_2(P) \quad , \tag{5.49}$$

which acts naturally on P_0 and exchanges the two SU_3 factors in $SU_3 \times SU_3$. To distinguish such SU(3) factors, let us denote them as $SU_3^{(+)} \times SU_3^{(-)}$; they are called in the physics literature the $\pm$ chirality group. The group (5.49) is a good frame for understanding the relation of P (parity operator) with the different interactions. This will become clear in the following. The diagonal subgroup $SU(3)^d \subset SU_3^{(+)} \times SU_3^{(-)}$ is the SU(3) group of invariance of 5.1, 5.2, and 5.3.

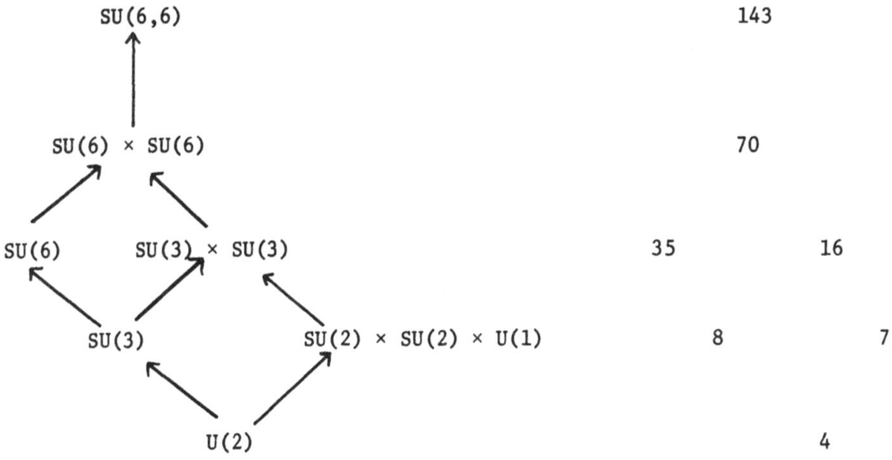

FIGURE 5.3. SYMMETRY GROUPS AND THEIR DIMENSIONS

Lattice of symmetry groups used in hadronic physics. → means injection as subgroup.

We will denote a vector of the 16-dimensional vector space E_{16} of the SU(3) × SU(3) Lie algebra by a direct sum of two vectors

$$\tilde{a} = a_+ \oplus a_- \quad , \tag{5.50}$$

$a_\pm$ belongs to the $SU_3^{(\pm)}$ octet.

The invariant Euclidean scalar product (given by the Cartan-Killing form) is, in terms of the octet scalar product

$$(\tilde{a},\tilde{b}) = (a_+ \oplus a_-, b_+ \oplus b_-) = \frac{1}{2}(a_+,b_+) + \frac{1}{2}(a_-,b_-) \quad . \tag{5.51}$$

The Lie algebra law is (we use $\wedge$ for it)

$$\tilde{a} \wedge \tilde{b} = (a_+ \wedge b_+) \oplus (a_- \wedge b_-) \quad ,$$

and since $\dim \mathrm{Hom}(E_{16} \vee E_{16}, E_{16})^{SU3 \times SU3} = 1$, there is a unique canonical symmetrical symmetrical algebra

$$\tilde{a} \vee \tilde{b} = (a_+ \vee b_+) \oplus (a_- \vee b_-) \quad . \tag{5.52}$$

The covariance property of the electromagnetic and weak interactions are most naturally extended to $SU(3) \times SU(3)$ by the following hypothesis: the electric current $j_\mu(x)$, the vector part $v_\mu^{(\varepsilon)}(x)$, the axial vector part $a_\mu^{(\varepsilon)}(x)$ of the (charged $\varepsilon = \pm 1$) weak hadronic current $h_\mu^{(\varepsilon)}(x) = v_\mu^{(\varepsilon)}(x) - a_\mu^{(\varepsilon)}(x)$ are images of the same E_{16} tensor operator, which we will denote $h_\mu(x;\tilde{a})$; the vector currents correspond to $SU(3)^d$ and the axial vector current to the anti-diagonal. The weak current has pure "-" chirality. Explicitly: electromagnetic current

$$\frac{2}{\sqrt{3}}\, e h_\mu(x;-(q \oplus q)) \quad , \tag{5.53}$$

(charged) weak currents

$$\frac{G}{\sqrt{2}}(h_\mu(x;0 \oplus c_1) \pm i h_\mu(x;0 \oplus c_2)) \tag{5.53$'$}$$

and the Radicati form of the (non leptonic) weak hadronic interaction is

$$\frac{G^2}{2}\int h_\mu(x) \vee h^\mu(x))(c) d^3\vec{x} = \frac{G^2}{2}\int (h_\mu(x) \,\dot{\vee}\, h^\mu(x)(0 \oplus c) d^3x \quad . \tag{5.54}$$

The generators of $SU(3) \times SU(3)$ are the space integral of the current, i.e.,

$$a \rightsquigarrow F(\tilde{a}) = \int h_0(x;\tilde{a}) d^3\vec{x} \quad , \tag{5.55}$$

is the representation (up to i) of the $SU(3) \times SU(3)$ Lie algebra on the Hilbert space of physics

$$[F(\tilde{a}), F(\tilde{b})] = i F(\tilde{a} \wedge \tilde{b}) \quad , \tag{5.56}$$

and for the particular case of the E_{16}-tensor operator $(h_\mu(x)\tilde{a})$

$$[F(\tilde{a}), h_\mu(x,\tilde{b})] = i h_\mu(x, \tilde{a} \wedge \tilde{b}) \tag{5.57}$$

as we saw in Equation (1.9).

In the approximation where $SU(3) \times SU(3)$ is an exact symmetry $\partial_\mu h^\mu(x,\tilde{a}) = 0$ and the $F(\tilde{a})$ are well defined. Since $SU(3) \times SU(3)$ is a broken symmetry, the usual difficulty to define the self-adjoint operator $F(\tilde{a})$ arises. (See O'Raifertaigh's lectures, this Volume.)

The equation†

$$\tilde{a} \,\dot{\vee}\, \tilde{a} = \lambda \tilde{a} \quad , \tag{5.58}$$

for unit vectors $\in S_{15} \subset E_{16}$ has two sets of solutions. One is the set of $1/\sqrt{2}\,\tilde{a} = \pm c \oplus 0$ or $\pm 0 \oplus c$, where c is a (normalized positive) pseudo-root and $\lambda = \mp\sqrt{2/3}$. This set is made up of two minimal strata, each consisting of two pieces of one orbit each. So each of the four orbits is a critical orbit for <u>every</u> smooth $SU(3) \times SU(3)$ invariant function in S_{15} the unit sphere of E_{16}. The stabilizers are, up to a conjugation, $SU_3^{(+)} \times U_c^{(-)}(2)$ and $U_c^{(+)}(2) \times SU^{(-)}(3)$ for the two strata.

† See L. Michel and L. Radicati, preprint, *Breaking of the SU(3) × SU(3) Symmetry in Hadronic Physics*.

The other type of solution is the set of vectors,

$$\pm(q_1 \oplus q_2) \quad ,$$

which form two orbits of a four separated orbit stratum ($\pm q_1 \oplus \mp q_2$ for the two other orbits) whose stabilizer is $(U_{q_1}(2) \times U_{q_1}(2))_{\square} Z_2$. The pseudo-roots $\pm(q \oplus q)$ of the diagonal $SU^{(d)}(3)$ are on the orbits of solutions while those of the anti-diagonal $(\pm q \oplus \mp q)$ are not. This has a bearing on parity.

It seems to us remarkable that the electromagnetic charge direction $-(q \oplus q)$ and the weak hypercharge direction $(0 \oplus c)$ give two solutions, one of each type, of Equation (5.58).

$SU(3) \times SU(3)$ is not only broken by electromagnetic and weak interaction, but also by semi-strong U_2-invariant interaction. There are two different interesting intermediate approximations of symmetry of strong interactions between U_2 and $SU(3) \times SU(3)$; those of the fourth line of Figure 5.3, $SU(3)$ already studied, and $SU(2) \times SU(2) \times U_1$, which implies the Adler-Weissberger sum rule, and more recently emphasized by Gell-Mann, Oakes and Renner. In both cases H_{strong} is, to a good approximation, the sum

$$H_{strong} = H_0 + H_1(\underset{\sim}{m}) \quad , \tag{5.59}$$

of H_0 invariant under $SU(3) \times SU(3)$ and of $H_1(\underset{\sim}{m})$ which is the image of $\underset{\sim}{m}$ by a $SU(3) \times SU(3)$ tensor operator for the (irreducible over reals) $(3,\bar{3}) \oplus (\bar{3},3)$ representation. The two corresponding directions $\underset{\sim}{m}$ for these two approximations are again idempotent or nilpotents of the canonical symmetric algebra. I refer to my preprint with Radicati for details. This 18-dimensional irreducible real representation of $SU(3) \times SU(3)$ on E_{18} (which is the one which naturally arises in a quark model) is such that $\dim \mathrm{Hom}(E_{18} \vee E_{18}, E_{18})^{SU(3) \times SU(3)} = 1$ so there is a unique canonical symmetrical (real) algebra on E_{18} which has $SU(3) \times SU(3)$ as group of automorphisms. We denote this algebra law by $\underset{\sim}{m}_1 \top \underset{\sim}{m}_2$.

The equation

$$\underset{\sim}{m} \top \underset{\sim}{m} = \lambda m \quad , \tag{5.60}$$

has only two types of solutions (for vectors on the invariant unit sphere $S_{17} \subset E_{18}$) belonging to two minimal strata, the one for $|\lambda| = 2/3$ corresponds to $SU^d(3)$ as stabilizer. The other, for $\lambda = 0$ corresponds to $SU_y^{(+)}(2) \times SU_y^{-}(2) \times U_y^d(1)$.

Theorem 1 shows that this latter case $(\lambda = 0)$ corresponds to a critical orbit for all $SU(3) \times SU(3)$ invariant functions on S_{17} (unit vectors of the $3\bar{3} + \bar{3}3$ irrep); this orbit is also a minimal stratum of dimension 9. The stratum corresponding to $SU(3)^d$ is also minimal; it is a nine-dimensional connected submanifold (of S_{17}) made up of eight-dimensional orbits. So from Theorem 2, each invariant function has at least two critical orbits in this stratum. For all functions of $(x, x \top x)$ these two orbits are $x \top x = \pm\, 2/3\, x$.

Note Added After the Seattle Rencontres.† I do not understand why I have not used in Seattle, as emphasized by Equation (5.49), $(SU(3) \times SU(3))_{\Box}Z_2$ instead of $SU(3) \times SU(3)$. Then, the two orbits on S_{17}, $x \top x = \pm 2/3\, x$ are critical for all functions. Radicati and I also wonder why we have not considered before the groups $(SU(3) \times SU(3))_{\Box}(Z_2 \times Z_2)$ where the discrete group $Z_2 \times Z_2 = \{I,P,C,PC\}$ is generated by the parity and the charge conjugation operators. Among the strata of S_{15} for this group, there are four which contain only one orbit. These orbits are the critical ones of S_{15}. Typical points (~ unit vectors up to a sign) of these orbits are

$\pm\tilde{q} = \pm(q \oplus q)$	the direction of electromagnetic interaction,	(5.53)
$\pm\tilde{c}_i = 0 \oplus \pm c_i (i = 1,2)$	the Cabibbo direction of weak coupling,	(5.53′)
$\pm\tilde{c} = 0 \oplus \pm c$	the "weak hypercharge" direction proposed by Radicati,	(5.54)
$\pm\tilde{r} = \pm(r \oplus \varepsilon r)$, $\varepsilon = \pm 1$ (root vectors)	a direction which some authors (for instance M. L. Good, L. Michel, and E. de Rafael, *Phys. Rev.*, 151, 1199 (1966), have used in their proposed theory of the CP violating interaction.	

Radicati and I have also included Theorem 1 into a more complete:

Theorem 1′

Let G be a compact Lie group acting smoothly on the real manifold M, $p \in M$. The three following propositions are equivalent.

a) the orbit of p is critical (for every G-invariant real valued smooth function f on M, $df_p = 0$),

b) the orbit of p is isolated in its stratum, i.e., $\exists$ a neighborhood V_p of p such that ($x \in V_p$ and $x \notin G_p$) $\Rightarrow G_x$ is not conjugate to G_p,

c) the canonical linear representation of G_p on the slice N_p does not contain the trivial representation.

Theorem 1 is simply $c \Rightarrow a$.

5.6. SU(6), Quarks, Current Algebra, Boot-Strap, Etc.

The title of this section is a statistical sample of key words found these last years in papers on fundamental particle physics.†† This last section is not a

† After the Seattle Recontre, L. Radicati and I collected the above results to present them in a lecture on September 19, 1969 in Rome (see preprint, *Geometrical Properties of the Fundamental Interactions*). The following improvements were then obvious to us.

†† For the last year, the passwords are Veneziano and duality. It is a sociological fact that there are fads in fundamental particle physics.

conclusion but an open-end to the description of a very rapidly changing situation; the view that physics gives us of the hadron world.

SU(6) Symmetry. SU(6) Symmetry was introduced independently[†] by Gürsey and Radicati (*Phys. Rev. Lett.*, 13, 299 (1964)) and by B. Sakita (*Phys. Rev.*, 136 B, 1756 (1964), for mesons only). It was noticeable that mass-differences between SU(3) multiplets were not larger than those inside multiplets.

Both groups of authors, inspired by the SU(4) = supermultiplet Wigner theory for nuclei (3.3) extended it to fundamental particles by enlarging the SU(2) isospin to SU(3). So in the non-relativistic version, the space of the one particle hadron states is the tensor product

$$\mathcal{H}^{(1)} = L_2(\mathbb{R}^3, t) \otimes K_\sigma \otimes K_\lambda \quad ;$$

here K_σ and K_λ are respectively, two- and three-dimensional Hilbert spaces and the action of $\bar{G}$, the central extension of the Galilee group, and of SU(6) on $\mathcal{H}^{(1)}$ are respectively, (with $\bar{G} \overset{\psi}{\twoheadrightarrow} SU(2)$ also (2.9) and Equations (2.57) and (2.58)[††]

$$\mathcal{H}^{(1)} = L_2(\mathbb{R}^3, t) \otimes \quad K_\sigma \otimes K_\lambda$$
$$\bar{g} \in \bar{G} \to \pi(\bar{g}) \otimes \psi(\bar{g}) \otimes I \qquad (5.61)$$
$$u \in SU(6) \to I \otimes u$$

The lowest two multiplets of SU(6) are given in Figure 5.4. For the baryon, it belongs to the irrep ▭▭▭ of dimension 56; for the meson, to the ▮,

the 35-dimensional adjoint irrep of SU(6). The χ^0 (not discovered in 1964!) is a singlet. We give here the decomposition of these irrep into $SU_2 \times SU_3$ irrep

$$\left. \Box\Box\Box \right|_{SU_2 \times SU_3} = \Box \times \text{⊞} \oplus \Box\Box\Box \times \Box\Box\Box$$
$$2 \times 8 \; + \; 4 \; \times \; 10 \quad = 56$$

$$\left. \text{▮} \right|_{SU_2 \times SU_3} = \bullet \times \text{⊞} \oplus \Box\Box \times \text{⊞} \oplus \Box\Box \times \bullet$$
$$(1 \times 8) + (3 \times 8) + (3 \times 1) \quad = 35$$

The mass formula for each SU(6) multiplet becomes

$$m = m_0 + m_1 y + m_2 \left(t(t+1) - \tfrac{1}{4} y^2 \right) + m_3 j(j+1) + m_4 q + m_5 \left(u(u+1) - \tfrac{1}{4} q^2 \right) \quad .$$

[†] In fact, Gell-Mann in, *Physics*, 1, 63 (1964), page 74 (reproduced in, *The Eightfold Way*, anthology, p. 203), was the first to introduce SU(6) in the physics of elementary particles but, for once, he did not work out its physical applications.

[††] For more details, see L. Michel, "The Problem of Group Extensions of the Poincaré Group and SU(6) Symmetry", p. 331; 2nd Coral Gables Conferences, *Symmetry Principles at High Energy*, Freeman and Co., San Francisco (1965).

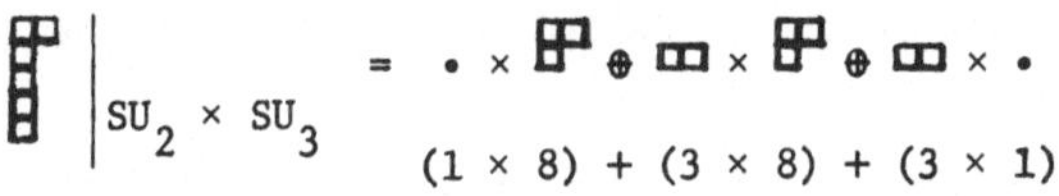

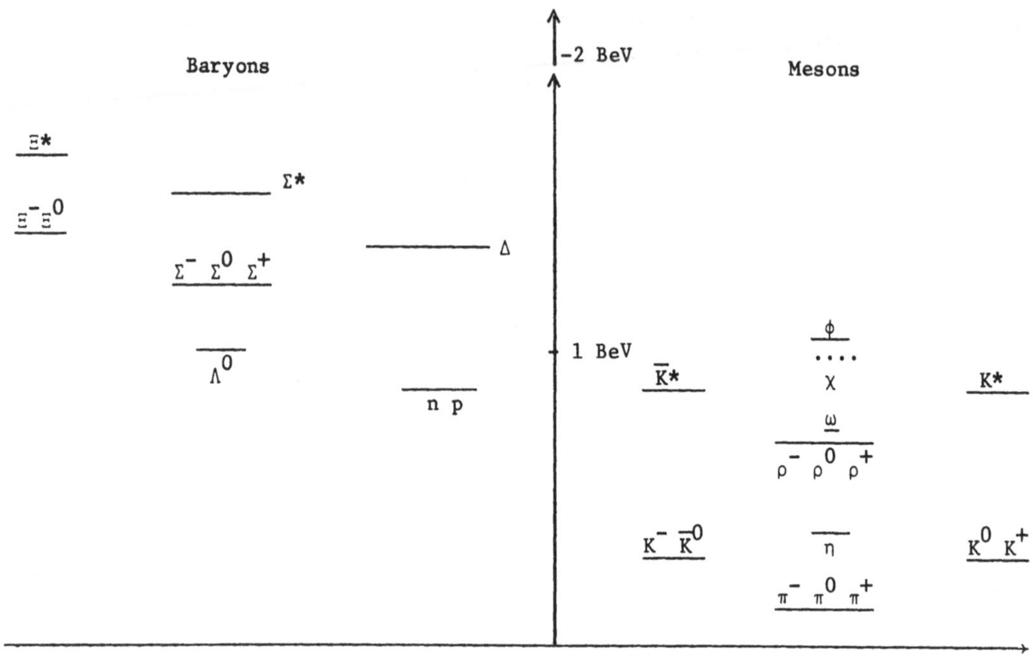

FIGURE 5.4. THE $(8 \times 2) + (10 \times 4) = 56$-PLET OF BARYONS AND THE $(8 \times 1) + (9 \times 3) = 35$-PLET OF MESONS IN THE SU-6 CLASSIFICATION OF HADRONS.

Neglecting the electromagnetic mass difference $(m_4 = m_5 = 0)$ the four-parameter formula predicts well the masses of the eight lowest $U_{(2)}$- multiplets of baryons. The magnetic moment of baryons depends on only one parameter μ_p so we have the relation

$$\mu_n = -\frac{2}{3}\mu_p \quad , \tag{5.62}$$

which is within 3 percent of reality (this is too good!).

How should one apply SU(6) invariance to particle reactions? Some physics and empirical rules (e.g., so called $SU(6)_W$) have to be injected, and the symmetry is still useful.

However, the drawback is the difficulty in reconciling SU(6) with relativistic invariance.†

Quarks. It is a natural tendency in science to try to explain the universe by the smallest number of different types of building blocks, such as the four elements of the Greeks, which at the end of the XIXth century had reached nearly ninety chemical elements. From 1910 to 1929 (measurement of the spin and statistics

† This will be dealt with by O'Raifeartaigh, when he studies the two upper lines of the diagram of Figure 5.3.

of the N^{14} nucleus, see 2.10) only three particles p^+, e^-, γ, were known and needed to build the universe again. But one had to add ν in 1931, n and e^+ in 1932, etc., so now we have the Table 3.2 of 3.5 = spectroscopy of hadrons.

Is it possible to return to "simplicity"? The hoped for building blocks have been called quarks by Gell-Mann: the 3 spin 1/2 quarks for the multiplet 3 (= fundamental irrep □) of SU(3) and 6 (= irrep □) of SU(6). There are also 3 antiquarks belonging to the contragredient irrep

$$\bar{3} = \begin{smallmatrix}\square\\\square\end{smallmatrix} \text{ of } SU(3) \quad \text{or} \quad \bar{6} = \begin{smallmatrix}\square\\\square\\\square\\\square\\\square\end{smallmatrix} \text{ of } SU(6) \quad .$$

Mesons of Table 3.2 are formed of one quark and one antiquark $\bar{q}$. Lowest bound states of $q + \bar{q}$ yield all expected meson states. Baryons of the same table are made of 3 quarks, which are, for the lowest state, in the SU(6) state □□□, so they must have a space symmetry $\begin{smallmatrix}\square\\\square\\\square\end{smallmatrix}$ to obey Fermi statistics; this from our experience acquired in Chapter 2 and 3 does not seem compatible with attractive forces. And how to explain the saturation by 3; why should 2-quark or 4- or 5-quark states not also be stable?†

Forgetting these difficulties one can search for quarks. (They should be very heavy, stable, have fractional quantum numbers $b = 1/3$, $q = 2/3$ or $-1/3$) and compute with them (good prediction of the "quarks model", e.g., by Dalitz, Lipkin.) They have not been found experimentally, and quarks can simply be looked at as the physicists' name for an orthonormal basis of the fundamental □ irrep of SU(6), used in their computations!

<u>Current Algebra.</u> Let $\tilde{a} \rightsquigarrow D(\tilde{a})$ be the SU(3) × SU(3) Lie algebra adjoint irrep E_{16}. Any E-tensor operator function on space time $f(\vec{y},\underset{\sim}{m})$ will satisfy Equation (1.9) at any <u>fixed</u> time

$$[F(\tilde{a}),f(\vec{y},\underset{\sim}{m})] = if(\vec{y},D(\tilde{a})\underset{\sim}{m}) \quad , \tag{5.63}$$

where $\underset{\sim}{m} \in E$. Equation (5.57) is a particular case for $f(\vec{x},\underset{\sim}{m}) = h^\mu(\vec{x},\tilde{b})$.

Replace $F(\tilde{a})$ by its expression (5.55). After commuting the symbols [and $\int$, Equation (5.63) reads (use $(1 = \int\delta(\vec{x} - \vec{y})d^3\vec{x})$),

$$\int d^3\vec{x}[h^0(\vec{x},\tilde{a}),f(\vec{y},\underset{\sim}{m})] = i\int d^3\vec{x}\delta(\vec{x} - \vec{y})f(\vec{y},D(\tilde{a})\underset{\sim}{m}) \quad ,$$

for <u>any</u> tensor operator function of $\vec{x}$. It is very suggestive to write the equality for the integrands

$$[h^0(\vec{x},\tilde{a}),f(\vec{y},\underset{\sim}{m})] = i\delta(\vec{x} - \vec{y})f(\vec{x},D(\tilde{a})\underline{m}) \quad . \tag{5.64}$$

---

† There are several ways out of these difficulties, but the most efficient seems to me that of O. W. Greenberg and collaborators who have introduced three types of 3(q and $\bar{q}$). They obtain a remarkable hadronic spectrum.

Equation (5.56) is written in this local form

$$[h^0(\vec{x},\tilde{a}),h^\mu(\vec{y},\tilde{b})] = i\delta(\vec{x}-\vec{y})h^\mu(\vec{x},\tilde{a}\wedge\tilde{b}) \quad . \tag{5.65}$$

This is called current algebra in the literature. For the time component $\mu = 0$, one speaks of the current algebra of charges. For a space-component one has to introduce in the second member a distribution (usually called Schwinger terms, see O'Raifeartaigh lectures).

Very few physical results require the local form of current algebra and cannot be deduced from the form (5.63). However, physicists prefer to consider current algebra as an hypothesis. They like the analogy with quantum mechanics which is expressed by the algebra (= Lie algebra of the Heisenberg group) of p's and q's at a given time. Let us note also that in this frame B. W. Lee (*Phys. Rev. Lett.*, 17, 145 (1965)) has given a meaning to SU(6) symmetry. There is an anthology on "current algebra" physics (see below).

Boot-Strap. When there are so many particles, one hesitates to distinguish which ones are elementary. Boot-strap is a physical concept† which deals with particles on a more democratic basis. Boot-strap is expressed by non-linear (simply quadratic) equations, invariant under the hadronic symmetry group G (no larger group than SU(3) has been used). Such equations yield solutions which break the symmetry of G. Indeed, from the abstract point of view of group invariance, these equations are of the form

$$a \vee a = \lambda a \quad ,$$

and we have already shown how this yields the directions in nature which break the SU(3) × SU(3) symmetry.

For the readers who wish to read the physics literature, we recall the existence of the anthologies (with commentaries) of original papers, that we have already mentioned.

- *Quantum Theory of Angular Momentum*, Biedenharn, L. C., and van Dam, H., Academic Press, New York (1965).
- *Symmetry Group in Nuclear and Particle Physics*, Dyson, F. J., Benjamin, New York (1966) (which also contains three lectures by Dyson).
- *The Eightfold Way*, Gell-Mann, M., and Ne'eman, Y., Benjamin, New York (1964).
- *Current Algebras*, Adler, S. L., and Dashen, R. F., Benjamin, New York (1968).

---

† Although its father, G. F. Chew has written recently a paper entitled "Boot-strap, a scientific concept?", and given an ambiguous answer!

ACKNOWLEDGMENTS

It was very exciting to prepare these lectures, and discuss some points with my colleagues in or near Bures (Deligne, Fotiadi, Lascoux, Radicati, Stora, Thom, et al.). For the preparation of these notes, I benefited from many discussions with the participants of the Rencontres, among them B. Kostant, G. Mackey, C. Moore, L. O'Rafeartaigh, and more especially the Rencontres Director, V. Bargmann. His friendly advice helped me to improve many points of the original draft. By their careful readings, Dr. Abellanas and Professor Bargmann suppressed most of the misprints of the original draft.

Unhappily, these notes do not convey the lively interruptions during the lectures. They are incomplete (no time to deal with molecular and solid state physics!) and written much too hastily. I apologize to the reader, asking him to remember that he is not reading a book, but perishable lecture notes. I still hope they will incite some readers to better learn this fascinating part of physics.

I acknowledge the wonderful hospitality offered by the Battelle Memorial Institute, to the participants (and their families!), and the perfect organization of this fruitful Rencontres. The only sad point was the absence of E. P. Wigner, the most, and yet not enough, quoted scientist in these notes.

UNITARY REPRESENTATIONS OF LIE GROUPS IN QUANTUM MECHANICS

by

L. O'Raifeartaigh*

1. NON-RELATIVISTIC CLASSICAL MECHANICS AND THE GALILEAN GROUP**

Let $S(3)$ denote Euclidean 3-space. A Cartesian observer of $S(3)$ is a mapping $s \in S(3) \to \vec{x} \equiv (x_1 x_2 x_3) \in R^3$ for which the metric $\rho(s,s')$ of $S(3)$ may be written as

$$\rho(s,s') = \{(x_1 - x_1')^2 + (x_2 - x_2')^2 + (x_3 - x_3')^2\}^{\frac{1}{2}} \quad . \tag{1.1}$$

The group of transformations between Cartesian observers is the Euclidean group $E(3)$

$$x_a' = R_{ab} x_b + C_a \quad , \quad a = 1,2,3 \quad , \tag{1.2}$$

where R_{ab} is any real orthogonal matrix and C_a any real vector (independent of x).

Let t denote Newtonian time, which is simply a parameter assumed to be the same, up to a change of origin $t \to t' = t + t_0$, for all Cartesian observers. Note that in general R_{ab} and C_a are functions of t, i.e., Cartesian observers may be accelerating relative to each other.

Newtonian physics assumes that physical objects occupy volumes in $S(3)$ and vary their positions continuously with time, the variation of any body being determined by the others. The business of physics is to determine the laws of variation.

We shall be concerned mainly with a simplifying limiting case of physical objects, namely, Newtonian particles. A Newtonian particle is a physical object to which is attached an intrinsic label called its mass m (which will be

---

* School of Theoretical Physics, Dublin Institute for Advanced Studies, 64-65 Merrion Square, Dublin 2, Ireland.

** Throughout this paper an asterisk (*) used in a mathematical expression denotes complex conjugation and a dagger (†) passing to the adjoint operator.

discussed in more detail in a moment) and whose volume is so small (relative to its distance from other particles) that for practical purposes it can be neglected and shrunk to a point in $S(3)$. Thus, a Newtonian particle is characterized at any time t by a point in $S(3)$ and its mass.

In view of the importance of the mass of a particle for our later discussion, we consider in a little detail how it enters in Newtonian theory. Its existence is, of course, empirical and may, in principle at least, be established as follows: If any 2 particles interact in isolation (in practice, sufficiently far from other objects), then there exists a set of Cartesian observers such that the quantity

$$m_{12} = -\frac{d^2x_a^{(1)}}{dt^2}\Big/\frac{d^2x_a^{(2)}}{dt^2} \qquad (1.3)$$

(the ratio of the acceleration of the particles) is positive and is independent of a, t, $x^{(1)}$, $x^{(2)}$ and the nature of the interaction. In other words, m_{12} is an intrinsic property of the pair of particles 1 and 2. Furthermore, if α, β, γ are any 3 particles then (again empirically)

$$m_{\alpha\beta} = m_{\alpha\gamma} \cdot m_{\gamma\beta} \quad . \qquad (1.4)$$

Equation (1.4), however, implies the existence of a set of intrinsic masses m_α, one for each particle, and unique up to a common scale factor, such that

$$m_{\alpha\beta} = m_\beta/m_\alpha \quad . \qquad (1.5)$$

As the masses m_α are relatively positive, they are chosen by convention to be positive.

The result that $m_{\alpha\beta}$ is constant already lays the foundations for the law of variation of the positions of the particles with respect to time. The general law (Newton's law) is a linear generalization, namely, given a set of n isolated particles $(m_\alpha, x^\alpha, \alpha = 1,\ldots,n)$, there exists a set of Cartesian observers such that

$$\sum_{\alpha=1}^{n} m_\alpha \frac{d^2x^\alpha}{dt^2} = 0 \quad . \qquad (1.6)$$

This law, in turn, brings out the importance of the force, defined by

$$F_\alpha = m_\alpha \frac{d^2x^\alpha}{dt^2} \quad , \qquad (1.7)$$

as a basic physical concept. Forces are additive, from (1.6), and have additional good properties, which might be described as follows:

What we are looking for is a description of the interaction of particles which is as simple and as universal as possible. Now a description would be provided by simply stating what each $\vec{x}^{\alpha}$ is as a function of t for each ensemble of particles, (this is what Kepler actually did for the planets), but such a description would be neither simple nor universal (as Kepler found to his cost). What Newton discovered is that there exists a quantity that is simple and universal, namely, F_α. The classic example of a simple universal F_α is in the Newtonian theory of gravitation, for which the simple inverse square law $F = m_1 m_2/r^2$ is sufficient to explain all (non-relativistic) effects. (Of course, one can reverse the logic and define gravitational effects to be those for which $F = m_1 m_2/r^2$. However, the point is that gravitational effects so defined cover a huge class of observed phenomena--falling bodies, projectiles, planetary motion, etc.)

From the group theoretical point of view, the interesting aspect of Newton's Equation (1.6) is its invariance group. Equation (1.6) does not hold for all Cartesian observers, but only for a subclass. Let us call the subclass Galilean observers. By noting that any Cartesian observer is related to a Galilean observer by a transformation of the form

$$x'_a = R_{ab}(t)x_b + C_a(t) \quad , \quad t' = t + t_0 \quad , \tag{1.8}$$

and inserting this result in (1.6), we see that the Galilean observers are those, and only those, for whom

$$R_{ab}(t) = R_{ab} \quad , \quad C_a(t) = C_a + V_a t \quad , \tag{1.9}$$

where R_{ab}, C_a, and V_a are independent of t. The subgroup G of (1.8) for which (1.9) holds is called the Galilean group G.

The geometrical significance of the Galilean group becomes clear if we note that it is formed exhaustively from the four subgroups:

$$\begin{array}{lll} 1) & \text{Time-translations} & t' = t + t_0 \\ 2) & \text{Space-translations} & x'_a = x_a + C_a \\ 3) & \text{Rotations} & x'_a = R_{ab}x_b \\ 4) & \text{Accelerations} & x'_a = x_a + V_a t \end{array} \tag{1.10}$$

The invariance of (1.6) under (1.10), 1) to 3), means that (1.6) does not prefer any origin in space or time or any direction in space, which is understandable. The invariance under 4) means that observers with different but constant velocities are equivalent. This is far less obvious, and was first discovered by Galileo. The invariance under 4) does have, however, a geometrical significance, namely, in the 4-space spanned by $S(3)$ and t, (1.6) does not prefer any slope for the t-axis.*

---

* I am indebted to Henri Bacry for this remark.

The force defined by Equation (1.7) is clearly Galilean invariant, provided that the Galilean transformation is universal, i.e., it is a transformation of the coordinates of all the particles. Thus, in guessing the forces for any problem, one can restrict oneself to those that are Galilean invariant.

Let us now consider the Galilean group by itself. By definition, it is a 10-parameter Lie group, which is the semi-direct product of its connected part ($\det R_{ab} = +1$) and the 2-element space reflexion (parity) group. Its Lie algebra dG has the basis:

1) Time-translations E
2) Rotations L_a
3) Space-translations P_a
4) Accelerations K_a

with commutation relations

$$
\begin{array}{lll}
[E,M_a] = 0 & [E,P_a] = 0 & [E,K_a] = P_a \\
[M_a,M_b] = \varepsilon_{abc}M_c & [M_a,P_b] = \varepsilon_{abc}P_c & [M_a,K_b] = \varepsilon_{abc}K_c \\
[P_a,P_b] = 0 & [P_a,K_b] = 0 & \\
[K_a,K_b] = 0 \ , & &
\end{array}
\tag{1.11}
$$

where $a,b,c = 1,2,3$ and ε_{abc} is the Levi-Civita symbol. In words, dG is the semi-direct sum of the rotation algebra L and a 7-dimensional solvable algebra made up of the two abelian commuting vectors P and K and a scalar E which projects K onto P and commutes with P.

One of the most important properties of Galilean transformations is that they are a special case of contact transformations [1], namely, transformations $x \to x'(x,p)$, $p \to p'(x,p)$ which leave the symplectic form

$$\{A,B\} = \sum \left(\frac{\partial A}{\partial p}\frac{\partial B}{\partial x} - \frac{\partial A}{\partial x}\frac{\partial B}{\partial p}\right) \ , \tag{1.12}$$

where $p_\alpha = m_\alpha \dfrac{dx_\alpha}{dt}$, invariant.

Now a property of the group of contact transformations[2] is that if σ is the parameter of any 1-parameter simply connected Lie subgroup, then there exists a function $G(p,q)$ such that

$$\frac{\delta F}{\delta\sigma} = \{G,F\} \ , \tag{1.13}$$

where F is any regular function of p and q, and $\frac{\delta F}{\delta\sigma}$ is its rate of variation with respect to the group parameter σ. The function G is called the generator function for the 1-parameter subgroup.

Furthermore, for an n-parameter Lie subgroup of contact transformations with parameters α, β,...

$$\left(\frac{\delta}{\delta\alpha}\frac{\delta}{\delta\beta} - \frac{\delta}{\delta\beta}\frac{\delta}{\delta\alpha}\right) F = C^{\gamma}_{\alpha\beta}\frac{\delta F}{\delta\gamma} \quad , \tag{1.14}$$

where $C^{\gamma}_{\alpha\beta}$ are the structure constants of the group. Hence, inserting (1.13) into (1.14) and using the Jacobi relation for $\{A,B\}$, we obtain

$$\left\{\{G_\alpha,G_\beta\}F\right\} = C^{\gamma}_{\alpha\beta}\{G_\gamma,F\} \quad , \tag{1.15}$$

whence,

$$\{G_\gamma,G_\beta\} = C^{\gamma}_{\alpha\beta}G_\gamma + \lambda_{\alpha\beta} \quad , \tag{1.16}$$

where the $\lambda_{\alpha\beta}$ have zero bracket with all F and hence are constants. Thus, under the bracket operation, the generator functions G_α of a Lie group of contact transformations form a representation (up to the constants $\lambda_{\alpha\beta}$) of the Lie algebra. The number of constants $\lambda_{\alpha\beta}$ can be minimized by transformations of the form $G_\alpha \to G_\alpha + \lambda_\alpha$, where the λ_α are constants, but whether the $\lambda_{\alpha\beta}$ can be eliminated entirely depends on the group structure.

The above results hold for <u>any</u> Lie group of contact transformations. Let us now return to the connected Galilean group G. For G, the generator functions corresponding to the generators in (1.11) can be seen to be

$$\begin{aligned} L &= \sum_\alpha x_\alpha \times p_\alpha \\ P &= \sum_\alpha p_\alpha \\ K &= \sum_\alpha m_\alpha x_\alpha - Pt \\ E &= \sum \frac{1}{2m_\alpha} p_\alpha^2 + \varphi \quad , \end{aligned} \tag{1.17}$$

where $p_\alpha = m_\alpha \frac{dx_\alpha}{dt}$, and φ is the potential from which the F_α can be derived, i.e.,

$$F_\alpha = -\frac{\partial\varphi}{\partial x_\alpha} \quad .$$

If we compute the brackets $\{L,E\}$, etc., for the generator functions (1.17), we obtain, as expected, the Lie algebra (1.11) up to constants. In fact, there is only one constant; namely, the relations (1.11) hold as they stand except that

$$[P_a,K_b] = 0 \to \{P_a,K_b\} = \delta_{ab}M \quad , \tag{1.18}$$

where $M = \sum_\alpha M_\alpha$ is the total mass. Further, the structure of the Galilean group is such that M cannot be eliminated (we shall be discussing this question again later).

Note that the generator for the time translations is just the Hamiltonian H for the system. Note further that $[H,K] \neq 0$, although $[H[H,K]] = 0$. Thus, although the Galilean group is an invariance group of Newton's Equations (1.6), it is not quite an invariance group of the Hamiltonian, or of Hamilton's equations of motion,

$$\frac{dp_\alpha}{dt} = -\frac{\partial H}{\partial x_\alpha} \qquad \frac{dx_\alpha}{dt} = \frac{\partial H}{\partial p_\alpha} \quad . \tag{1.19}$$

This is understandable since a choice of Hamiltonian forces a choice of direction for the t-axis in $S(3) \otimes R$ and thus destroys the Galilean invariance. Incidentally, the term $-Pt$, which is explicitly time-dependent, is inserted in the definition of K, so that in spite of the fact that $[H,K] \neq 0$, K can be a constant of the motion, i.e., so that

$$\frac{dK}{dt} = \frac{\partial K}{\partial t} + \{H,K\} = -P + P = 0 \quad . \tag{1.20}$$

2. NON-RELATIVISTIC QUANTUM MECHANICS

As is well known, Newton's laws, or the more general and sophisticated versions of them, such as Hamilton's, sufficed to explain all physical phenomena until the end of the last century. But after the turn of the century, the Newtonian framework was shattered both by the theory of relativity and by the quantum theory. In this lecture, we shall be concerned only with quantum theory. As is also well-known, the crux of the quantum theory is to replace the functions x and $p = m\frac{dx}{dt}$ needed to describe particles, by linear operators X and P on a Hilbert space, satisfying the relation

$$[X,P] = i\hbar \quad . \tag{2.1}$$

(This relation will be made mathematically more precise later.) For the moment, we shall only emphasize that the assumption (2.1) is the <u>only</u> new assumption made in the quantum theory. The old equations of motion

$$\frac{dX}{dt} = \frac{\partial H}{\partial P} \quad , \quad \frac{dP}{dt} = -\frac{\partial H}{\partial X}$$

are retained with $x \to X$, $p \to P$ (which is unambiguous since $H = \frac{p^2}{2m} + \varphi(x)$). There are four questions which we wish to discuss briefly:

1) How one arrives at the particular Ansatz (2.1)
2) How to make it mathematically precise
3) How to relate it to experiment
4) How the group structure of Newtonian theory is affected.

Let us begin with 1). The decision to replace x and p by operators was based on a large number of empirical observations and on partial theories formed from these observations [1]. Since we could not even begin to describe the general picture in a part of one lecture, let us concentrate on one experimental result, namely, the discrete frequency of the light emitted from atoms, and try to sketch the motivation from that result. It was known at the time the quantum theory was founded that the atom consisted of a positively charged kernel of very small radius with negatively charged electrons circling it, about 10^{-8} cms out.

For such a system Newton's laws (extended to include Maxwell's) would predict a continuous emission of radiation from the circling (and therefore accelerating) electrons, leading to a continuous loss of energy on the part of the electrons (so that the atom would finally run down) and a continuous change in the frequency of the emitted radiation. The experimental situation, however, was quite the opposite. First, the atoms were quite stable (otherwise, our universe would not exist). Second, from spectroscopy it was known that the frequency of the radiation emitted from atoms, far from being continuous, could only have special sharp values (spectral lines) characteristic of the atom (yellow for sodium, green for copper, and so on). Hence, Newton's laws were incompatible with experiment on the atomic level. The question was: how to change them?

One worked backwards. If one assumes

1) Einstein's empirical law $E = h\nu$, where h is Planck's constant, ν the frequency of the emitted light, and E its energy, and
2) conservation of energy, i.e., energy lost by electron in the atom = energy of emitted radiation,

it follows from the discreteness of the frequency of the emitted radiation that the energy levels of the electron in the atom must be discrete. It follows that the Hamiltonian

$$H = \frac{1}{2m} p^2 - \frac{Ze^2}{r} \quad , \tag{2.2}$$

for an electron in an atom with nucleus of charge Ze, cannot take continuous values. This leaves one with three options:

1) Abandon the Hamiltonian (2.2)
2) Impose some conditions on it from outside
3) Change it so that it can naturally take only discrete values.

1) has the difficulty that it is almost impossible to think of a classical Hamiltonian which would take discrete values. 2) is what was done in the so-called "old quantum theory" (1900-25), and is very *ad hoc*. 3) is the option chosen by Schrödinger and Heisenberg. The choice they made was to interpret H as a linear operator, since H could then take discrete values naturally. This means interpreting x and p as linear operators X and P. To determine the kind of operators P and X should be, one must do more. Heisenberg analyzed the atomic

spectra in detail and concluded that P and X should be the matrices

$$P = \frac{\hbar}{i\sqrt{2}} \begin{bmatrix} 0 & \sqrt{1} & 0 & 0 & \cdot \\ -\sqrt{1} & 0 & \sqrt{2} & 0 & \cdot \\ 0 & -\sqrt{2} & 0 & \sqrt{3} & \cdot \\ 0 & 0 & -\sqrt{3} & 0 & \cdot \\ \cdot & \cdot & \cdot & \cdot & \cdot \end{bmatrix} \qquad X = \frac{1}{\sqrt{2}} \begin{bmatrix} 0 & \sqrt{1} & 0 & 0 & \cdot \\ \sqrt{1} & 0 & \sqrt{2} & 0 & \cdot \\ 0 & \sqrt{2} & 0 & \sqrt{3} & \cdot \\ 0 & 0 & \sqrt{3} & 0 & \cdot \\ \cdot & \cdot & \cdot & \cdot & \cdot \end{bmatrix}, \tag{2.3}$$

where $\hbar = \frac{h}{2\pi}$. Schrödinger, on the other hand, built on a partial theory due to de Broglie. According to de Broglie, free particles should diffract like light from sufficiently small gratings and should therefore satisfy, in the relativistic case, a wave equation of the form

$$\left(\frac{1}{c^2}\frac{\partial^2}{\partial t^2} - \nabla^2 - m^2\right)\psi(x) = 0 \quad . \tag{2.4}$$

Comparing this with the classical energy moment relation,

$$\varepsilon^2 - p^2 - m^2 = 0 \quad , \tag{2.5}$$

Schrödinger concluded that P should be the operator

$$\frac{\hbar}{i}\frac{\partial}{\partial x} \quad , \tag{2.6}$$

on $L^3_2(-\infty,\infty)$, and went on to postulate that this identification should persist in the non-relativistic limit and in the presence of a potential.

One sees that the Schrödinger and Heisenberg Ansatz are equivalent by noting that they are special realizations of the Ansatz (2.1). Note, incidentally, that the Ansatz (2.1) need only be made at a single (initial) instant of time and it is therefore a <u>kinematical</u> Ansatz. Newton's laws then guarantee it for all times.

It might be wondered if the Ansatz (2.1) is absolutely necessary to obtain agreement with experiment, or whether one could get away with less. Wigner [2], for example, has proposed that (2.1) might be replaced by the weaker commutation relations

$$[H,P] = i\,\frac{\partial H}{\partial X} \quad , \qquad [H,X] = -i\,\frac{\partial H}{\partial P} \quad , \tag{2.7}$$

where H is the Hamiltonian, which would seem to be necessary from Heisenberg's analyses of the spectral lines. However, except in the case (2.1), the Ansatz (2.7) would make the commutation relations depend on H, i.e., on the dynamics.

Let us now turn to question 2), namely the question of putting the Ansatz $[X,P] = i\hbar$ on a better mathematical footing. For this we proceed as follows:

Let $\mathcal{H}$ be a Hilbert space, and let X and P be operators on it such that there exists for them a common invariant dense domain $\mathcal{D}$ on which

a) X and P are symmetric

b) $X^2 + P^2$ is essentially self-adjoint

c) $[X,P] = i\hbar$

d) the only bounded operator which commutes with X and P is a multiple of the unit operator.

Then X and P are uniquely and rigorously defined [3] on $\mathcal{H}$ up to a unitary transformation (which may depend on the time). They are essentially self-adjoint on $\mathcal{D}$. A realization of X and P, is the Schrödinger realization x and $\frac{\hbar}{i}\frac{\partial}{\partial x}$ on $L_2^3(-\infty,\infty)$, where the domain $\mathcal{D}$ could be, for example, the space K of all infinitely differentiable functions of compact support, or the space S of all infinitely differentiable functions of fast decrease (i.e., which decrease faster than any inverse power of x as $|x| \to \infty$). We shall see later (from Nelson's theorem) that conditions a) to d) are precisely the necessary and sufficient conditions, that X and P can be exponentiated to form a unique unitary irreducible representation of the Weyl-Heisenberg group W, i.e., that

$$e^{i\sigma X}e^{i\tau P} = e^{i\tau P}e^{i\sigma X}e^{i\sigma\tau\hbar} \quad , \text{ on } \mathcal{H} . \tag{2.8}$$

Thus, an alternative definition of X and P is that they satisfy (2.8), i.e., that they are the generators of the unitary irreducible representation (UIR) of W, [4]. In fact, this definition of X and P was the starting point for von Neumann's celebrated proof [5] of the uniqueness of X and P up to a unitary transformation.

Having disposed of these mathematical points, we come to the experimental numbers. To extract the experimental numbers, we first put the self-adjoint operators A on $\mathcal{H}$ into a 1-1 correspondence with the measurable quantities (observables) which we shall then also denote by A. In practice, for the self-adjoint operators for which it is meaningful, the correspondence is [4]

$$A = f(P,X) = \frac{1}{(2\pi)^2}\int e^{i(P\nu+X\zeta)}d\nu d\zeta \int e^{-i(p\nu+x\zeta)}f(p,x)dpdx \quad , \tag{2.9}$$

where $f(p,x)$ are the corresponding classical functions. (The bounded subset of the operators for which (2.9) is meaningful form a dense set in the ring of bounded self-adjoint operators.)

Now let $P_\lambda(A)$ be the projection operator on the eigenspace of A belonging to the eigenvalue λ, where for the moment we assume λ to be discrete and the eigenspace finite dimensional. The numbers to be extracted are then

$$\text{trace } (P_\lambda(A)P_\mu(B)) \quad , \tag{2.10}$$

with appropriate modifications in the case that λ and μ are not discrete and that both eigenspaces are infinite dimensional. The meaning of the numbers (2.10) is that they are probabilities; namely, trace $(P_\lambda(A)P_\mu(B))$ is the probability of finding the value μ from a measurement of B, having just previously found λ from a measurement of A, except points in spectra of self-adjoint operators, the probabilities are the only experimental numbers that quantum mechanics can predict.

In the particular case that the eigenvalues λ and μ are simple, i.e., that $P_\lambda(A)$ and $P_\mu(B)$ project onto 1-dimensional subspaces, (2.10) reduces to

$$|(f_\lambda(A),\, f_\mu(B))| \quad , \tag{2.11}$$

where $f_\lambda(A)$ and $f_\mu(B)$ are any unit vectors in the respective subspaces. This is the case which will be of most interest to us. (For future reference, we shall need for this case the concept of a quantum mechanical <u>state</u>. The state of a system after a measurement of A with result λ, where λ is simple, is defined to be the set of unit vectors in the 1-dimensional eigenspace. Such a set of unit vectors $\ell^{i\alpha}f_\lambda(A)$, $0 \leq \alpha < 2\pi$ is often called a <u>ray</u>. Thus, the states of a system are in 1-to-1 correspondence with the rays.)

Let us turn now to question 4), the group theoretical properties of non-relativistic quantum mechanics, and first consider the Hamiltonian

$$H = \frac{1}{2m} P^2 + \varphi(X) \quad , \tag{2.12}$$

for a single particle in an external potential.

In most cases of interest, H is essentially self-adjoint on the domain $\mathcal{D}$ above. Hence, by Stone's theorem [6], there exists a unique continuous 1-parameter group of unitary transformations U(t) on $\mathcal{H}$, such that

$$\frac{dU(t)}{dt} = HU(t) \quad \text{on} \quad \mathcal{D} \quad . \tag{2.13}$$

We now show that U(t) is the group of time translations. Since the Newtonian equations of motion are the same in classical and quantum theory, we have in both cases

$$\frac{dX}{dt} = \frac{1}{m} P \quad , \quad \frac{dP}{dt} = - \frac{\partial\varphi(X)}{\partial X} \quad . \tag{2.14}$$

In the quantum mechanical case, however, we have the extra condition

$$[X,P] = i\hbar \quad .$$

Inserting this equation into (2.14) and (2.12), we see that in the quantum mechanical case we have

$$\frac{dX}{dt} = \frac{i}{\hbar} [H,X] \quad , \quad \frac{dP}{dt} = \frac{i}{\hbar} [H,P] \quad \text{on} \quad \mathcal{D} \quad . \tag{2.15}$$

If we assume that the domain $\mathcal{D}$ is invariant with respect to U(t), it follows at once that

$$X(t) = U(t)X(0)U^{-1}(t) \quad , \quad P(t) = U(t)P(0)U^{-1}(t) \quad \text{on} \quad \mathcal{D} \quad , \tag{2.16}$$

and, in general, for suitably defined $F(P,X)$ in (2.9)

$$F(P(t)X(t)) = U(t)F(P(0)X(0))U^{-1}(t) \quad . \qquad (2.17)$$

Thus, $U(t)$ is the group of time translations. In quantum mechanics, therefore, the Hamiltonian H, like P and X, plays a dual role. It is a physical observable (energy) and it generates the group of time translations.

It may happen that H is not essentially self-adjoint on $\mathcal{D}$. In this case, there is usually a good physical reason, and the corresponding classical Hamiltonian also has bad properties, e.g., sends the particle off to infinity in a finite time [7].

Turning now to the Galilean group for a system of interacting particles, we find that, in analogy to P, X, and H, if we replace the classical generator functions of the Galilean group by their quantum mechanical counterparts to obtain

$$\begin{aligned} E = H &= \sum \frac{1}{2m} P_\alpha^2 + \varphi \\ L &= \sum X_\alpha \times P_\alpha \\ P &= \sum P_\alpha \\ K &= \sum m_\alpha X_\alpha - Pt \quad , \end{aligned} \qquad (2.18)$$

then, in analogy to P, X and H, these ten operators (2.18) play a dual role. They are physical observables and at the same time they are the generators of unitary representations of the 1-parameter subgroups of the Galilean group G on $\mathcal{H}$, i.e., if σ is a parameter,

$$\frac{dF}{d\sigma} = \frac{i}{\hbar} [G_\sigma, F] \quad , \quad \sigma = 1 \ldots 10 \quad . \qquad (2.19)$$

This is the quantum-mechanical analogue of the classical Poisson bracket relation

$$\frac{dF}{d\sigma} = \{\tilde{G}_\sigma, F\} \quad . \qquad (2.20)$$

Using the quantum mechanical relation $[X,P] = i\hbar$, we can easily compute the commutators of the operators (2.18) amongst themselves. We obtain

$$\begin{array}{lll} [M_a,M_b] = i\varepsilon_{abc}M_c & [P_a,P_b] = 0 & [K_a,K_b] = 0 \\ [M_a,P_b] = i\varepsilon_{abc}P_c & [P_a,K_b] = i\delta_{ab}M & [K_a,H] = 0 \quad . \\ [M_a,K_b] = i\varepsilon_{abc}K_c & [P_a,H] = 0 \quad , & \\ [M_a,H] = 0 \quad . & & \end{array} \qquad (2.21)$$

These relations are the analogue of the classical Poisson bracket relations for the generator functions amongst themselves. Note that (2.21) even contains the term $\delta_{ab}M$ which occurs in the classical Poisson bracket relations, but not in the Lie algebra of G.

Apart from the term $\delta_{ab}M$, (2.21) is just the Lie algebra of G. Hence, if the term $\delta_{ab}M$ were absent, the 1-parameter subgroups of G, generated by the G_α, would mesh together to form a unitary representation of G on $\mathcal{H}$ (modulo some domain restrictions which will be discussed later and which are normally satisfied). Thus, in quantum mechanics the generators G_α play the dual role of observables and generators (modulo $\delta_{ab}M$) of a unitary representation [8] of G on $\mathcal{H}$. This is true, of course, in classical mechanics also, where the generator functions are observables and generators of group transformations in the sense of Poisson brackets. But the relationship in quantum mechanics is more direct. In particular, the operation of commutation is simpler and more direct than the operation of forming Poisson brackets. In this sense, group theory, which plays a background role in classical theory, may be said to come into its own and play a central role in quantum mechanics.

Let us now consider the term $\delta_{ab}M$. Since it commutes with all the G_α, it cannot make a big difference to the representation of G on $\mathcal{H}$. It is easily checked that the difference it makes is that the 1-parameter subgroups of G, instead of meshing together to form a true unitary representation of G on $\mathcal{H}$, mesh together to form a unitary ray representation of G on $\mathcal{H}$, i.e., a representation by unitary operators $U(g)$ satisfying

$$U(g)U(g') = U(gg')e^{i\omega(g,g')} \quad , \tag{2.22}$$

where $g,g' \in G$ and ω is real. The reason for the name ray representation is that the factor $\exp i\omega(g,g')$ is irrelevant for rays, (where rays are defined as above to be sets of unit vectors related to a given unit vector f by $\exp(i\alpha)f$, where $0 \le \alpha < 2\pi$). If we now recall that the experimental numbers which can be extracted from quantum theory are

$$|(f,g)| \quad , \tag{2.23}$$

where f and g are unit vectors, we see at once that they do not distinguish between vectors in the same ray. Thus, the experimental numbers do not distinguish between unitary ray representations and true unitary representations. We shall be returning in more detail to this point later, but for the moment we merely note that the failure of the experimental numbers to distinguish between true and ray representations means that the appearance of ray representations and hence, in particular, of the term $\delta_{ab}M$ in the Lie algebra (2.21), is quite natural in quantum mechanics.

In the case of a single free particle, the generators reduce to

$$\begin{aligned} M_a &= \frac{1}{2}\,\varepsilon_{abc}P_aX_c \\ P_a &= P_a \\ K_a &= mX_a - P_at \\ E &= \frac{1}{2m}\,P^2 \quad , \end{aligned} \tag{2.24}$$

where m is now the mass of the particle and E is both a generator of the Galilean group and the generator of time translations. Thus, a free particle "carries" a unitary ray representation of G. Furthermore, if the quantum mechanical commutation relation

$$[X,P] = i\hbar \quad ,$$

is irreducible on $\mathcal{H}$, then so is the representation (2.21) of G. A non-relativistic free particle may, therefore, be said to carry an irreducible unitary ray representation of G.

An interesting question is what would happen if we reversed our line of approach and demanded that a free non-relativistic particle carry a true unitary representation of G. This question has been investigated by Inönü and Wigner [9]. They showed that in a true irreducible unitary representation of G the quantum mechanical relation

$$[X,P] = i\hbar \quad ,$$

cannot be realized, which has the unpleasant physical consequence that X cannot be localized. The crucial point is that P^2 is a Casimir operator for G. Hence, in any unitary irreducible representation, it is a number, and the Fourier transform $\tilde{f}(X)$ of any $f(P)$ must therefore have a spread in X.

In a ray representation, the situation is saved by the ray relation

$$i[K_a,P_b] = \delta_{ab}m \quad , \tag{2.25}$$

or

$$i[K_a,P^2] = 2mP_a \quad . \tag{2.26}$$

The latter relation implies that P^2 assumes all values in the range $0 \leq P^2 < \infty$, which together with

$$[M_a,P_b] = i\varepsilon_{abc}P_c \quad , \tag{2.27}$$

implies that $\vec{P}$ takes all values in R^3, in which case the Fourier transform $\vec{X}$ is localizable.

In conclusion it might be worth remarking that the twin postulates of quantum mechanics, $|(f,g)|^2$ = probability, and $[X,P] = i\hbar$ are not entirely independent. The second can be deduced from the first, using group theoretical and other general arguments of a more or less plausible nature (see ref. 4, lecture 6).

3. INVARIANCE GROUPS IN NON-RELATIVISTIC QUANTUM MECHANICS

In the last two sections, we saw that the Galilean group G was the group of invariance of the non-relativistic equations of motion of an isolated system of n particles. Let us now consider a 2-particle system and "factor-off" the Galilean invariance by introducing center of mass and relative coordinates.

$$X = \frac{1}{M}(m_1x_1 + m_2x_2) \quad , \quad P = p_1 + p_2 \quad , \quad M = m_1 + m_2 \quad ,$$

and

$$y = x_1 - x_2 \ , \quad \pi = \frac{1}{M}(m_2p_1 - m_1p_2) \quad , \tag{3.1}$$

respectively. Because of Galilean invariance the Hamiltonian splits into $H = H_{CM} + H_r$, where

$$H_{CM} = \frac{P^2}{2M} \quad , \quad [X,P] = i\hbar \quad ,$$

and

$$H_r = \frac{\pi^2}{2\mu} + \phi(y) \quad , \quad [y,\pi] = i\hbar \quad , \tag{3.2}$$

where $\mu = m_1m_2/M$ is called the reduced mass.

Clearly H_{CM} describes the motion of the centre of mass and H_r the relative motion of the particles.

The equations of motion derived from the "relative" Hamiltonian (3.2) will not, in general, retain any of the original Galilean invariance. However, in particular cases (i.e., for particular potentials $\phi(y)$) they may retain invariance under a subgroup of the Galilean group (e.g., the rotation group) or they may happen to be invariant under special groups which have nothing to do with Galilean invariance. In this lecture we wish to consider such cases. For this purpose, we define an invariance group.

Definition: An invariance group is defined to be any group of transformations on $\mathcal{H}$, the Hilbert space of y, π, which leaves invariant

a) the Hamiltonian H

b) the absolute values of the inner products $|(f,g)|$.

We first discuss the motivation for this definition. That the group should leave the Hamiltonian invariant is practically self-explanatory since this is true of an invariance group even in classical mechanics. We only note that (in both classical and quantum mechanics) the invariance of H is slightly stronger than the demand that the group leave the equations of motion invariant. (For example, as we saw for an isolated system, the Galilean group left the equations of motion invariant but not the Hamiltonian.) However, for invariance groups of the relative Hamiltonian, the distinction between H and the equations of motion usually does not arise, and the invariance of H is used as the simplest and most compact was of defining invariance.

The more interesting question concerns b), namely the invariance of the inner products $|(f,g)|$ which are peculiar to quantum mechanics. The question is whether this demand is necessary, or at least reasonable.

For a group of transformations which have a passive interpretation, as is the case for the Galilean group $\mathcal{G}$, the answer is yes. For if we change the observer of a system, without changing the system itself, the probability of the system making any particular transition $g \to f$ cannot change (since the system "does not know who is looking at it") and this is just another way of saying that $|(f,g)|$ is invariant.

For transformations which do not have a passive interpretation, i.e., for which we must change the system itself to implement them (these are usually transformations which have no geometrical interpretation), the argument is not so easy to establish. However, it is usual to demand the invariance of the probabilities in this case also, if only for simplicity and to preserve the analogy with the active case.

Demanding that the probabilities $|(f,g)|^2$ remain invariant, we come to a second question: Are unitary ray representations the most general group representations which leave the probabilities invariant?

To answer this, one first concentrates on a single transformation T and asks: What is the most general T such that

$$|(Tg,Tf)| = |(g,f)| \quad , \quad g,f \in \mathcal{H} \quad . \tag{3.3}$$

If T is linear, then the answer is simple: T must be unitary. In general, however, there is no need for T to be linear. In that case, we fall back on the following remarkable theorem due to Wigner [1].

Theorem

Let T be a transformation satisfying (3.3). Then there exists a unitary or anti-unitary transformation U such that for all $f \in \mathcal{H}$

$$(U^{-1}T)f = e^{i\delta(f)}f \quad . \tag{3.4}$$

Note that U is then unique up to a phase-factor, $\exp(i\delta)$, which is independent of f. [An anti-unitary transformation is defined to be a transformation such that

$$(Uf,Ug) = (g,f) = (f,g)^* \quad] \quad . \tag{3.5}$$

This theorem means that, for rays, T is equivalent to, and may be replaced by, a unitary or anti-unitary transformation.

This theorem was first stated by Wigner in his book on group theory in 1931. [1] However, the proof given in the book is not complete, and since then many papers [2] have been devoted to completing, simplifying and generalizing the proof.

The most definitive proof is that given by Bargmann [3] in 1964. This proof has the advantage of being basis-free and hence valid for non-separable as well as separable Hilbert spaces.

Wigner's theorem applies to any fixed transformation T. Consider now a group of transformations $T(g)$. For each fixed g, $T(g)$ can be replaced by a unitary or anti-unitary transformation $U(g)$, unique up to a phase-factor $\exp i\delta(g)$. Using the group relation

$$T(g)T(g') = T(gg') \quad , \tag{3.6}$$

Equation (3.4), and the unitarity (or anti-unitarity) of $U(g)$, one sees that

$$U(g)U(g') = U(gg')e^{i\omega(g,g')} \quad , \tag{3.7}$$

where $\omega(g,g')$ is a real number. It follows that any group of transformations $T(g)$ preserving the probabilities (3.2) is equivalent to a set of unitary or anti-unitary transformations $U(g)$ forming a ray representation of the group. In this sense, unitary of anti-unitary ray representations are the most general group representations preserving the probabilities.

In practice, only one anti-unitary transformation is used in physics. This is the time-reversal transformation. To keep the quantum mechanical equations of motion

$$\frac{dF}{dt} = \frac{i}{\hbar}[H,F] \quad , \tag{3.8}$$

invariant under time-reversal, it is necessary to let either $H \to -H$ or $i \to -i$ when $t \to -t$ · $H \to -H$ is ruled out because $H \geq 0$. Hence, $i \to -i$, and this leads to an anti-unitary transformation.

We turn now to some examples of invariance groups in quantum mechanics. For this purpose, it is usual to consider the relative motion Hamiltonian

$$H = \frac{\pi^2}{2\mu} + \phi(y) \quad . \tag{3.9}$$

The problem is, given $\varphi(y)$, to find unitary groups of operators which commute with this H, and have a direct physical meaning. Indeed, in practice, it is usually the physical meaning that enables us to find the groups. The advantages of finding such groups are:

1) Since for the group generators G,

$$[H,G] = 0 \quad , \tag{3.10}$$

the group provides in the G's at least some of the constants of the motion.

2) At the same time, the G's are natural operators to diagonalize simultaneously with H.

3) The group can be used to reduce enormously the labor involved in making a calculation with the Hamiltonian, e.g., calculating an energy level, an emission probability, or a scattering amplitude.

Note that Equation (3.10) can be looked at from two points of view: The group generated by G leaves H invariant (is an invariance group of the equations of motion). Conversely, the group generated by H leaves G invariant (G is conserved).

Let us illustrate points 1), 2), and 3) above with the most important special case of an invariance group; namely, the case when $\varphi(y)$ in (3.9) is central, i.e., depends only on r where $r^2 = y_1^2 + y_2^2 + y_3^2$. In this case, H commutes with the rotation group generated by the three operators $L = y \times \pi$, with Lie algebra $[L,L] = iL$, and which are at the same time identified with the relative angular momenta of the particles in the 1, 2, 3 directions. [The transition from the group to the algebra and back will be justified in the next section.] Now with respect to 1) above it is clear that L_1, L_2, and L_3 are conserved. With respect to 2) it is not difficult to show that the so-called total relative angular momentum $L^2 = L_1^2 + L_2^2 + L_3^2$ and any one of L_1, L_2, L_3 (usually L_3) can be added to H to form a complete set on $\mathcal{H}$ ($\mathcal{H}$ being assumed irreducible with respect to $[y,\pi] = i\hbar$). Thus, a convenient and physically relevant basis in $\mathcal{H}$ is $f(\varepsilon,\ell,m)$ where

$$\begin{aligned} Hf(\varepsilon\ell m) &= \varepsilon f(\varepsilon\ell m) \quad , \\ L^2 f(\varepsilon\ell m) &= \ell(\ell+1) f(\varepsilon\ell m) \quad , \\ L_3 f(\varepsilon\ell m) &= m f(\varepsilon\ell m) \quad , \end{aligned} \tag{3.11}$$

where, because the rotation group is compact, ℓ is a non-negative integer and $-\ell \le m \le \ell$.

With respect to 3), we see at once that in calculating the eigenvalues of H, which are the eigenvalues of the differential operator

$$-\frac{\hbar^2}{2m}\nabla^2 + V(r) \quad , \tag{3.12}$$

on L_2, the use of (3.11) reduces the partial differential operator (3.12) to the simple differential operator

$$-\frac{\hbar^2}{2m}\left(\frac{1}{r^2}\frac{d}{dr}r^2\frac{d}{dr} - \frac{\ell(\ell+1)}{r^2}\right) + V(r) \quad , \tag{3.13}$$

and so simplifies the calculation.

But the group does much more for us than that. For example, if we wish to calculate the probability of a particle in the state $f(\varepsilon,\ell,m)$ emitting a photon with momentum $\vec{k}$ and ending up in a state $f(\varepsilon',\ell',m')$, then, to lowest order in the EM coupling constant e, and provided the wavelength of the emitted photon is large compared with the size of the atom [1,4], the relevant inner products to compute are the multipole moments of the particle. A typical one of these is the dipole moment,

$$E_a = \int (f_{\varepsilon'\ell'm'}, y_a f_{\varepsilon\ell m}) \quad , \quad a = 1,2,3 \quad . \tag{3.14}$$

Now for even quite low value of ℓ and ℓ', the number of quantities (3.14) to be computed is quite large, since $-\ell' \le m' \le \ell'$, $-\ell \le m \le \ell$. But thanks to the group properties of y (it is a polar vector with respect to rotations and space reflexions), we can

a) show that the E_a vanish unless $\ell' = \ell \pm 1$, $m' = m, m \pm 1$,

b) for $\ell' = \ell \pm 1$, reduce the calculations in each case to <u>one</u> calculation. In fact, the group invariance implies that

$$E_a = \varphi(\ell'm',\ell m)\int_0^\infty r^2 dr F^*_{\varepsilon'\ell'}(r)\, r F_{\varepsilon\ell}(r) \quad , \quad \ell' = \ell \pm 1 \quad , \tag{3.15}$$

where $m' - m = 0, \pm 1$ for $a = 3,\ 1 \pm i2$ respectively and the $F_{\varepsilon\ell}$ are the eigenfunctions of the simple differential operator (3.13). The crucial point about (3.15) is that the m' and m dependence appears only in the coefficients (Clebsch-Gordon coefficients) <u>which are independent of</u> $V(r)$. Thus, these coefficients need only be calculated once and for all (Figure 3.1), and then they can be used for <u>any</u> central potential. (The functions in the integral will, of course, depend on $V(r)$.)

$\ell' =$	$m' = m + 1$	$m' = m$	$m' = m - 1$
$\ell + 1$	$\sqrt{\frac{(\ell+m')(\ell+m'+1)}{(2\ell+1)(2\ell+2)}}$	$\sqrt{\frac{(\ell-m'+1)(\ell+m+1)}{(2\ell+1)(\ell+1)}}$	$\sqrt{\frac{(\ell-m')(\ell-m'+1)}{(2\ell+1)(2\ell+2)}}$
ℓ	$-\sqrt{\frac{(\ell+m')(\ell+m'+1)}{2\ell(\ell+1)}}$	$\frac{m'}{\sqrt{\ell(\ell+1)}}$	$\sqrt{\frac{(\ell-m')(\ell+m'+1)}{2\ell(\ell+1)}}$
$\ell - 1$	$\sqrt{\frac{(\ell-m')(\ell-m'+1)}{2\ell(2\ell+1)}}$	$-\sqrt{\frac{(\ell-m')(\ell+m')}{\ell(2\ell+1)}}$	$\sqrt{\frac{(\ell+m'+1)(\ell+m')}{2\ell(2\ell+1)}}$

FIGURE 3.1. VALUES OF $\phi(\ell'm',\ell m)$

The labor saved by using one group to obtain the results a) and b) in this example is obviously immense. Furthermore, the use of the group gives a much deeper insight into what is going on. It isolates the group properties of a central potential (independence of the potential of the angular variables θ, φ) from the dynamical properties (form of the dependence of $V(r)$ on r). The results a) and b) for this example are, of course, a special case of the Wigner-Eckart theorem, which has already been mentioned by Louis Michel and will be formulated for completeness in the next chapter.

We conclude by considering two Hamiltonians which have special invariance groups. The first is the harmonic oscillator Hamiltonian

$$H = \frac{1}{2m}\pi^2 + \frac{1}{2\kappa}y^2 \quad , \tag{3.16}$$

where κ is a constant. This is centrally symmetric and has the angular momentum invariance group generated by L discussed above. But, in addition, H commutes with the six operators

$$M_{ab} = X_a X_b + P_a P_b \quad , \tag{3.17}$$

where $X = \alpha y$, $P = \alpha^{-1}\pi$, $\alpha^4 = m\kappa$, and these six operators, together with the corresponding L_a, form the Lie algebra

$$\begin{aligned} [L,L] &= iL \quad , \\ [L,M] &= iM \quad , \\ [M,M] &= iL \quad , \end{aligned} \tag{3.18}$$

of the compact, connected Lie group $U(3)$. Thus, the Hamiltonian (3.16) is $U(3)$ invariant and, in fact, is just $M_{aa}(4m\kappa)^{-1/2}$.

For 1 particle, this result is not particularly exciting because the Hamiltonian (3.16) is so simple that we can calculate its properties directly anyway. However, in nuclear physics, in the nuclear shell model, it is much more interesting. [5] In the nuclear shell model, it is assumed that the particles in the nucleus interact with each other in such a way that, for each particle, the total effect is the same as if it were in a strong <u>central</u> potential due to <u>all</u> the other particles, together with somewhat weaker potentials due to the effects of other individual particles. A special case of this model is the Elliott model, in which one assumes that

a) the central potential is the harmonic oscillator potential.
b) the smaller potentials, while not $U(3)$-invariant, have definite $U(3)$ tensor properties (like X in the dipole moment). (These

properties are guessed from the general nature of the individual potentials, e.g. that they are 2-body interactions.)

From a) and b), one can go ahead, apply the Wigner-Eckart theorem, and deduce some general properties of the nuclei (e.g., the spacing of the energy levels) without having specified the potential in detail.

The second Hamiltonian we consider is the more spectacular

$$H = \frac{P^2}{2m} - \frac{Ze^2}{R} \quad , \tag{3.19}$$

of a particle in an attractive $1/R$ potential, e.g., of an electron in a hydrogen atom, considered already by Louis Michel. As he points out, using the SO(4) invariance with generators L and the Lenz vector

$$A = \frac{1}{\sqrt{-2H}} \left\{ \frac{\vec{R}}{R} + \frac{1}{2mZe^2} (L \times P - P \times L) \right\} \quad , \tag{3.20}$$

one can predict [6]

1) the SO(3) (angular momentum) content of each energy level (Figure 3.2), and
2) the value of the energy for each level.

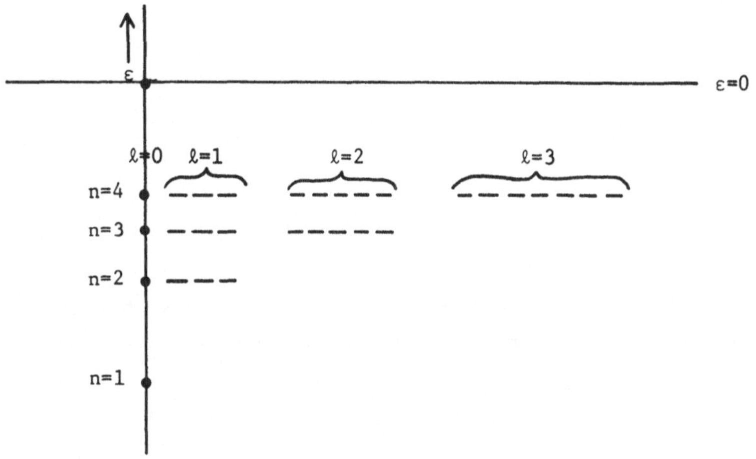

FIGURE 3.2. ANGULAR MOMENTUM CONTENT OF H-ATOM ENERGY LEVELS

The only thing one cannot predict is the multiplicity of the SO(4) representation for each level. I should like to add just two comments to Michel's remarks.

a) The Lenz vector also has a meaning in classical physics; namely, for the planets in the gravitational field of the sun, it is a vector directed along the major axis of the ellipse with length equal to the eccentricity. The fact that it is a constant of the motion is reflected in the fact that the ellipse does not precess. It is perhaps amusing to see that the absence of planetary precession and the degeneracy of the spectrum of the hydrogen atom have the same origin!

b) The second point is just a remark in defense of the groups $SO(4) \times T_H$ or $SO(4,1)$ which contain $SO(4)$ and have representations that can be used to describe all the bound states of the H-atom with the correct multiplicity. The remark is that these two groups can be used to simplify many spectroscopic calculations, and have even been used for calculations which were not feasible by direct methods [7].

4. GENERAL RESULTS ON REPRESENTATIONS OF LIE GROUPS

In this section, we will fill in some of the mathematical gaps which were left in the previous discussion. In particular, we wish to establish the connection between representations of the Lie algebras and the corresponding representations of Lie groups, to define unbounded tensor operators, and finally to formulate the Wigner-Eckart (WE) theorem.[1]

We begin with the case of 1-parameter continuous groups. From Stone's theorem, we know that to any 1-parameter continuous group of transformations $U(t)$ on $\mathcal{H}$, there corresponds a unique skew-adjoint generator G with a dense domain $\mathcal{D}$ on which

$$\frac{dU(t)}{dt} = GU(t) \quad , \tag{4.1}$$

and, conversely, to any skew-adjoint operator G with dense domain $\mathcal{D}$ there corresponds a unique continuous group of unitary transformations such that (4.1) is true on $\mathcal{D}$.

Furthermore, from the spectral resolution [2],

$$iG = \int \lambda dE(\lambda) \quad , \tag{4.2}$$

of G, we see at once that the vectors

$$[E(a) - E(b)]h \quad , \tag{4.3}$$

for all finite intervals $[a,b]$ and all $h \in \mathcal{H}$, form a dense domain A on which

$$\sum_{n=0}^{\infty} \frac{t^n}{n!} G^n \quad , \tag{4.4}$$

converges to U(t). A vector f for which (4.4) converges is said to be an <u>analytic</u> vector for G.

The question now is: What are the analogues, if any, of these results for general groups of continuous unitary transformations on $\mathcal{H}$? The answer is that for completely general groups, definitive results are not available. But for the important special case of simply connected finite parameter Lie groups, almost exact analogues of the above results have been established. We shall restrict ourselves to this case and let U(g) denote henceforth a continuous unitary representation of a simply connected Lie group G on $\mathcal{H}$.

Since the Lie algebra of G contains r elements, where r is the number of independent parameters of G, the question in this case concerns the existence of <u>common</u> domains analogous to $\mathcal{D}$ and A above, for the r elements

$$G_\alpha = \left(\frac{dU(g)}{dx_\alpha}\right)_{x=0} \quad , \quad \alpha = 1 \ldots r \quad , \tag{4.5}$$

of the Lie algebra of G.

The existence of a common dense domain for the G_α was first established by Gårding [3] who in 1947 exhibited the domain $\mathcal{D}_g$ consisting of the vectors

$$\int d\mu(g')U(g')f(g')h \quad , \tag{4.6}$$

where $\mu(g')$ is the group invariant measure, $f(g')$ is any infinitely differentiable function of compact support over the group, and h is any vector in $\mathcal{H}$. (Note that $\mathcal{D}_g$ is not only a common dense domain G_α, but is invariant with respect to both G_α and U(g).)

It was soon shown by Segal [4] that the G_α are actually essentially skew-adjoint on $\mathcal{D}_g$, i.e., their restrictions to $\mathcal{D}_g$ have <u>unique</u> skew-adjoint extensions. This is actually a special case of the following lemma which was later proved by Nelson. [5]

<u>Lemma</u>

The G_α are essentially skew-adjoint on any dense domain $\mathcal{D}$ which is invariant with respect to U(g).

<u>Proof</u>. Let f be an eigenvector of $G_\alpha^\dagger$ with complex eigenvalue ζ. Then the function $\psi(g) = (f,U(g)d)$, $d \in \mathcal{D}$ both satisfies the differential equation

$\frac{\partial\psi(g)}{\partial x_\sigma} = \zeta\psi(g)$ and is bounded. Hence it is zero, in which case, since $\mathcal{D}$ is dense, f = 0. Thus, the deficiency indices of G_α are zero, i.e., G_α is essentially skew-adjoint [2].

The next question is whether there exists a common dense domain of analytic vectors for the G_α, i.e., a dense domain on which

$$\sum_{n=0}^{N} \frac{(t)^n}{n!} (G)^n \tag{4.7}$$

converges where G is any linear combination of the G_α. Here the spectral theorem does not help since, in general, the closures $\overline{G}_\alpha$ and $\overline{G}_\beta$ do not commute and so cannot be simultaneously resolved. Furthermore, the Gårding domain $\mathcal{D}_g$ does not help since that is not in general analytic. However, it has been shown by Cartier and Dixmier [6], Nelson [7] and Gårding [8] that for a unitary representation of a Lie group, a common dense analytic domain for the Lie algebra does in fact exist. Here we describe briefly a simplification of Nelson's proof due to Gårding. The point is to replace the infinitely differentiable functions of compact support f(g) in the Gårding integral by a dense set of analytic functions a(g) of sufficiently fast decrease to counter the (at most exponential) growth of the Haar measure and make the integral converge. Such a dense set of functions is given by

$$a(g) = e^{t\overline{\Delta}} f(g) \quad , \quad t > 0 \quad , \tag{4.8}$$

where $\overline{\Delta}$ is the unique self-adjoint extension of the operator

$$\Delta = 1 - G_1^2 - G_2^2 - \dots - G_r^2 \quad , \tag{4.9}$$

on the Gårding domain $\mathcal{D}_g$ for the regular representation. The functions a(g) have Gaussian decrease for $t > 0$.

It is interesting to note that the above results concerning the existence of a Gårding and analytic dense domain are not confined to unitary representations. They hold for any continuous representation by bounded operators. This is clear for the Gårding domain and follows for the analytic domain because, for a continuous representation, the growth of U(g), like the Haar measure, is at most exponential. Even the result that the G_α are skew-adjoint on any group invariant domain $\mathcal{D}$ generalizes; namely, if superscript c denotes contragredient quantities, we have

$$(G_\alpha^c/\mathcal{D}_c)^\dagger = -\overline{(G_\alpha/\mathcal{D})}$$

So far, we have given the group representation U(g) and asked questions about the Lie algebra. Now we ask the converse question: What is a necessary and sufficient condition that a Lie algebra G_α generate a unique unitary group U(g) on $\mathcal{H}$? An answer was given by Nelson [7] in 1959, who established the following theorem:

<u>Theorem</u>

A necessary and sufficient condition that a Lie algebra of symmetric operators iG_α be the Lie algebra of a unique unitary Lie group U(g) on $\mathcal{H}$ is that there exist in $\mathcal{H}$ a common dense invariant domain $\mathcal{D}$ for the G_α on which the iG_α are symmetric, and the operator

$$\Delta = -\sum_{\alpha=1}^{r} G_\alpha^2 + 1 \quad ,$$

is essentially self-adjoint.

In the course of the proof, Nelson has shown that the analytic domain for the self adjoint extension $\bar{\Delta}$ of the operator Δ is a common analytic domain for the Lie Algebra, and thus furnished an alternative proof of the existence of a dense analytic domain for the unitary representations. The essence of Nelson's proof is to obtain, from the general form of the commutation relations and the obvious bounds $\|G\| < \|\bar{\Delta}\|$, $\|G^2\| < \|\bar{\Delta}\|$, a bound $\|G^n\| < C_n\|\bar{\Delta}^n\|$, where $C_n \leq n!$. Then, if $\sum \frac{(t\bar{\Delta})^n}{n!} < \infty$ for all t, $\sum \frac{(tG)^n}{n!} < \infty$ for $t < t_0$, where $t_0 > 0$. Note that, in general, the <u>entire</u> vectors for $\bar{\Delta}$, i.e., $\sum \frac{(t\bar{\Delta})^n}{n!} < \infty$ all t, are not necessarily entire vectors for $\bar{G}$. Indeed, in general there do not exist any entire vectors for the Lie algebra of a unitary Lie group. The unitary representations of SL(2,C) already provide a counter-example. Recently it has been shown by R. Goodman [10] that the analytic domain for the Lie algebra is exactly the analytic domain for the operator $\Delta^{1/2}$. Goodman has also discussed the question of the existence of entire vectors [11].

From the above results, namely the existence of an analytic domain for any continuous representation, and the existence of a unique continuous unitary representation when Δ is essentially self-adjoint, it is evident that for continuous Lie groups the relationship between Lie algebra and Lie group representations is all that could be required. We can operate relatively freely with the algebra in spite of the unbounded nature of the operators, a circumstance we had anticipated earlier. We close with a few incidental remarks:

First, in the case of UIR's of <u>semisimple</u> Lie groups, there are some stronger results due to Harish-Chandra.[9] For example, the vectors in the (necessarily finite dimensional) subspaces, which are invariant with respect to the maximal compact subgroup of the group, are analytic vectors for the whole group. Furthermore, the linear span of such vectors, which is dense in $\mathcal{H}$, can be generated from any one such vector using the enveloping algebra of the Lie algebra.

Second, there are still some outstanding problems. One is to find an analogue of Nelson's results (Δ essentially self-adjoint) for non-unitary representations. Another is to ask for statements concerning the analytic continuation of the functions $(h,U(g)a)$ to complex values of the group parameters. How close are the singularities? Are they poles or cuts? And so on.

We next consider briefly the domain question for tensor operators. For a set of operators T_a, $a = 1...s$ to transform as a tensor under a unitary group $U(g)$, we need only a dense domain $\mathcal{D}$ with

1) the T_a essentially self-adjoint on $\mathcal{D}$,
2) $\mathcal{D}$ stable with respect to $U(g)$,
3) $U(g)T_aU^{-1}(g) = \hat{D}_{ba}(g)T_b$ on $\mathcal{D}$, where $\hat{D}(g)$ is a representation of $U(g)$.

$\hat{D}(g)$ is usually finite-dimensional $(r < \infty)$, but the definition can be extended to cover infinite dimensional representations as well.

If the group $U(g)$ is compact, one is usually interested not in the full (generally unbounded) tensor components T_a, but only in the restrictions $P'T_aP$, where P,P' are the projections onto finite dimensional subspaces of $\mathcal{H}$ which are invariant with respect to $U(g)$. For the restrictions $P'T_aP$ to exist, one needs only the weaker condition that there exist a dense domain $\mathcal{D}$ for the T_a such that $P\mathcal{D} \subset \mathcal{D}(\overline{T}_a)$, where $\overline{T}_a$ is the unique self-adjoint extension of T_a. The physical conditions are usually enough to guarantee this.

For example, in the dipole radiation example of the last section, the relevant matrix elements were $(f_{\varepsilon'\ell'm'}, y_a f_{\varepsilon\ell m})$, i.e., they were the matrix elements of the restrictions of y_a to the finite spaces $f_{\varepsilon\ell m}$. One can see that these restrictions must exist from the physical point of view as follows: The dipole radiation is actually just the first coefficient in the expansion of $(f_{\varepsilon'\ell'm'}, e^{iy_a/\lambda} f_{\varepsilon\ell m})$ in powers of $1|\lambda$, where λ is the wavelength of the emitted radiation. Now the restriction $P' \exp iy_a/\lambda P$ certainly exists since $\exp iy_a/\lambda$ is a bounded operator, so the only question is the validity of the subsequent expansion in powers of $1|\lambda$. This expansion is justified on the physical grounds that the wavelength λ can be (and in practice usually is) large compared with the mean value of $|y|$ for the wavefunction $f_{\varepsilon\ell m}$, i.e., compared with the "size" of the atom.

Finally, we consider the WE theorem. Let $U(g)$ be a unitary representation of G on $\mathcal{H}$ and T_a a tensor component belonging to the representation $\hat{D}(g)$. Let $\mathcal{H}_1,\mathcal{H}_2$ be irreducible subspaces of $\mathcal{H}$ with respect to $U(g)$, let

$\mathcal{H}_\Lambda$ be the Hilbert space for $D^\Lambda(g)$, and let the product space $\mathcal{H}_\Lambda \otimes \mathcal{H}_2$ decompose into

$$\mathcal{H}_\Lambda \otimes \mathcal{H}_2 = \sum_\lambda \oplus \mathcal{H}_\lambda \quad , \tag{4.10}$$

with respect to U(g). The WE theorem states that

$$(f_1, T_a f_2) = \sum_\lambda (f_\lambda, f_a f_2) \varphi(\mathcal{H}_1 T \mathcal{H}_2)_\lambda \quad , \tag{4.11}$$

where the sum is taken over all λ such that the representations $D^\lambda(g)$ and $(U(g)/\mathcal{H}_1)$ are equivalent and f_λ, f_a are vectors in the directions f_1 and T_a, respectively. In other words, the T-dependent tensor $(f_1, T_a f_2)$ can be expanded linearly in terms of the T-independent tensors $(f_\lambda, f_a f_2)$ with scalar coefficients $\varphi(\mathcal{H}_1 T \mathcal{H}_2)_\lambda$. In particular, if $U(g)/\mathcal{H}_1$ occurs only once in the decomposition (4.10), then

$$(f_1, T_a f_2) = (f_1, f_a f_2) \varphi(\mathcal{H}_1 T \mathcal{H}_2) \quad , \tag{4.12}$$

i.e., $(f_1, T_a f_2)$ is parallel to $(f_1, f_a f_2)$.

The coefficients $\varphi(\mathcal{H}_1 T \mathcal{H}_2)_\lambda$ are usually called reduced matrix elements, and the T-independent tensors $(f_\lambda, f_a f_2)$ are called Clebsch-Gordon coefficients. Note that the $(f_\lambda, f_a f_2)$ are just the matrix elements of the unitary (intertwining) operator which transforms the direct product basis in $\mathcal{H}_\Lambda \otimes \mathcal{H}_2$ into the basis in which U(g) is diagonal.

5. SURVEY OF EXPERIMENTAL AND THEORETICAL BACKGROUND TO ELEMENTARY PARTICLE PHYSICS

The rest of these chapters will be devoted to the group theory of elementary particle physics. But before going on to the group theory proper, it might be worthwhile to fill in a little of the experimental and theoretical background. This we shall do in the present chapter.

First we consider the experimental background [1].

The non-relativistic quantum mechanics discussed up to now suffices to describe completely the greater part of modern physics--atomic, molecular, plasma, solid state, low temperature, etc., physics. It is built on the twin postulates of Newton's laws and $[X,P] = i\hbar$. The basic constituents of matter for all these branches of physics are the protons, neutrons, and electrons which form the atoms, and the photons, which carry the EM (electromagnetic) field. These constituents of matter, or particles, are regarded as elementary. In particular, the protons, neutrons, and electrons are regarded as indestructible.

As soon, however, as one wishes to inquire into the finer features of atomic phenomena or wishes to investigate the structure of the atomic nucleus or the structure of the protons, neutrons, and electrons themselves, then the situation changes drastically. First, the energies necessary for the investigation are relativistic. Second, the electrons, protons, and neutrons are found to be far from indestructible. They can be destroyed and created almost at will. Third, not only can these particles be destroyed and created, but new particles are created and destroyed along with them. The new particles include the anti-particles of the proton, neutron, and electron, the π-meson which keeps the protons and neutrons bound in the nucleus, and many other particles (along with their anti-particles). To date, the number of new particles which have been produced is of the order of 100.

It should, perhaps, be emphasized that the particles referred to here differ in some fundamental ways from the Newtonian particles defined in the first lecture; namely,

a) they can be created and destroyed.

b) Although they can be created and destroyed, their masses are not arbitrary but are fixed by nature to have definite values outside our control. For example, the electron has a mass 9.11×10^{-28} grams.

c) As well as an intrinsic mass, the particles have an intrinsic angular momentum. The Casimir operator of the intrinsic angular momentum group takes the values $J(J + 1)$, where J (the spin of the particle) is half-integer.

Thus, the particles appear to be particles in the sense of Democritus (fixed, ultimate constituents of matter) rather than of Newton (fictitious limits of small bodies). For this reason they are called elementary particles. Of course, it is difficult to believe that 100 particles can be elementary, but until something more elementary is discovered, they are regarded as such. (An analogy is provided by the chemical elements, all 92 of which were regarded as elementary until the advent of atomic theory.)

In Figure 5.1, a list of the particles is presented. They are grouped together into multiplets (so-called isospin multiplets) of particles with approximately the same mass and spin. Even so, the number of multiplets is very large and it might help to clarify the situation a little if we briefly classify them by word.

The broadest classification of the particles is in terms of their interactions. Apart from the gravitational interactions, in which all the particles participate, but which are so weak as to be negligible, the particles can interact in only three ways:

a) By electromagnetic interactions, with coupling constant $e^2/\hbar c \sim 1/137$

b) By weak interactions, with coupling constant $g^2 \ll e^2/\hbar c$

c) By strong interactions, with coupling constant $G^2 \gg e^2/\hbar c$.

Name	$I^G(J^P)C_n$ estab. ?=guess
$\pi^{\pm}(140)$ $\pi^0(135)$	$1^-(0^-)+$
$\eta(549)$	$0^+(0^-)+$
$\eta_{0^{++}}(700)$ "ε"$\to\pi\pi$	$0^+(0^+)+$
$\rho(765)$	$1^+(1^-)-$
$\omega(784)$	$0^-(1^-)-$
$\eta'(958)$ or X^0	$0^+(0^-)\pm$
$\delta(692)$ $\pi_N(1016)$ $\to K\bar{K}$	$\geq 1(\ )$
$\phi(1019)$	$0^-(1^-)-$
$\eta_{0^{++}}(1060)$ "S*"$\to K_S K_S$	$0^+(0^+)+$
A1(1070)	$1^-(1^+)-$
$\rho_N(1660)$ "g"$\to 2\pi$	$1^+(N)-$
$\rho(1710)$? $\to 4\pi$	$1^+(\ )-$
U(2375)	$1^+(\ )-$
$K^+(494)$ $K^0(498)$	$1/2(0^-)$
$K^*(892)$	$1/2(1^-)$
$K_A(1240)$ or C	$1/2(1^+)$
K_A(1280 to 1360)?	$1/2(1^+)$
$K_N(1420)$	$1/2(2^+)$
$K_A(1775)$ or L	$1/2(A)$
B(1235)	$1^+(1^+)-$
f(1260)	$0^+(2^+)+$
D(1285)	$0^+(A)+$
$A2_L(1280)$	$1^-(2^+)+$
$A2_H(1320)$	$1^-(2^+)+$
E(1422)	$0^+(0^-)\pm$
f'(1514)	$0^+(2^+)+$
$\pi/\rho(1540)$ "F_1"$\to K^*\bar{K}$?	1(A)
$\pi_A(1640)$ $\to 3\pi$)	$1^-(A)+$

Isospin 0

$\eta(550)$	0^-
$\omega(780)$	1^-
$\eta^*(960)$	0^- (?)
$\phi(1020)$	1^-
$\eta(1070)$	0^+
f(1260)	2^+
D(1285)	$P=(-1)^{J+1}$
E(1420)	0^- (?)
f*(1515)	2^+
$\eta(700)$	0^+

Isospin 1/2

K(490)	0^-
K*(890)	1^-
K(1320)	1^+
K**(1420)	2^+
K(1780)	$P=(-1)^{J+1}$

Isospin 1

$\pi(140)$	0^-
$\rho(760)$	1^-
$\pi(1016)$	0^+
$A_1(1070)$	1^+ (?)
$A_2(1270)$	$P=(-1)^J$
$A_2(1315)$	2^+ (?)
$\pi(1640)$	$P=(-1)^{J+1}$
$\rho(1650)$	$P=(-1)^J$
B(1235)	1^+ (?)

SU(3) multiplets

$\pi(140)$, K(490), $\eta(550)$, $\eta^*(960)$	0^-
$\rho(760)$, K*(890), $\omega(780)$, $\phi(1020)$	1^-
$A_2(1315)$, K**(1420), f(1260), f*(1515)	2^+

FIGURE 5.1b MESONS

The following bumps have also been observed, but their spins and parities are not yet known; $\sigma(410)$; H(990); $\eta_\nu(1080)$; A1.5(1170); $A2_2(1320)$; $\rho\rho(1410)$; $K_S K_S(1440)$; $\phi(1650)$, R(1750); η or $\rho(1830) \to 4\pi$; ϕ or $\pi(1830) \to \omega\pi\pi$; S(1930); $\rho(2100)$; T(2200); $\rho(2275)$ $N\bar{N}_{I=0}(2380)$; $\kappa(725)$; $K_N(1080\text{-}1260)$; $K_{A(I=3/2)}(1175)$: $K_{A(I=3/2)}(1265)$; $K_{N(I=1/2)}(1660)$; $K^*(2240) \to \bar{Y}N$; $X^-(2500)$; $X^-(2620$; $X^-(2880)$.

Particle or resonance[2]	I (J^P)
p n	$1/2(1/2^+)$
N′(1470)	$1/2(1/2^+)$
N′(1520)	$1/2(3/2^-)$
N′(1535)	$1/2(1/2^-)$
N(1670)	$1/2(5/2^-)$
N(1688)	$1/2(5/2^+)$
N′′(1700)	$1/2(1/2^-)$
N′′(1780)	$1/2(1/2^+)$
N(1860)	$1/2(3/2^+)$
N(1990)	$1/2(7/2^+)$
N′′′(2040)	$1/2(3/2^-)$
N(2190)	$1/2(7/2^-)$
N(2650)	$1/2(?^-)$
N(3030)	$1/2(?)$
Δ(1236)	$3/2(3/2^+)$
Δ(1650)	$3/2(1/2^-)$
Δ(1670)	$3/2(3/2^-)$
Δ(1890)	$3/2(5/2^+)$
Δ(1910)	$3/2(1/2^+)$
Δ(1950)	$3/2(7/2^+)$
Δ(2420)	$3/2(11/2^+)$
Δ(2850)	$3/2(?^+)$
Δ(3230)	$3/2(?)$

N

Δ

Λ	$0(1/2^+)$
Λ(1405)	$0(1/2^-)$
Λ′(1520)	$0(3/2^-)$
Λ′(1670)	$0(1/2^-)$
Λ′′(1690)	$0(3/2^-)$
Λ(1815)	$0(5/2^+)$
Λ(1830)	$0(5/2^-)$
Λ(2100)	$0(7/2^-)$
Λ(2350)	$0(?)$

Λ

Σ	$1(1/2^+)$	
Σ(1385)	$1(3/2^+)$	P_{13}
Σ(1670)	$1(3/2^-)$	D_{13}
Σ(1750)	$1(1/2^-)$	S_{11}
Σ(1765)	$1(5/2^-)$	D_{15}
Σ(1915)	$1(5/2^+)$	F_{15}
Σ(2030)	$1(7/2^+)$	F_{17}
Σ(2250)	$1(?)$	
Σ(2455)	$1(?)$	
Σ(2595)	$1(?)$	

Σ

Ξ	$1/2(1/2^+)$
Ξ(1530)	$1/2(3/2^+)$
Ξ(1820)	$1/2(?)$
Ξ(1930)	$1/2(?)$
Ξ(2030)	$1/2(?)$
Ξ(2250)	$1/2(?)$
Ξ(2500)	$1/2(?)$
Ω^-	$0(3/2^+)$

Ξ

SU(3) multiplets

p	$\frac{1}{2}^+$
n	"
Λ	"
Σ	"
Ξ	"
Δ(1236)	$\frac{3}{2}^+$
Σ(1385)	"
Ξ(1530)	"
Ω^-(1686)	" (?)
N(1525)	$\frac{3}{2}^-$
Λ(1520)	"
Σ(1660)	"
Ξ(1815)	" (?)
N(1688)	$\frac{5}{2}^+$
Λ(1820)	"
Σ(1910)	"
Ξ(1930)	" (?)

Regge Recurrences

$\frac{1}{2}^+$-multiplet	N(1525)	$\frac{3}{2}^-$	Λ(1520)	$\frac{3}{2}^-$
$\frac{3}{2}^+$-multiplet	N(2190)	$\frac{7}{2}^-$	Λ(2100)	$\frac{7}{2}^-$

FIGURE 5.1a. BARYONS

Data are taken from A. Rosenfeld *et al.*, *Rev. Mod. Phys.* (January, 1970). The numbers in brackets are masses in millions of electron volts. J is the spin (half-odd-integer and integer for baryons and mesons respectively), and P is the parity.

Apart from the photon, which carries the EM field and interacts only electromagnetically, there are three main classes of particles:

1) The leptons: These do not interact strongly. There are four of them; the electron e, the μ-meson, and the two neutrinos ν_e, ν_μ. All have spin 1/2.

2) The baryons: The particles which interact strongly and obey Fermi-Dirac statistics (i.e., have half-odd integer spin).

3) The mesons: The strongly interacting particles which obey Bose-Einstein statistics (have integer spin).

The mesons and baryons can, of course, also interact weakly and electromagnetically, both with each other and with the leptons. The collective name for all strongly interacting particles is hadrons.

Anti-particles are omitted in Figure 5.1 because they have the same masses and spins as the particles. Further subdivisions of the particles have already been considered by Michel and will be touched on again in later lectures. An important property of the particles is their stability, or lack of it, (when left alone). The only really stable ones are the photon, neutrinos, electron, and proton. However, many others are metastable, i.e., have relatively long (10^{-13} sec) lifetimes. These include the leptons, n, Σ, Λ, Ξ, and Ω in class 2), and π, K, η in class 3). The rest of the particles are unstable. They have lifetimes of $\sim 10^{-23}$ secs and are usually not observed directly but as resonances in the scattering cross-sections for metastable particles.

It should, perhaps, be emphasized at this point that the experimental information that we can get on the elementary particles is very limited. The particles are so tiny and so unstable that essentially all one can do is scatter them and watch them decay.

In particular, one can only build particles with masses up to the energies available in the accelerators. Figure 5.1 is based on the present energies (pending the building of the 200 Gev Weston machine and Super-Cern). This table may not be, and probably is not, sufficient to let us see the true picture. For example, ten years ago only the part of Figure 5.1 above the Ω-line was available, and it is now clear that this would have been insufficient to predict today's picture.

Further, one gets information for weak and electromagnetic interactions only when these interactions are not swamped by the strong ones and, for the weak interactions in particular, the information is limited to decay.

For the strong interactions themselves, the information is limited not only by the energies available, but by the particles which are available as targets and projectiles for the scattering. Essentially the only available ones are:

Target: Protons, neutrons (and electrons)

Projectiles: Protons, neutrons (and electrons), photons and the metastable mesons π and K, together with their anti-particles.

What is actually measured in the strong collisions is the scattering amplitude $A(p_A; p_B; p_{C_1} \ldots p_{C_N})$ for the processes

$$A + B \to C_1 + C_2 + \ldots + C_N \quad ,$$

(Figure 5.2), which is a function of the momenta $p_A \ldots p_{C_N}$ of the particles and whose absolute value squared is the probability for A and B to scatter into particles $C_1 \ldots C_N$ with these momenta.

Similarly, what is measured in electromagnetic interactions is the form-factor $F_{AB}(t)$ whose square is the probability for the particle A with momentum p_A to interact with the EM field, lose momentum k, and emerge as particle B (possibly the same as A) with momentum $p_B = p_A - k$ (Figure 5.3). On account of Lorentz invariance, $F_{AB}(t)$ is essentially a function of

$$t = k^2 = (p_B - p_A)^2 \quad ,$$

only. (It may have some polynomial dependence on P_A and P_B through the spins of A and B.) Actually, at present $F_{AB}(t)$ is known reasonably well (up to $t \simeq$ proton-mass) only for the electron (for which it is trivial), the proton, and the neutron. For some other metastable particles, notably π, Σ, Λ, K, a little is known about it for $t \to 0$.

Thus, to sum up, what has been established experimentally is the existence of a large number (~100) of particles of definite masses and spins and various life-times, most of them short. What can be measured, essentially, are their electro-magnetic form-factors $F_{AB}(t)$, their strong scattering amplitudes, and their weak decays, all subject to strong experimental limitations.[1]

The business of elementary particle physics is to construct a theory which will

1) explain the interactions (form-factors, scattering amplitudes, decays) of the particles, and
2) predict their masses and spins.

This is a tall order since it combines 1) solving Newton's problem at a subnuclear level with solving 2) the problem of the structure of matter.

Not surprisingly, one has at present nothing like a complete theory of the elementary particles, though one does have some ideas and a workable, if not yet mathematically rigorous, theory of electromagnetic interactions. Almost all the ideas one has can be traced back to the theory of quantized fields introduced by Pauli, Heisenberg, and Dirac [2] in the heroic days of quantum mechanics, 1925-28. Because they lie at the root of most later developments and because they are necessary later as background for relativistic group theory, we conclude this lecture with a

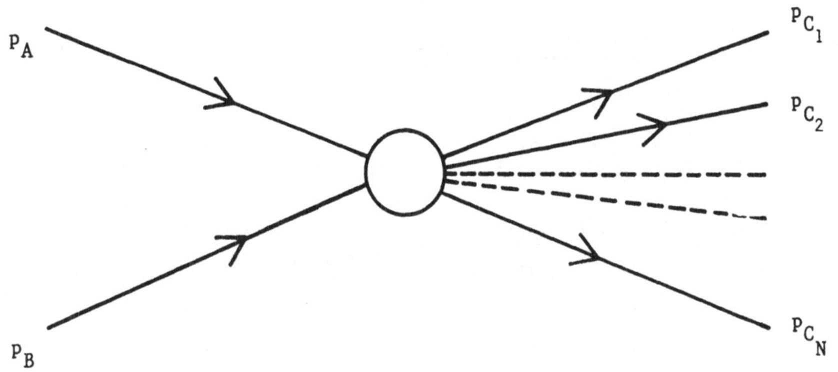

FIGURE 5.2

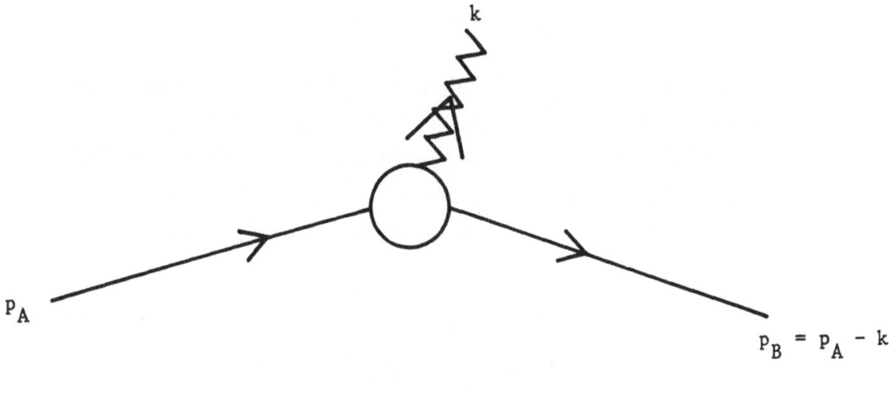

FIGURE 5.3

brief review of the ideas underlying the théory of quantized fields.

To begin with, we return to the Hamiltonian, which describes a non-relativistic classical particle in a potential

$$H = \frac{p^2}{2m} + \varphi(x) \quad . \tag{5.1}$$

Generalizing to describe interactions with the EM field and the field equations for the EM field itself, we have

$$H = \frac{p^2}{2m} + e[\varphi(x) + \frac{1}{c}\vec{\nabla}\cdot\vec{A}(x)] + \frac{1}{2}\int d^3y[\pi(y)^2 + (\nabla A(y))^2] \quad , \tag{5.2}$$

where $A = (\varphi,\vec{A})$ is the EM potential and the integral term is the Hamiltonian for the free EM field. (It is equivalent to $\frac{1}{2}\int d^2y[E(y)^2 + H(y)^2]$ where $(E,H) = F_{\mu\gamma} = \partial_\mu A_\gamma - \partial_\gamma A_\mu$, but the form (5.2) is better for later quantization.) We can also write the interaction term (with coupling coefficient, or charge, e) as

$$e\int d^3y j_\mu(y)A_\mu(y) \quad , \tag{5.3}$$

where

$$j_\mu(y) = \delta(x - y)[1,v/c] \quad . \tag{5.4}$$

If we now quantize the particle according to non-relativistic quantum mechanics, wé obtain

$$H = \frac{p^2}{2m} + e\int d^3y j_\mu(y)A_\mu(y) + \frac{1}{2}\int d^3y[\pi^2 + (\nabla A)^2] \quad , \tag{5.5}$$

where

$$2j_\mu(y) = \delta(X - y)[1, \frac{P}{mc}] + [1, \frac{P}{mc}]\delta(X - y) \quad , \tag{5.6}$$

and P and X are now the usual quantum mechanical operators, satisfying $[X,P] = i\hbar$. This Hamiltonian is only

a) semirelativistic because the EM field is relativistic but the particle is not.

b) semi-quantized because the particle is quantized but the EM field is not.

To remedy these defects, one quantizes the EM field by the Ansatz

$$[A_\mu(x),A_\gamma(x')] = i\hbar g_{\mu v}D(x - x') \quad , \tag{5.7}$$

where D(x) is a numerical function (or, more precisely, distribution) to be discussed in a moment, and one makes the particle relativistic by the substitution

$$\frac{1}{2m}P^2 \to \vec{\alpha}\cdot\vec{P} + \beta m \quad , \tag{5.8}$$

$$j_\mu(x) \to \delta(x - X)\gamma_\mu \quad , \tag{5.9}$$

where the γ_μ are the 4 × 4 Dirac matrices defined by

$$[\gamma_\mu,\gamma_\nu]_\pm = 2g_{\mu\nu} \quad , \tag{5.10}$$

β is γ_0, $\vec{\alpha}$ is $\gamma_0\vec{\gamma}$, $g_{\mu\nu}$ is the metric tensor, and, for simplicity, we have assumed that the particle in question has spin $\frac{1}{2}$ (e.g. is an electron). For other spins we use an appropriate generalization of the γ_0 (see Section 7).

The Ansatz (5.7) for the EM field is the analogue of $[X,P] = i\hbar$ for the particle. Indeed, one can expand the purely EM part of the Hamiltonian as a sum of formal harmonic oscillators

$$\frac{1}{2}\int d^3x[\pi(X)^2 + (\nabla A)^2] = \frac{1}{2}\int d^3k[P(k)^2 + \omega(k)^2Q(k)^2] \quad ,$$

where

$$\vec{Q}(k) = \int d^3k \sin kx \, \vec{A}(x) \quad ,$$

$$\vec{P}(k) = \dot{\vec{Q}}(k) \quad ,$$

and

$$\ddot{Q}(k) + \omega^2(k)Q(k) = 0 \quad ,$$

and then (5.7) amounts to the Ansatz

$$[Q(k),P(k')] = i\hbar\delta(k - k') \quad ,$$

for the formal oscillators. The important properties of the distribution $D(x,t)$ are that $D(x)$ is Lorentz invariant,

$$\begin{aligned} D(x) &= 0 \quad , \quad x^2 < 0 \quad , \\ D(\vec{x},0) &= 0 \quad , \\ \dot{D}(\vec{x},0) &= \delta^3(\vec{x}) \quad , \\ \left(\frac{1}{c^2}\frac{\partial^2}{\partial t^2} - \nabla^2\right)D(x) &= 0 \quad . \end{aligned} \tag{5.11}$$

The Ansatz (5.8)(5.9) for the particle means that it is no longer described in the Hilbert space $L_2(-\infty,\infty)$ for $[X,P] = i\hbar$, but in a Hilbert space $L_2[-\infty,\infty) \times R_4$ where R_4 is the 4-dimensional Dirac space.

It turns out, however, that while the relativistic quantized Hamiltonian (5.5)(5.10) is sufficient to describe processes in which the relativistic particle is conserved, it cannot take account of the experimental fact that when the relativistic energies are large enough, the particle can be created or destroyed. To allow for this possibility, one must go further and <u>second quantize</u> the Hamiltonian. This means introducing for the <u>particle</u> a field $\psi_\alpha(x)$, which is quantized according to the rule

$$[\psi_\alpha(x),\psi_\beta(x')]_\pm = i\hbar D_{\alpha\beta}(x - x') \quad , \tag{5.12}$$

where $D_{\alpha\beta}(x)$ is a function analogous to $D(x)$, the $\pm$ commutator is taken according as to whether the particle obeys Fermi-Dirac statistics (has half-odd-integer spin) or Bose-Einstein statistics (integer spin), and α,β are spin indices. (In the case of the electron, which is a spin $\frac{1}{2}$ particle, the + sign is taken and the indices α,β are the Dirac indices.) Using the field $\psi_\alpha(x)$, one makes the substitutions

$$\vec{\alpha}\cdot\vec{P} + \beta m \rightarrow \psi^+(x)(\vec{\alpha}\cdot\vec{\partial} + \beta m)\psi(x) \quad , \tag{5.13}$$

$$j_\mu(x) \rightarrow \psi^+(x)\gamma_0\gamma_\mu\psi(x) \quad , \tag{5.14}$$

in the relativistic first-quantized Hamiltonian (5.5)(5.8) and (5.9) and obtains finally

$$H = \psi^+(\vec{\alpha}\cdot\vec{\partial} + \beta m)\psi + e\int d^3x\psi^+(x)\gamma_0\gamma_\mu\psi(x)A_\mu(x)$$

$$+ \frac{1}{2}\int d^3x[\pi(x)^2 + (\nabla A(x))^2] \quad . \tag{5.15}$$

This is the fully quantized, relativistic, Hamiltonian of Dirac, Heisenberg, and Pauli. Note that in this theory the particles and the EM field are on the same footing. Each is described by a field and the field has a particle interpretation (photon interpretation in the case of the EM field), which is obtained by analyzing the quantization Ansätze (5.7) and (5.12).

Without accepting the Hamiltonian H (and its generalization to include interactions between particle-fields other than the electron $\psi(x)$ and photon $A(x)$) too literally, one can extract from it most of the ideas which are used in the later theories. Let us summarize briefly the most important and relevant ideas: [3]

1) The particles are described in some way by fields $\varphi(x)$ ($\psi(x)$ and $A(x)$ above) which are <u>quantized locally</u>, i.e., whatever quantization rules are adopted for the interacting fields, they should at least satisfy the conditions

$$[\varphi(x),\varphi(x')]_\pm = 0 \quad , \quad (x - x')^2 < 0 \quad . \tag{5.16}$$

These conditions are dictated by the principle of strong microscopic causality; measurements which are separated by spacelike distances should not interfere. (The + sign in (5.16) is taken for fermion fields for which only <u>bilinears</u> in the field are observables.) The locality assumption is usually strengthened by the demand that the fields, which, to make sense both mathematically and physically, are not operators but operator-valued distributions, should not be too wild in the sense of distributions.

2) The fields interact locally. For example, if a Hamiltonian exists, the interaction term in it would be of the form

$$H_{int} = g \int d^3x \psi^+(x)\gamma_0\psi(x)\varphi(x) \quad ,$$

$$H_{int} = g \int d^3x \psi^+(x)\gamma_0\gamma_\mu\psi(x)\partial_\mu\varphi(x) \quad , \tag{5.17}$$

etc., but not of the form

$$H_{int} = g \int d^3x \int d^4x' d^4x'' \psi^+(x'')\gamma_0$$

$$f(x - x', x - x'')\psi(x')\varphi(x) \quad , \tag{5.18}$$

where f is some Lorentz invariant function which does not vanish for $x \neq x'$, $x \neq x''$.

3) Under Lorentz transformations, the fields transform according to the law

$$\varphi_\alpha(x) \overset{\Lambda, a}{\rightarrow} S_{\alpha\beta}(\Lambda)\varphi_\beta(\Lambda^{-1}(x - a)) \quad , \tag{5.19}$$

where Λ is a homogeneous Lorentz transformation, a is a translation, and $S_{\alpha\beta}(\Lambda)$ is a representation of Λ. The choice of representation $S_{\alpha\beta}(\Lambda)$ is determined by the masses and spins of the particles. For free fields, or in the free field limit of interacting fields, the above description can be made a little more exact. The fields can be expanded in the form

$$\varphi(x) = \int d\mu(p)[\varphi(p)a(p)e^{ipx} + \bar{\varphi}(p)b^\dagger(p)e^{-ipx}] \quad , \tag{5.20}$$

where the unquantized "wavefunctions" $\varphi(p)$, $\bar{\varphi}(p)$ carry the Lorentz properties of $\varphi(x)$, and the operators $a(p)$ and $b^\dagger(p)$, which satisfy quantization relations of the form

$$[a(p), a^+(p')]_\pm = \hbar\delta(p - p') \quad ,$$

$$[a(p), b(p')] = 0 \quad , \quad \text{etc.} \quad , \tag{5.21}$$

carry the quantization properties. An analysis of the algebra (5.21) in Hilbert space shows that the operators $a(p)$ and $b^\dagger(p)$ can be considered as creation and destruction operators for states which have the right properties to be identified with free particle states. Thus, the particle description of the field may be said to be embodied in the quantization relations.

To sum up, one is confronted with a huge number of elementary particles experimentally and one is looking for a theory which will explain the elementary particles and their interactions. For want of better alternatives, one tries to find such a theory by using general ideas derived from local field theory. The fields in local field theory have particle properties in the free field limit, have definite transformation properties with respect to the Lorentz group, and they interact and are quantized locally.

6. REPRESENTATIONS OF THE POINCARE GROUP IN HILBERT SPACE

In the last lecture, we sketched briefly the experimental background to elementary particle physics and the basic theoretical tool, namely the theory of quantized fields, which is used to attack it. We saw that one of the most important properties of the fields was that they transformed in a manifestly covariant manner,

$$\psi_\alpha(x) \xrightarrow{\Lambda, a} S_{\alpha\beta}(\Lambda)\psi_\beta(\Lambda^{-1}(x - a)) \quad , \tag{6.1}$$

under inhomogeneous Lorentz, or Poincaré, transformations. In this lecture we wish to consider the question of Poincaré covariance in a more general way, that is, divorced from any particular theory such as field theory, and using nothing but the most fundamental quantum mechanical ideas. Later we shall try to establish the link with field theory.

We begin, as usual, with the probabilities

$$|(f,h)| \quad , \tag{6.2}$$

where f and h are vectors in the Hilbert space $\mathcal{H}$. The assumption that apart from the spectra these, and only these, are the physical numbers to be extracted from the theory is made not only in non-relativistic but in relativistic quantum theory, and underlies all other assumptions. (For simplicity, we assume that all vectors in $\mathcal{H}$ represent physical states (no super-selection rules), but the argument can easily be generalized to the case where this is not so.)

Let us now suppose that, due to the geometry of space-time, we wish to impose an invariance principle on the quantum mechanical system--we wish to demand that the system be invariant under some group G of space-time transformations. Let us for the moment not specify the group although, in practice, it will be the Galilean group or the Poincaré group. How are we to impose the invariance principle? Following the arguments used earlier, namely that under a change of observer the probability of a system making a given transition remains unchanged (the old argument that "the system does not care who is looking at it"), we impose the invariance principle by demanding that, under the transformations of the group, the inner products (6.2) remain invariant, i.e.,

$$|(T(g)f,T(g)h)| = |(f,h)| \quad , \tag{6.3}$$

f, h $\in \mathcal{H}$, g $\in G$. We also demand that the Hamiltonian transform under the group in a way appropriate for the energy. The latter demand generalizes the idea of invariance groups used in non-relativistic theory.

Using Wigner's theorem, it follows that the invariance group can be implemented on $\mathcal{H}$ by a set of unitary or anti-unitary operators U(g), forming a ray representation

$$U(g)U(g') = e^{i\omega(g,g')}U(g,g') \quad , \tag{6.4}$$

of the group.

If the group is continuous, physical continuity demands that as $g \to 1$ in the group topology, T(g)f should represent the same state as f, i.e.,

$$T(g)f \to e^{i\alpha}f \quad , \tag{6.5}$$

whence

$$U(g)f \to e^{i\gamma(g,f)}f \quad , \tag{6.6}$$

i.e., physical continuity demands that U(g) be ray-continuous in the sense of (6.6).

We see, therefore, that from quite general principles the invariance of a quantum mechanical system under a geometrical group demands that the Hilbert space $\mathcal{H}$ of the system carry a unitary or anti-unitary ray representation of the group. If the group is continuous, the representation must be ray-continuous.

For connected Lie groups, such a representation can be shown [1] to be equivalent to (or can be "lifted" to) a true continuous unitary representation of the covering group of either the group itself or some continuous central extension of it.

Thus, without loss of generality, we can confine ourselves to continuous unitary group representations. Whether we can use continuous unitary representations of the geometrical group itself or of some central extension depends on the geometrical group in question.

To proceed further, we must therefore specify the geometrical group more precisely. We shall specify finally to the Galilean and Poincaré group, in particular to the Poincaré group, but before doing so it might be interesting to point out that we could first limit ourselves to <u>kinematical</u> groups, i.e., 10-parameter, continuous, connected space-time Lie groups with rotations, a scalar time translation, vector space translations, and vector accelerations, with the commutation relationships not mentioned left open. Under general conditions [2], it can be shown that there are, in fact, only eight such groups, four non-relativistic ($t' = t + t_0$) groups including the Galilean group, and four relativistic groups including the Poincaré group. For the four relativistic groups, the phase-factors exp $i\omega(gg')$ can be lifted completely. For the four non-relativistic groups, the lifting requires a 1-parameter central extension. We have already seen this in the case of the Galilean group for which the central extension is generated by the total mass M.

Let us now concentrate on the relativistic case and in particular on the connected Poincaré group. From what we have just said, the Hilbert space $\mathcal{H}$ must carry a true continuous unitary representation of its covering group, which we denote by

$$P_+^\uparrow = T_4 \textcircled{s} SL(2,C) \quad , \tag{6.7}$$

where T_4 is the 4-dimensional translation group, s denotes semi-direct product, and $\uparrow+$ mean that time-and space-inversions are not included. Group multiplication is to the left. In particular, if we use the conventional paramatrization (Λ,a) for $P_+^\uparrow$, we have $(\Lambda,a)(\Lambda',b) = (\Lambda\Lambda',a + \Lambda b)$.

Needless to say, the representation of $P_+^\uparrow$ carried by $\mathcal{H}$ will not, in general, be irreducible. However, $P_+^\uparrow$ is a type 1 group, which means that any continuous unitary representation decomposes uniquely into a direct sum and/or a direct integral of continuous unitary irreducible representations (CUIR's). It follows that, from the group theoretical point of view, the elementary objects to study are the CUIR's of $P_+^\uparrow$. Some of the CUIR's will, in fact, be identified directly (i.e., without summation or integration) with elementary particles. This point will be discussed in more detail later. For the moment, we merely remark that for the case of non-relativistic quantum mechanics, we have already seen that a free Newtonian particle carries a CUIR of the extended Galilean group.

The CUIR's of $P_+^\uparrow$ were first classified by Wigner [3] in 1939. However, they are most simply classified by Mackey's method [4] of induced representations, which generalizes and simplifies Wigner's approach. We, therefore, proceed using Mackey's method. We first describe the method for a general group G, and then specialize to $P_+^\uparrow$.

Let G be any separable locally compact group, H any closed subgroup, G/H the right coset space, and $\mu(s)$ the left invariant (or left quasi-invariant) measure on G/H. Let $W(h)$, $h \in H$ be any unitary representation of H on a Hilbert space N, and $f(g)$ the set of vector functions over G with values in N satisfying the

1) subsidiary condition

$$f(hg) = W(h)f(g) \quad , \tag{6.8}$$

2) square integrability condition

$$\int d\mu(s)(f(g),f(g)) < \infty \quad , \tag{6.9}$$

where the inner product in the integrand is with respect to W and, on account of 1), is a function over G/H only.

The representation U(g) of G defined by letting G act transitively on f(g), i.e.,

$$f(g) \overset{g'}{\rightarrow} f(gg') \quad , \tag{6.10}$$

is unitary and is called the unitary representation of G induced by the representation W of H.

Note that if H = 1, W = 1, U is just the regular representation. At the other extreme, if H = G, then U = W. Note also that to induce U, two choices are necessary: a choice of subgroup H and a choice of representation U(H) of H. In general, there is no guarantee that U will be irreducible or that the set of all induced representations will be exhaustive.

Let us turn now to the special case of $P_+^\uparrow$. The question is how, in this case, we make our choice of H and W. To answer it, we first have to introduce the concept of optics.

Orbits. Consider T_4. Every unitary irreducible representation of T_4 is 1-dimensional and of the form exp ipa, where $a_\mu \in R$, $\mu = 1\ldots4$, are the group parameters, and $p \in R_4$ is the character. Now let $g \in P_+^\uparrow$ act on a. We have

$$\exp(ip.a) \longrightarrow \exp(ip.ga) = \exp(ipg.a) \quad . \tag{6.11}$$

where $pg \in R_4$, i.e., we have an associated action of $P_+^\uparrow$ on p. The orbit of p is defined to be the subset pg of R_4, $g \in P_+^\uparrow$. Clearly, R_4 breaks up disjointly into orbits, and there are six kinds:

a) $p^2 = m^2$	$p_0 > 0,\ p_0 < 0$	(timelike)	SU(2)
b) $p^2 = -m^2$		(spacelike)	SU(1,1)
c) $p^2 = 0$	$p_0 < 0,\ p_0 > 0$	(lightlike)	E(2)
d) $p = 0$		(trivial)	SL(2,C)

where m^2 is any fixed positive number.

We are now in a position to choose the subgroup H and its representation W. The rules are as follows:

1) Choose an orbit (e.g., $p^2 = -m^2$),
2) Choose any point $p = \alpha$ on the orbit,
3) Determine the stability (little) group of α, i.e., the maximal subgroup K of SL(2,C), leaving α fixed,
4) Choose $H = T_4 \circledS K$,
5) Choose $W(H) = \exp i\alpha a \otimes V(K)$, where V(K) is any unitary irreducible representation of K,
6) Induce with H and W(H).

With this choice of H, the representations of $P_+^\uparrow$ obtained are irreducible and (using all possible V(K)) exhaustive.

One can gain an intuitive feeling why this is so by noting that the following three things coincide: the coset space G/H, the orbit 0, and the simultaneous spectrum S of the infinitesimal generators of T_4. Thus

$$G/H = 0 = S \quad .$$

The irreducibility can then be seen intuitively as follows. From the subsidiary condition 1), f(g) is essentially a function over G/H and the Hilbert space of W(H) only. But $P_+^\uparrow$ acts irreducibly on 0 by definition. Hence, $P_+^\uparrow$ acts irreducibly on G/H = 0. And V(K) acts irreducibly on N. Hence, $P_+^\uparrow$ acts irreducibly on both G/H and N. Hence, $P_+^\uparrow$ acts irreducibly on f(g), as required. To see why the induced representations should be exhaustive, we note that given any representation $P_+^\uparrow$, the infinitesimal generators of T_4 can be simultaneously diagonalized and hence the vectors in the representation space can be written as functions over S. Hence, these vectors can be written as functions over 0 = S. For a fixed point in $s \in S$, the only remaining freedom is to transform according to some representation of the group leaving S invariant. But since S = 0, the group leaving $s \in S$ invariant is just the stability group for a point $p = \alpha$ in 0. Thus, the representation of $P_+^\uparrow$ corresponds to an induced representation.

The little group corresponding to the orbits a) to d) above are written beside them. The invariant differential form is

$$d\mu(p) = \frac{d^3\underline{p}}{p_0} \quad , \qquad d\mu(p) = \frac{d^3\underline{p}}{p_1} \quad , \tag{6.12}$$

for a), c), and b), respectively. The continuous unitary irreducible representations of SU(2), E(2), SU(1,1), and SL(2,C) are all known. We are thus in a position to determine explicitly all the CUIR's of $P_+^\uparrow$. In the next section, we shall do this in some detail, at least for the physically relevant representations. In particular, we shall try to express the induced representations in a form which is immediately useful for physics. For the rest of the present lecture, we turn to the more general question of the physical interpretation of the CUR's of $P_+^\uparrow$ carried by $\mathcal{H}$.

First, according to the theorems of Nelson *et al.*, there exists in $\mathcal{H}$ a domain $\mathcal{D}$ on which it is permissible to work with the Lie algebra of $P_+^\uparrow$. A canonical basis for the Lie algebra is

$$\begin{array}{lll} [P_0,L] = 0 & [P_0,P] = 0 & [P_0,K] = P \\ [L,L] = iL & [L,P] = iP & [L,K] = iK \\ [P,P] = 0 & [P,K] = iP_0 & \\ [K,K] = -iL \quad , & & \end{array} \tag{6.13}$$

on $\mathcal{D}$.

Following non-relativistic quantum mechanics, we <u>identify</u> P_0,P and L with the physical energy, 3-momentum, and angular momentum, respectively, and call K, by analogy, the relativistic angular momentum. Thus, once again the operators play a dual role--group generators and physical observables.

Note that the relations (6.13) differ from the Galilean relations in only two respects,

$$[P,K] = M \rightarrow [P,K] = iP_0$$
$$[K,K] = 0 \rightarrow [K,K] = -iL \quad , \tag{6.14}$$

the first of which means that [P,K] maps back onto the algebra itself instead of onto a central extension.

We have already seen that the spectrum S of the generators of T_4 can be identified with the orbit 0. More precisely, if we denote the four generators of T_4 by P_μ, $\mu = 0,1,2,3$, they take values p_μ, $\mu = 0,1,2,3$, in the orbit ($p^2 = \pm m^2$,0). The orbit in a unitary irreducible representation of $P_+^\uparrow$ is, therefore, precisely the energy momentum spectrum. (Note that it makes sense to talk about the simultaneous spectrum of the P_μ since they commute on a domain $\mathcal{D}$ on which they are essentially self-adjoint.)

The identification of the orbits with the energy-momentum spectrum means that we can use direct physical arguments to decide which orbits and hence which CUIR's $\mathcal{H}$ should carry. Since physical mass-squared and energy are not negative, one usually makes the following assumptions about the energy-momentum spectrum and, hence, about the orbits:

1) $\mathcal{H}$ contains a unique normalizable ray (the vacuum state), which is invariant under $P_+^\uparrow$.
2) If there are no massless particles, the energy-momentum spectrum contains at least one isolated hyperbola (Figure 6.1) plus a continuum beginning at twice the height of the lowest hyperbola.
3) If there are massless particles present, the energy momentum spectrum fills the closed forward light cone.

On the isolated hyperboloids

$$P^2 = p^2 = \text{constant} \quad . \tag{6.15}$$

Furthermore, for each CUIR on such a hyperboloid, $\vec{P}$ takes all values in R_3. Hence, in contrast to the case of true unitary representations of the Galilean group, a position operator (Newton-Wigner operator) [5] satisfying $[X,P] = i\hbar$ can be defined. Hence, the CUIR's on the isolated hyperboloids can be identified with stable 1-particle states. The CUIR of the little group K(= SU(2) in the case when $p^2 > 0$) used to induce the CUIR of $P_+^\uparrow$ is then identified with the spin group of the particle. Thus, the spin group, which in non-relativistic quantum mechanics is introduced empirically and forms a direct

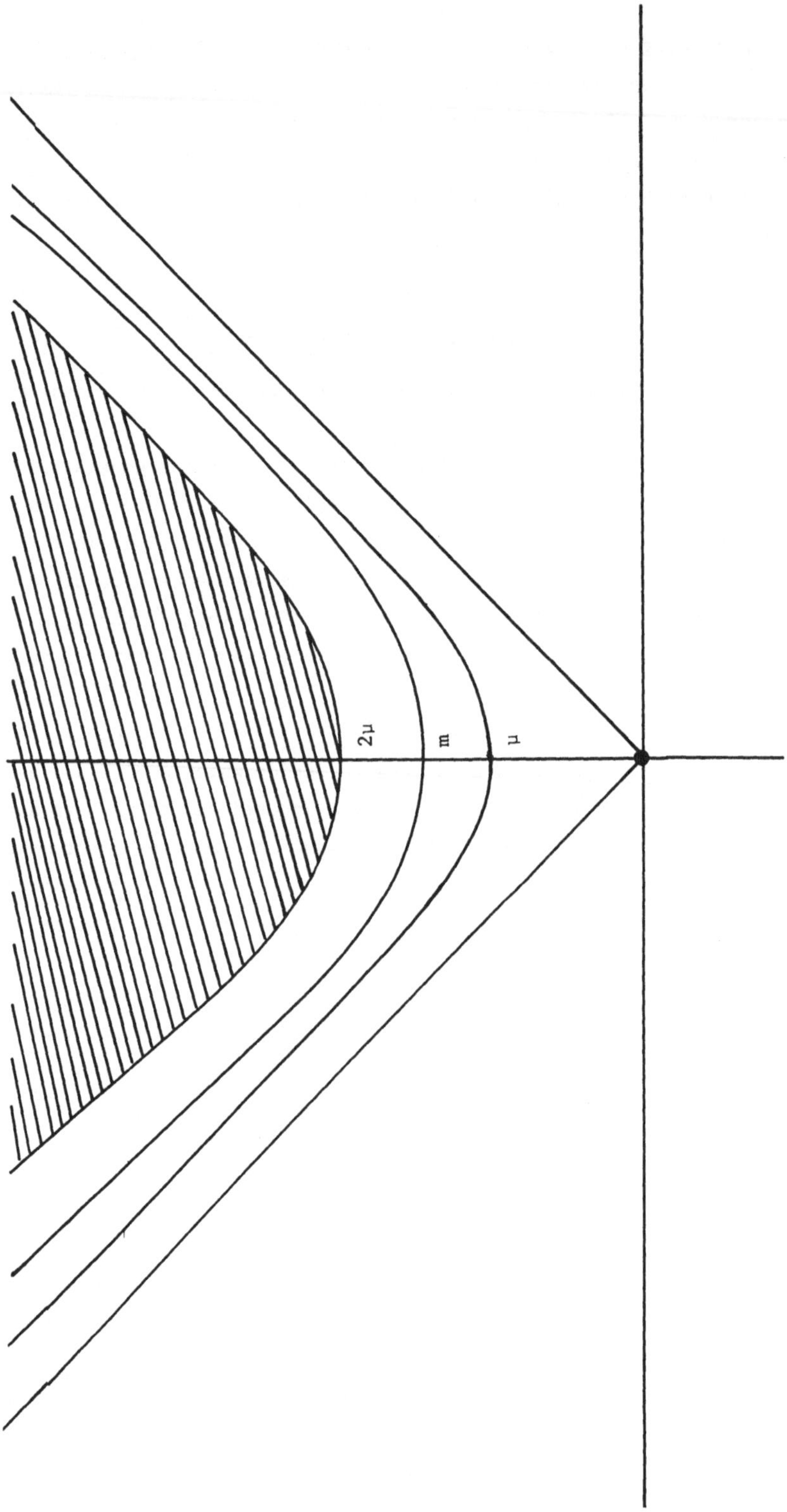

FIGURE 6.1. ENERGY-MOMENTUM SPECTRUM FOR MASSIVE PARTICLES

product with the Galilean group, is included automatically in the relativistic case.

Empirically, it is found that for any given mass there are only a finite number of elementary particles. Hence, the isolated hyperbolas are assumed to be finitely degenerate, i.e., to carry only a finite number of CUIR's of $P_+^\dagger$.

The continuum states in the energy-momentum spectrum represent, in general, two or more particle states, but may include 1-particle states which happen to have a higher mass than the lowest two particle states. In general, the continuum states are infinitely degenerate. In the case that zero mass particles are present, the continuum is everywhere in and on the forward light-cone and there is a serious problem as to how one should identify the 1-particle states, including the zero-mass particle states themselves. One possibility would be to identify them with normalizable, non-isolated, eigenvalues of P^2. But this is by no means the only possibility and, within some of the postulated frameworks, it is even impossible. [6]

From the point of view of the orbits of $P_+^\dagger$ allowed on $\mathcal{H}$, the energy-momentum spectrum conditions imposed are very strong. They reduce the six possible kinds of orbit which could be carried by $\mathcal{H}$ to the two kinds $p^2 \geq 0$, $p_0 \geq 0$. These orbits we shall call physical orbits. (They are actually characterized by $p_0 \geq 0$.)

The corresponding little groups are SU(2) and E(2). The CUIR's of SU(2) are well-known and require no comment. Those of E(2) are not so well-known, perhaps, but are actually simpler, as can be seen in the following way. The Lie algebra of E(2) is

$$[L_3, E_\alpha] = i\varepsilon_{\alpha\beta} E_\beta \quad ,$$
$$[E_\alpha, E_\beta] = 0 \quad , \qquad (6.16)$$

where $\alpha, \beta = 1,2$. It follows at once that $\exp(2i\pi L_3)$ and E^2 are the Casimir operators of E(2). Assuming that $\exp(2\pi i L_3) = \pm 1$ (i.e., integer or half-odd-integer values for L_3), it is then easy to see that there are only two possibilities:

a) $E^2 \neq 0$. The CUIR is infinite-dimensional and L_3 takes all integer or half-odd-integer values.

b) $E^2 = 0$. The CUIR is 1-dimensional and L_3 takes one integer or half-odd-integer value.

Case a), the so-called continuous spin case, does not seem to be realized in nature. Case b) is realized (it describes the photons and neutrinos for $L_3 = \pm 1$ and $L_3 = \pm \frac{1}{2}$, respectively). When Case b) does occur, it is usual to use a 2-dimensional reducible CUR of E(2) with $L_3 = \pm m$, rather than the 1-dimensional CUIR. This is to accommodate the parity operator.

Since $\mathcal{H}$ can carry only the physical orbits $p^2 \geq 0$, $p_0 \geq 0$, it follows that only the CUIR's of $P_+^\dagger$ corresponding to these orbits are directly related to physics. This does not mean that the other orbits are completely irrelevant. As we shall see later, they play an important role in the analyses of scattering amplitudes. The reason is that, in practice, one uses not only the matrix elements of operators on $\mathcal{H}$ themselves, but also the analytic continuation of these matrix elements, considered as functions of p_μ, to points other than those in the physical spectrum.

7. REDUCTION OF REPRESENTATIONS OF $P_+^\uparrow$ TO MANIFESTLY COVARIANT FORM

In the last section, it was shown that on quite general grounds the Hilbert space $\mathcal{H}$ of a relativistic quantum mechanical system must carry a CUR of $P_+^\uparrow$, and the CUIR's which this CUR could contain were described from the point of view of Mackey's theory. For a complete description of the elementary particles (origin of the masses and spins, nature of the interactions, etc.), however, much more is needed. For example, in a field theory, as we saw in Lecture 5, we need not only the Poincaré transformation properties of the field, but its commutation relations and interaction laws as well. The next step, therefore, is to try to relate the CUIR's of $P_+^\uparrow$ to other aspects of relativistic particle physics.

The question is: How is the contact between the group theoretical properties and the other physical properties to be made?

Traditionally, following non-relativistic quantum mechanics, Maxwell's theory, and Dirac's (non-second-quantized) relativistic quantum mechanics, the contact is made through wave functions $\psi(p)$ or fields $\psi(x)$ which transform in a manifestly covariant way, i.e.,

$$\psi(p) \xrightarrow{g} S(\Lambda)\psi(p')e^{ip\cdot a} \quad , \quad p = p\Lambda \quad , \tag{7.1}$$

where $g = (\Lambda,a)$, $\Lambda \in SL(2,C)$, $a \in T_4$, and $S(\Lambda)$ is a finite-dimensional representation of the homogeneous part $SL(2,C)$ of $P_+^\uparrow$. In the second-quantized theory of free particles of Dirac, Heisenberg, and Pauli, we have, as mentioned in Lecture 5, also the relation

$$\psi(x) = \int \frac{d^3p}{p_0} \{e^{ip\cdot x}a(p)\,\psi(p) + e^{-ip\cdot x}b^\dagger(p)\bar{\psi}(p)\} \quad , \tag{7.2}$$

between $\psi(x)$ and $\psi(p)$ where $a(p)$ and $b^\dagger(p)$ are the creation and destruction operators and where the fields $\psi(x)$ have local commutation relations and, when interactions are introduced, local interactions.

We shall follow the above tradition to the extent that we shall try to relate Mackey's method to manifestly covariant wavefunctions. [1] As we shall see for the physical orbits, this can always be done, and so it implies no restrictions. (Restrictions come when we try to relate the manifestly covariant wavefunctions to local fields, but that shall concern us only peripherally.)

We first recall Mackey's prescription for $P_+^\uparrow$ on $\mathcal{H}$:

a) Choose an orbit $p^2 = m^2 \geq 0$, $p_0 > 0$.

b) Choose a point $p = \alpha$ in the orbit.

c) Determine the little group K of α.

d) Let $H = T_4 \circledS K$.

e) Induce with $W(H) = e^{i\alpha a}V(K)$.

The induction procedure, we recall, is to choose the functions $f(g)$ over the group satisfying

1) $f(hg) = V(h)f(g)$

2) $\int d\mu(p)(f(g),g(g)) < \infty$

and letting the group act transitively on these functions,

3) $f(g_1) \xrightarrow{g_2} f(g_1 g_2)$.

We now make the transition from $f(g)$ to manifestly covariant wavefunctions in two steps.

<u>First Step</u>. We have a natural α-mapping $P_+^\uparrow \rightarrow$ Orbit given by $\alpha g = p$. We now define an inverse mapping $p \rightarrow P_+^\uparrow$ by introducing a <u>representative</u> Lorentz transform $\Lambda_0(\alpha,p) \in SL(2) \subset P_+^\uparrow$ for each p. The choice of $\Lambda_0(\alpha,p)$ is arbitrary but two standard ways of defining it are:

1) The canonical method: [2] Λ_0 is defined to be the unique Lorentz transformation in the 2-flat spanned by p and α.

2) The helicity method: [3] An arbitrary direction is chosen for the ζ-axis and Λ_0 is defined to be a pure Lorentz transformation in the ζ-direction to momentum $|\vec{p}|$, <u>followed</u> by a rotation from $(\varepsilon\ 0\ 0\ |\vec{p}|)$ to $(\varepsilon, \vec{p})$.

We then make the transformation

$$f(g) \rightarrow \phi(g) = V(\Lambda_0\Lambda^{-1})f(g) \quad , \tag{7.3}$$

where $g = (\Lambda,a)$, $\alpha\Lambda = \alpha\Lambda_0 = p$ and $V(\Lambda_0\Lambda^{-1})$ makes sense because $\Lambda_0\Lambda^{-1} \in K$. The point of this transformation is that, as is easily verified from Condition (1) and the relation $(k\Lambda)_0 = \Lambda_0$, which follows from the definitions of k and Λ_0, $\phi(g)$ satisfies the simpler subsidiary condition

1′) $$\phi(hg) = \phi(g)e^{i\alpha a} \quad ,$$

$$h = (k,a) \quad , \quad k \in K \quad , \quad a \in T_4 \quad .$$

Recalling that group multiplication is to the left, one sees at once from 1′) that $\phi(g)$ must be of the form

$$\phi(g) = \phi(\Lambda,a) = \theta(\Lambda)e^{i\alpha\cdot a}$$

where

$$\theta(\Lambda) = \theta(k\Lambda) \quad .$$

It follows that $\theta(\Lambda)$ is a function of p only, i.e.

$$\phi(g) = \theta(p)e^{i\alpha\cdot a} \quad .$$

Since V is unitary, the inner product remains unchanged.

$$2') \qquad \int d\mu(p)(f(g),f(g)) = \int d\mu(p)(\theta(p),\theta(p)) \quad . \tag{7.4}$$

The group multiplication law changes, however. In place of the simple transitivity 3), we obtain

$$3') \qquad \theta(p) \xrightarrow{g} V(\Lambda_0\Lambda\Lambda_0'^{-1})\theta(p')e^{ip\cdot a} \tag{7.5}$$

where

$$g = (\Lambda,a) \quad , \quad \Lambda_0 = \Lambda_0(\alpha,p) \qquad \Lambda_0' = \Lambda_0(\alpha,p') \quad , \quad p' = p\Lambda \quad .$$

Note that $V(\Lambda_0\Lambda\Lambda_0'^{-1})$ makes sense since $\Lambda_0\Lambda\Lambda_0'^{-1} \in K$. The rotations $V(\Lambda_0\Lambda\Lambda_0'^{-1})$ are called Wigner rotations. We see that, in effect, what we have done essentially is to change the "twist" introduced by $V(k)$ from the subsidiary condition to the group transformation.

For many purposes, the wavefunctions $\theta(p)$ are the most convenient. For example, the standard analysis of scattering amplitudes for general spin carried out by Jacob and Wick [3] is done in terms of $\theta(p)$. However, if we wish for manifest covariance, we must go farther. The transformation law (7.5) is not manifestly covariant on two counts:

1) It depends explicitly on p.

2) V is a representation of the little group, not $SL(2,C)$.

This brings us to Step 2.

Second Step. Elimination of the p-dependence from the transformation (7.14).

The basic idea underlying Step 2 is to modify $V(\Lambda_0\Lambda\Lambda_0'^{-1})$ so that it can be split into $V(\Lambda_0)V(\Lambda)V(\Lambda_0')^{-1}$. At the moment, $V(\Lambda_0)$, etc., make no sense since Λ_0, Λ, and Λ_0' are not separately in K. The modification is achieved by embedding $V(k)$ in any representation $S(\Lambda)$ of $SL(2,C)$ which is unitary when restricted to K. Letting $V_\lambda(K)$ be the representations of K occurring in $S(\Lambda)$, we define a <u>set</u> of wavefunctions $\theta_\lambda(p)$ (including $\theta(p)$) with the transformation law

$$\theta_\lambda(p) \overset{\Lambda,a}{\rightarrow} V_\lambda(\Lambda_0\Lambda\Lambda_0'^{-1})\theta_\lambda(p')e^{ip\cdot a} \tag{7.6}$$

In other words, we induce with the <u>reducible</u> representation $\sum \oplus V_\lambda$ of K (all on the orbit $p^2 = m^2$).

Now by definition

$$S_{\lambda\mu}(K) = \delta_{\lambda\mu}V_\lambda(K) \quad . \tag{7.7}$$

Hence, (7.6) can be written as

$$\theta_\lambda(p) \xrightarrow{\Lambda,a} S_{\lambda\mu}(\Lambda_0\Lambda\Lambda_0'^{-1})\theta_\mu(p')e^{ip\cdot a} \quad . \tag{7.8}$$

But since $S(\Lambda)$ makes sense, we then have

$$\theta_\lambda(p) \xrightarrow{\Lambda,a} [S(\Lambda_0)S(\Lambda)S^{-1}(\Lambda_0')]_{\lambda\mu}\theta_\mu(p')e^{ip\cdot a} \quad , \tag{7.9}$$

or

$$S^{-1}(\Lambda_0)\varphi(p) \xrightarrow{\Lambda,a} S(\Lambda)S^{-1}(\Lambda_0')\varphi(p')e^{ip\cdot a} \quad , \tag{7.10}$$

where

$$\varphi(p) = \sum \oplus \, \theta_\lambda(p) \quad .$$

Remembering that Λ_0 depends only on α and p, we see that (7.10) is equivalent to

$$\psi(p) \xrightarrow{\Lambda,a} S(\Lambda)\psi(p')e^{ip\cdot a} \quad , \tag{7.11}$$

where

$$\psi(p) = S^{-1}(\Lambda_0(\alpha),(p))\varphi(p) \quad . \tag{7.12}$$

Equation (7.11) has the required manifestly covariant transformation properties. Note that in the manifestly covariant formulation the Lie algebra of $P_+^\uparrow$ takes the simple form

$$P_\mu = p_\mu, \; L_{\mu\nu} = \frac{1}{i}\left(p_\mu \frac{\partial}{\partial p_\nu} - p_\nu \frac{\partial}{\partial p_\mu}\right) + S_{\mu\nu} \quad , \tag{7.13}$$

where $S_{\mu\nu}$ are the generators of $S(\Lambda)$ and $L_{\mu\nu} = (\vec{L},\vec{K})$.

Equation (7.13) shows that in the manifestly covariant formulation, $L_{\mu\nu}$ splits into the direct sum of an "orbital" part and a "spin" part $S_{\mu\nu}$.

For the manifest covariance, we have, however, to pay a heavy price:

1) The representation $S(\Lambda)$ of $SL(2,C)$ is arbitrary.

2) We have introduced the unwanted subsidiary fields

$$\theta_\lambda(p) \neq \theta(p) \quad .$$

3) Since $S(\Lambda)$ is, in general, not unitary, the inner product must be changed accordingly.

Let us discuss these points in turn:

1) The representation $S(\Lambda)$ in (7.10), which is usually called the spin group, is completely arbitrary. It is usually chosen to be a

finite-dimensional (non-unitary) representation of SL(2,C) and as we shall be considering infinite dimensional spin groups in the next section, let us concentrate on the finite dimensional case. Even for the finite-dimensional representations, there is much arbitrariness. All choices of $S(\Lambda)$ will, of course, be the same from the point of view of the original CUIR of $P_+^\dagger$. But they will not necessarily be the same from other points of view. For example, an interaction which involves no derivatives for one choice of $S(\Lambda)$ will have derivatives for another. Indeed for spin ≥ 1, the whole question of choosing the correct $S(\Lambda)$ is very much open. [4,5]

2) With regard to the subsidiary fields $\theta_\lambda(p)$, the point is that they should be eliminated in a manifestly covariant way. That this is possible for $P_+^\dagger$ and finite-dimensional $S(\Lambda)$ follows from the following two properties of SL(2,C): (a) The $\psi(p)$ for every irreducible finite-dimensional representation $D(n,m)$ of SL(2,C) is of the form

$$\psi_{\dot\alpha_1\ldots\dot\alpha_n;\,\beta_1\ldots\beta_m}(p) \quad , \tag{7.14}$$

where the β are 2-valued indices belonging to the fundamental 2-dimensional representation, the $\dot\alpha$ are similar 2-valued indices for the conjugate representation, and ψ is completely symmetric in the α and β, respectively [6]. (b) p_μ is of the form $p_{\dot\alpha\beta}$, and hence if $p_{\dot\alpha\beta}$ is contracted with $\psi_{\dot\alpha_1\ldots\dot\alpha_n;\,\beta_1\ldots\beta_m}(p)$ to remove either all the undotted or dotted indices, the remaining indices carry an <u>irreducible</u> representation of SL(2,C). These two properties canbe used in an obvious way to project out, with polynomials in p, the parts of $\psi_{\dot\alpha_1\ldots\dot\alpha_n;\,\beta_1\ldots\beta_m}(p)$ which are irreducible with respect to $P_+^\dagger$. The use of multispinors (7.14) is due originally to Fierz and Pauli [7].

3) With regard to the inner product, for finite dimensional-representations of SL(2,C), which carry a parity operator, the situation is saved by the fact that although $S(\Lambda)$ is not unitary, it is pseudo-unitary, i.e., there exists a metric η in $S(\Lambda)$-space such that

$$\begin{aligned} S^\dagger(\lambda)\eta S(\lambda) &= \eta \quad , \quad \lambda \in \Lambda \quad , \\ [S(k),\eta] &= 0 \quad , \quad k \in K \quad , \\ \eta &= \eta^\dagger = \eta^{-1} \quad , \end{aligned} \tag{7.15}$$

where the adjoint is with respect to the V(K) space. In fact, η is just the spinspace part of the parity operator. Hence the inner product

$$\overline{\psi}_1(p)\psi_2(p) = (\psi_1(p),\eta\psi_2(p)) \tag{7.16}$$

is SL(2,C) invariant and Mackey's inner product can be replaced by

$$\int d\mu(p)\overline{\psi}_1(p)\psi_2(p) \quad , \tag{7.17}$$

which is manifestly invariant i.e. invariant under $P_+^\uparrow$ and $S(\Lambda)$ separately. Note that

$$\int d\mu(p)\overline{\psi}(p)\psi(p) \quad , \tag{7.18}$$

is positive-definite on account of the subsidiary conditions.

We conclude this chapter with some examples of manifestly covariant fields.

a) On the orbit $p^2 = m^2$, we choose a $D(n,n)$ representation of $SL(2,C)$. The corresponding field $\psi_{\dot{\alpha}_1\ldots\dot{\alpha}_n;\,\beta_1\ldots\beta_n}(p)$ carries the spin $j = 0, 1, 2, \ldots n$ representations of $P_+^\uparrow$. We can eliminate the spins $j = 0, 1, 2, \ldots n-1$ by the manifestly covariant subsidiary conditions

$$p^{\dot{\alpha}_1\beta_1}\psi_{\dot{\alpha}_1\ldots\dot{\alpha}_n;\,\beta_1\ldots\beta_n}(p) = 0 \quad . \tag{7.19}$$

We usually see this field in its traceless symmetric tensor form $\psi_{\mu_1\ldots\mu_n}(p)$ with the subsidiary conditions $p^2 = m^2$,

$$p^{\mu_1}\psi_{\mu_1\ldots\mu_n}(p) = 0 \quad , \tag{7.20}$$

b) On $p^2 = m^2$ we choose a $D(n,n+1)$ representation

$$\psi_{\dot{\alpha}_1\ldots\dot{\alpha}_n\beta_1\ldots\beta_n\beta_{n+1}} \quad . \tag{7.21}$$

This carries the spins $j = \frac{1}{2}, \frac{3}{2} \ldots n + \frac{1}{2}$, and we can eliminate the lower spins by the subsidiary condition

$$p^{\dot{\alpha}_1\beta_1}\psi_{\dot{\alpha}_1\ldots\dot{\alpha}_n;\,\beta_1\ldots\beta_n\beta_{n+1}} = 0 \quad . \tag{7.22}$$

Again, one can use vector notation and replace (7.21) by the field

$$\psi_{\mu_1\ldots\mu_n\alpha}(p) \quad , \tag{7.23}$$

with the subsidiary conditions

$$p^{\mu_1}\psi_{\mu_1\ldots\mu_n\alpha}(p) = 0 \quad ,$$

and

$$\tau^{\alpha\beta}_{\mu_1}\psi_{\mu_1\ldots\mu_n\beta}(p) = 0 \quad , \tag{7.24}$$

where the τ_μ are the Pauli and unit 2×2 matrices.

c) Because the field $\psi_{\mu_1\ldots\mu_n\alpha}(p)$ for $\alpha = 1,2$ does not accommodate parity, it is customary to replace $\alpha = 1,2$ by a Dirac index $\alpha = 1,2,3,4$. The subsidiary conditions then become

$$(\gamma^\mu p_\mu + m)\psi_{\mu_1\ldots\mu_n\alpha} = 0 \quad , \quad \gamma^{\mu_1}\psi_{\mu_1\ldots\mu_n\alpha} = 0 \quad . \tag{7.25}$$

These equations are known as the Rarita-Schwinger equations, [8] and describe spin $j = n + \frac{1}{2}$.

d) One can similarly use $\psi_{\alpha_1 \ldots \alpha_n}(p)$ where $\alpha_r = 1,2,3,4$ are Dirac indices, with the subsidiary conditions

$$(\gamma^{\mu}_{(r)} p_{\mu} + m)\psi_{\alpha_1 \ldots \alpha_n}(p) = 0 \quad , \quad r = 1 \ldots n \quad . \tag{7.26}$$

These fields carry spin $\frac{1}{2}(n + 1)$ and the subsidiary conditions are known as the Bargmann-Wigner [9] equations. The Rarita-Schwinger and Bargmann-Wigner equations automatically include the orbit condition $p^2 = m^2$.*

A simpler and somewhat more general approach to the results of Section 7 will appear in the Proceedings of the 1970 Istanbul Nato Summer School in Mathematical Physics.

8. INFINITE COMPONENT WAVE FUNCTIONS

In the last section, we saw that <u>any</u> representation of the Lorentz spin group $SL(2,C)$ whose restriction to the little group was unitary could be used to product a manifestly covariant unitary representation of $P_+^{\uparrow}$. We then devoted our attention to the finite-dimensional (non-unitary) spin groups. In the literature also, attention has been devoted almost entirely to finite-dimensional spin groups. In this section we wish to discuss why this is so.

In the first place, there are good historical precedents for using finite-dimensional spin groups, since the classical fields of Newton, Maxwell, Einstein, and Dirac are of this form (they use the finite-dimensional $D(00)$, $D(10) + D(01)$, $D(1,1)$ and $D(\frac{1}{2}0) + D(0\frac{1}{2})$ representations of $SL(2,C)$, respectively).

Secondly, in particle physics, each of the particles one wishes to describe is known empirically to have finite spin. Hence, it is natural to use a finite-dimensional spin group to describe it.

On the other hand, one could legitimately ask the question:

1) Since in the spin group a number of superfluous representations of the little group appear anyway and are eliminated by subsidiary conditions, why not use an infinite dimensional spin group plus infinite dimensional subsidiary conditions?

---

* Note added in proof: In Step 1 of this chapter, if one wishes to avoid the explicit decomposition of g into (Λ,a) one can do so by defining $\theta(p)$ according to the equation

$$\theta(p) = \theta(g) = W(g_0 g^{-1})f(g) \quad , \quad g_0 = (\Lambda_0, 0) \quad .$$

Also, if one is interested only in the final manifestly covariant form (7.11) and wishes to eliminate Step 1, one can do so by letting $f(g) \to \sum_{\lambda} f_{\lambda}(g) = F(g)$, and defining $\psi(p) = \psi(g) = S^{-1}(\Lambda)e^{-i\alpha\Lambda.a}F(g)$.

2) Since what we observe experimentally is, in any case, not just one particle but the infinite family of particles suggested by Figure 5.1, why not go the whole hog and try to describe all of the particles, or at least large sub-families of them, by means of a single covariant field. This field, in order to carry an infinite number of UIR's of $P_+^\uparrow$, would have to correspond to an infinite dimensional representation of SL(2,C).

The possibility raised by Question 2) is even highly attractive. What we shall show, however, is that the attraction is deceptive and that infinite spin groups lead to difficulties which, at present at any rate, seem to be unsurmountable. We shall do this first for two special models, and then present a general no-go theorem which has been proved recently.

The difficulties come under two headings:

a) Violation of the spectral condition $p^2 \geq 0$

b) Violation of locality for quantized fields.

We first illustrate a) for two special models.

The first model we consider dates back to 1932 and was proposed by Majorana [1] as a possibility for avoiding the "negative energy" states of Dirac's theory, which were thought to be an embarrassment at that time. Majorana proposed that one use a wave function, with spin group corresponding to the $(j_0 = \frac{1}{2}, c = 0)$ or $(j_0 = 0, c = \frac{1}{2})$ UIR of SL(2,C) and satisfying the subsidiary condition

$$(\Gamma_\mu p_\mu - \kappa)\psi(p) = 0 \quad , \tag{8.1}$$

where κ is a positive number and Γ_μ is a p-independent SL(2,C) vector. (The "Majorana representations" $(j_0 = \frac{1}{2}, c = 0)$ and $(j_0 = 0, c = \frac{1}{2})$ are the only <u>irreducible</u> UR's of SL(2,C) to carry a vector operator.)

The question then is: What UIR's of $P_+^\uparrow$ does Majorana's $\psi(p)$ carry?

To answer it, consider an orbit $p^2 = m^2 > 0$, $p_0 > 0$. Such an orbit would contain the vector $\alpha = (m000)$ whence from (8.1)

$$(\Gamma_0 m - \kappa)\psi(\alpha) = 0 \quad , \tag{8.2}$$

which is possible if and only if m is equal to one of the eigenvalues of κ/Γ_0. The eigenvalues of κ/Γ_0 turn out [2] to be $\kappa(j_0 + \frac{1}{2} + n)^{-1}$, $n = 0,1,2,3\ldots$ Thus, Majorana's $\psi(p)$ carries the orbits $p^2 = m^2$, $m = \dfrac{\kappa}{j_0 + \frac{1}{2} + n}$, $p_0 > 0$.

Furthermore, the little group of such an orbit is SU(2) and, in the reduction of Majorana representations with respect to SU(2), each representation $j = j_0, j_0 + 1, \ldots$ of SU(2) occurs once and only once, with

$$\Gamma_0 = j + \frac{1}{2} = j_0 + \frac{1}{2} + n \quad . \tag{8.3}$$

Hence, each orbit $m = \kappa/\Gamma_0 = \kappa/(j + \frac{1}{2})$ carries exactly one UIR of $P_+^\uparrow$, and we have the mass-spin relationship

$$m = \frac{\kappa}{J + \frac{1}{2}} \quad . \tag{8.4}$$

Experimentally this mass-spin relationship is disastrous, but that is not a real problem as it could easily be modified. For example, by replacing κ by κp^2 in the subsidiary condition, it could be inverted, which would be very good experimentally.

The real difficulty comes from the non-physical orbits $p^2 = m^2 < 0$. These exist because they can be generated from vectors $\alpha = (000m)$ for which (8.1) is equivalent to

$$(\Gamma_3 m - \kappa)\psi(\alpha) = 0 \quad , \tag{8.5}$$

and this equation has non-trivial solutions since Γ_3 is self-adjoint. (Note that in this connection the unitarity of the Majorana representation of SL(2,C) is actually a liability, since it implies that if Γ_0 is self-adjoint, then so is Γ_3; the above argument would have broken down for the finite-dimensional non-unitary Dirac representation of SL(2,C), for which γ_0 is hermitian but γ_3 is not.)

The $p^2 < 0$ orbits are undesirable but are not an immediate catastrophe for the Majorana equation since they could simply be ignored. The trouble is that, in practice, one is interested not merely in the free Majorana particles, but also in their interactions. For example, if we try to introduce the EM interaction by means of the traditional minimum principle

$$(\Gamma_\mu p_\mu - \kappa)\psi(p) = 0 \rightarrow (\Gamma_\mu p_\mu - \kappa)\psi(p) = \Gamma_\mu A_\mu(k)\psi(p + k) \quad , \tag{8.6}$$

where A_μ is the vector potential and k the momentum transfer, one can show that for $k \neq 0$ the system makes transitions from $p^2 > 0$ to $p^2 < 0$ states, and similarly for any other interactions which are local in the Fourier transformed space. Now, of course, one might do better with some more complicated, non-local interaction. But since the purpose of the manifestly covariant wavefunctions is to provide a framework for introducing simple, local commutation relations and interactions, this would defeat the purpose. For this reason, the $p^2 < 0$ orbits are a real difficulty in Majorana's theory.

The second model we consider is a wavefunction $\psi(p)$ carrying a Dirac $\otimes$ unitary spin representation and satisfying the subsidiary condition

$$(\gamma \cdot p + M)\psi(p) = 0 \quad , \tag{8.7}$$

where M is a spin invariant, e.g.,

$$M = m_0 + m_1 \sigma_{\mu\nu}\Sigma_{\mu\nu} \quad ,$$

where m_0, m_1 are constants and $\sigma_{\mu\nu}$ and $\Sigma_{\mu\nu}$ are the generators of the Dirac and unitary representations, respectively. This equation was first studied by Abers Grodsky and Norton [3] (AGN) in 1965 and has since been used in current algebra theory. An analysis of the equation, similar to that described above for the Majorana equation, for the case in which the unitary representation is $(j_0, C = 0)$, shows that for the $p^2 > 0$ orbits there is a mass-spin relationship

$$\pm m = m_1(J + \tfrac{1}{2}) \pm \{(m_0 - m_1)^2 + m_1^2[J(J + 1) - j_0(j_0 + 1) - \tfrac{3}{4}]\} \quad , \qquad (8.8)$$

which can be drawn graphically as in Figure 8.1. The rising curve for $m > 0$ fits well with the observed particles (and with Regge theory, which we shall be describing later). However, the falling curve for $m > 0$ has no satisfactory interpretation. (The $m < 0$ curves can be identified with anti-particles.)

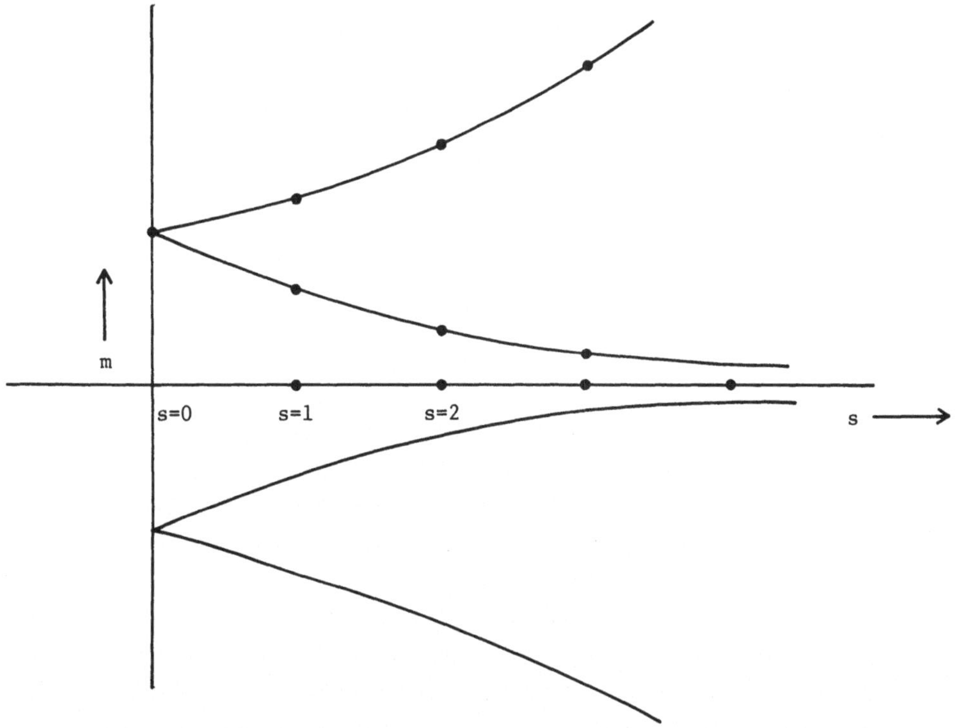

FIGURE 8.1. MASS-SPIN RELATIONSHIP FOR AGN EQUATION

Leaving aside the interpretation of the falling curve, we ask again whether $\psi(p)$ carries unphysical $p^2 < 0$ orbits. The answer is yes. The proof is perhaps worth giving.

Proof. Write the subsidiary condition (8.7) in the form

$$(\vec{\alpha} \cdot \vec{p} + \beta M)\psi(p) = p_0\psi(p) \quad , \qquad (8.9)$$

where $\vec{\alpha}$ and β are the self-adjoint Dirac matrices $\gamma_0\vec{\gamma}$ and γ_0, respectively. Now βM must be self-adjoint to provide a mass spectrum in the rest frame $\vec{p} = 0$, and $\vec{\alpha} \cdot \vec{p}$ is self-adjoint and bounded. Hence, for each $\vec{p}$, $\vec{\alpha} \cdot \vec{p} + \beta M$ is self-adjoint. Hence, (8.9) may be regarded as an eigenvalue equation for the self-adjoint operator $\vec{\alpha} \cdot \vec{p} + \beta M$, i.e., p_0 is any point in the spectrum of $\vec{\alpha} \cdot \vec{p} + \beta M$.

The condition that there be no $p^2 < 0$ orbits is that $p_0^2 \geq \vec{p}^2$. But since p_0 is any point in the spectrum of $\vec{\alpha} \cdot \vec{p} + \beta M$, this implies that

$$(\vec{\alpha} \cdot \vec{p} + \beta M)^2 \geq \vec{p}^2 \quad , \tag{8.10}$$

or, since $\alpha^2 = 1$,

$$\vec{p} \cdot [\vec{\alpha},\beta M]_+ + (\beta M)^2 \geq 0 \quad . \tag{8.11}$$

But since $\vec{p}$ varies over the whole Euclidean 3-space, this is possible if and only if

$$[\vec{\alpha},\beta M]_+ = 0 \quad , \tag{8.12}$$

which on account of the anti-commutativity of $\vec{\alpha}$ and β, reduces to

$$[\vec{\alpha},M] = 0 \quad . \tag{8.13}$$

But since

$$\sigma_{\mu\nu} = \vec{\alpha}, \vec{\alpha} \times \vec{\alpha} \quad ,$$

this means that $p^2 \geq 0$ is possible if and only if

$$[\sigma_{\mu\nu},M] = 0 \quad , \tag{8.14}$$

i.e., if and only if M is a Dirac invariant, in which case equation (8.7) can be reduced to a direct sum of Dirac equations with M = constant.

Thus, the AGN equation, like the Majorana, is either trivial or contains unphysical orbits $p^2 < 0$ and, once again, it can be checked that local interactions couple the physical orbits to unphysical ones.

Note that the $p^2 < 0$ difficulties arise whether or not the wavefunction $\psi(p)$ is quantized. If the field is quantized, then there are the further difficulties (b) concerning locality. To illustrate the point, consider an infinite component wavefunction $\psi(p)$ which has not yet been quantized, introduce a set of creation and destruction operators for particles satisfying Bose-Einstein or Fermi-Dirac statistics on a Hilbert space $\mathcal{H}$, i.e., satisfying

$$[a(p),a^+(p')]_\pm = \delta(p - p') \quad , \tag{8.15}$$

etc., and construct from $\psi(p)$ and $a(p)$ a quantized field in the standard way, namely,

$$\varphi(x) = \int d\mu(p)\{e^{ip\cdot x}a(p)\psi(p) + e^{-ip\cdot x}b^+(p)\tilde{\psi}(p)\} \quad , \tag{8.16}$$

where $\tilde{\psi}(p)$ transforms like $\psi(p)$. The locality difficulties can be subdivided into

a) locality proper

b) spin-statistics

c) CPT-invariance

d) analyticity.

Locality proper is the question whether the commutator

$$[\varphi(x),\varphi(x')] \quad , \tag{8.17}$$

vanishes for $(x - x')^2 < 0$. In the finite-dimensional case, the commutator does vanish for suitable choice of $\pm$ in (8.16). In the infinite dimensional case, however, in general no choice of sign in (8.15) and no simple modification will make (8.17) vanish. The possibilities for evading this difficulty have been investigated in some detail in the recent literature [4], but with no particularly attractive solution.

The spin-statistics difficulty is an extension of the problem: In the finite-dimensional cases, (8.17) vanishes for $\pm$ in (8.15), but the choice of $\pm$ is not arbitrary. It must be (+) (Fermi-Dirac statistics) if the field carries half-odd-integer spin and (-) (Bose-Einstein statistics) if the field carries integer spin, a correlation which is verified experimentally and is regarded as one of the most fundamental results of quantum field theory. But in the infinite dimensional case, since (8.17) does not vanish for either choice of sign, the spin-statistics correlation gets lost. (In the cases that (8.17) can be made to vanish, it can be made to vanish for <u>either</u> choice of sign, so the correlation becomes, at best, arbitrary.)

The other two difficulties, CPT invariance and analyticity, are special cases of the general result that for finite-dimensional spin groups, the Lorentz transformations can be continued to any complex values of the parameters whereas for infinite dimensional spin representations, this is not the case. (Infinite-dimensional representations of SL(2,C) have dense sets of analytic vectors, but no entire vectors.) As a result, the EM form factors and the scattering matrix S have different analytic properties (as functions of the inner products of the momenta) in the finite and infinite-dimensional cases, and the analytic properties in the infinite-dimensional case do not seem to be the most desirable.

All models so far constructed using infinite-dimensional representations of SL(2,C) have been found to be unsatisfactory in at least some of the above ways. This suggests that it might be possible to rule out infinite component fields on quite general grounds and, thus, restrict oneself to the finite-dimensional spin representations without any real loss in generality.

One such general set of conditions was found recently by Streater and Grodsky [5]. Their argument is as follows:

Let $\varphi(\sigma,x)$ be an infinite component field operating on a physical Hilbert space $\mathcal{H}$ with vacuum state h, and carrying a continuous bounded irreducible infinite dimensional spin group, $S_{\lambda\sigma}$. Rather than specify precisely how $\varphi(\sigma,x)$ is quantized, they assume only that it has been quantized in such a way that the vacuum expectation value

$$F(\sigma,\sigma',x,x') = (0,\varphi^{\dagger}(\sigma,x)\varphi(\sigma',x')0) \quad , \tag{8.18}$$

with unique vacuum state 0), has the following properties:

a) Translational invariance: $F(\sigma,\sigma',x,x') = F(\sigma,\sigma',x - x')$

b) Reasonable spectrum: $\tilde{F}(\sigma,\sigma',p) = 0$ for $p^2 < 0$, where $\tilde{F}$ denotes Fourier transform

c) Causality (locality): $F(\sigma,\sigma'x) = 0$ for $x^2 < 0$

d) Temperedness: $F(\sigma,\sigma',x)$ is a tempered distribution in x for all σ,σ'

e) Finite degeneracy of the lowest isolated mass-hyperboloid.

These are all assumptions that are made normally in quantum field theory. The temperedness assumption is a strengthening of locality (it implies that $f(\sigma,\sigma',x)$ is not too singular on the light cone) and, although this assumption can be relaxed, it cannot be relaxed very much if the correct analyticity properties are to be obtained for the S-matrix.

Grodsky and Streater now claim that these assumptions are incompatible. To prove this, they make use of a theorem due to Bogoliubov and Vladimirov [6] which states that if $f(x)$ is a tempered distribution with $f(x) = 0$ for $x^2 < 0$ and the Fourier transform $\tilde{f}(p) = 0$ for $p^2 < 0$, then $\tilde{f}(p)$ is a finite covariant, i.e., $\tilde{f}(p)$ has the representation

$$\tilde{f}(p) = \sum_{[n]} C_{[n]} p_0^{n_0} \cdots p_3^{n_3} \int dm^2 \rho_{[n]}(m^2)\delta(p^2 - m^2) \quad , \tag{8.19}$$

where the sum is finite, $[n] = [n_1 n_2 n_3 n_0]$ and $\rho_{[n]}$ is tempered. Applying this theorem to $F(\sigma,\sigma',x)$, which obviously satisfies the conditions, and smearing with a test function $f(x)\sim\tilde{f}(p)$ with support only in the neighborhood of the lowest mass-hyperboloid in p-space, one obtains

$$(0,\varphi^{\dagger}(\sigma,f)\varphi(\sigma',f)0) = \text{Const.}\ \delta^4(p' - p) \sum_{[n]} C_{[n]}(\sigma,\sigma') p_0^{n_0} \cdots p_3^{n_3} \quad .$$

But since the spin-representation is assumed to be continuous, $C_{[n]}(\sigma,\sigma')$ is continuous in σ and σ'. Hence, $C_{[n]}(\sigma,\sigma')$ is the matrix element of a bounded linear operator in spin space V. Hence, for fixed σ', $C_{[n]}(\sigma,\sigma')$ may be regarded as a vector in V and since there are only a finite number of $C_{[n]}$, the linear span

$$\sum_{[n]} C_{[n]}(\sigma,\sigma') p_0^{n_0} \cdots p_3^{n_3} \quad , \tag{8.20}$$

for all $p_0 \dots p_3$ and fixed σ', is finite dimensional. It follows that the expression (8.20) vanishes for an infinite number of values of σ. Referring back to (8.18), we see that there are, therefore, an infinite number of states $\varphi(\sigma,f)0)$ in $\mathcal{H}$, orthogonal to the state $\varphi(\sigma' f)0)$ for all $\vec{p}$ and $\vec{p}'$. Furthermore, since the spin group is irreducible, $\varphi(\sigma,f)0)$ vanishes if and only if $\varphi(\sigma',f)0)$ vanishes. It follows that the orthogonal states are not zero. Thus, the lowest mass-hyperboloid is infinitely degenerate. This is the result of Grodsky and Streater.

A corollary to their result, which has been pointed out by Grodsky and Streater, is that since any field $\psi(x)$ which is obtained by quantizing in the usual manner (8.16), a wavefunction $\psi(p)$ whose support is in $p^2 > 0$ and whose SL(2,C)-space projection on $p_0 > 0$ is polynomially bounded in p, will be automatically tempered and causal, it must belong to a finite dimensional representation of SL(2,C).

What does this result mean physically? It means that if we use infinite-dimensional representations of SL(2,C) one of two things must happen. Either the subsidiary conditions imposed on the wavefunctions are too weak, in which case there is an infinite number of spin states on each mass-hyperboloid (in gross contradiction to experiment), or else the subsidiary conditions are too strong (as in the Majorana and AGN cases discussed above). In that case, there is no spin degeneracy but the wavefunction cannot be quantized so as to describe a tempered local field with $p^2 > 0$.

Note that the temperedness of the distribution plays a critical role in the above arguments. It leads directly to the <u>finiteness</u> of the expansion (8.19), which leads in turn to the finiteness of the linear span (8.20) and hence to the infiniteness of the orthogonal complement. (Note added in proof: a generalization of the GS theorem which allows more general distributions, including Jaffe distributions, is now available [7].)

Perhaps the best way to summarize the results of this chapter is to say that while there are no <u>group-theoretical</u> reasons for excluding infinite spin groups, there appear to be other reasons to exclude them, namely, mass-spectrum, locality, and finite-spin degeneracy considerations. Thus, one can return, (with some relief!) to the finite dimensional spin representations.

9. LITTLE GROUP DECOMPOSITION OF THE SCATTERING AMPLITUDE

In the last couple of chapters we saw how the Poincaré group $P_+^\uparrow$ and its little group for $p^2 \geq 0$ could be used to characterize relativistic particles. In this chapter I should like to mention briefly how $P_+^\uparrow$ and its little group can be used to analyze scattering processes. One of the interesting features will be that, in spite of the spectral condition, the SU(1,1) little group for the orbits $p^2 < 0$ will also be relevant.

To put the role of the little groups into perspective, we consider the scattering amplitude (Figure 9.1) for 2-particles scattering into 2 particles (not necessarily the same), e.g. $\pi N \to \Sigma K$. The probability of the particles 1 and 2

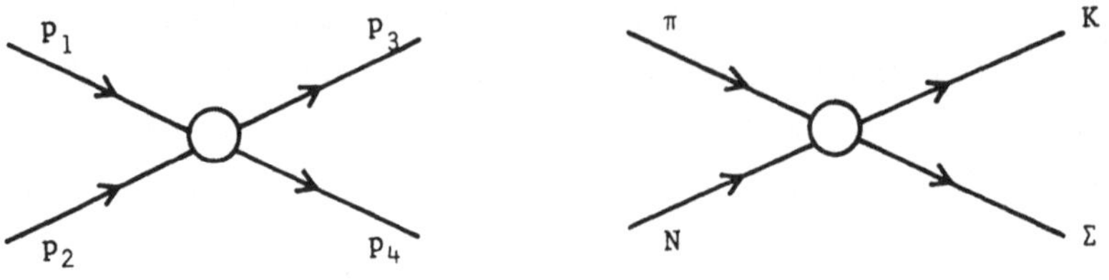

FIGURE 9.1. SCATTERING IN S-CHANNEL

with momenta p_1 and p_2 scattering into particles 3 and 4 with momenta p_3 and p_4 is given by

$$P(p_1p_2 \to p_3p_4) = |(p_3p_4, T\, p_1p_2)|^2 \quad , \tag{9.1}$$

where T is the scattering matrix. Because of Poincaré invariance, the scattering amplitude $(p_3p_4, T\, p_1p_2)$ is (apart from some kinematical factors, which we omit) a function of two invariant variables, s and t

$$(p_3p_4, T\, p_1p_2) = F(s,t) \quad , \tag{9.2}$$

where

$$s = (p_1 + p_2)^2 \quad , \quad t = (p_1 - p_3)^2 \quad . \tag{9.3}$$

For symmetry we can also define $u = (p_1 - p_4)^2$, but u is not an independent variable. In fact $u + s + t = \sum_{\alpha=1}^{4} m_\alpha^2$, where m_α are the masses. (In general, the scattering amplitude for 2 particles into n-2 particles depends on 3n - 10 invariant variables, the 3n variables being the n 3-momenta of the n particles involved, the ten constraints coming from the conservation of the ten generators of $P_+^\uparrow$.) If the four particles involved in the scattering of Figure 9.1 are spinless (as we shall assume for simplicity) then F is a scalar function.

Now consider the process of Figure 9.2, namely the scattering of particles

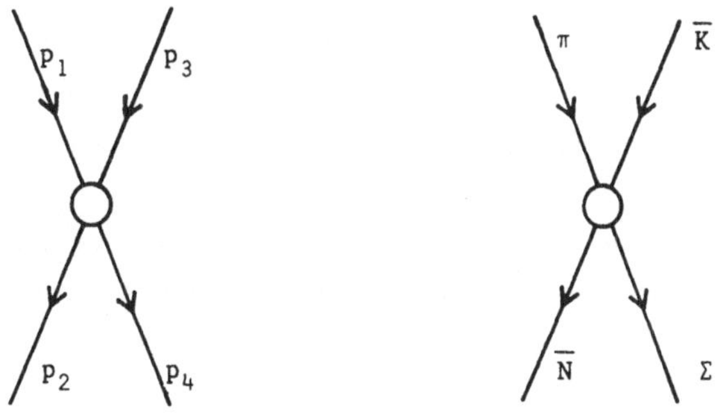

FIGURE 9.2. SCATTERING IN S-CHANNEL

1 and 3 with momenta p_1 and p_3 into particles 2 and 4 with momenta p_2 and p_4 (e.g. $\pi\overline{K} \to \overline{N}\Sigma$). The probability for this scattering is given by

$$P(p_1 p_3 \to p_2 p_4) = |(p_2 p_4, T\, p_1 p_3)|^2 \quad , \tag{9.4}$$

where

$$(p_2 p_4, T\, p_1 p_3) = F'(s',t') \quad , \tag{9.5}$$

and

$$s' = (p_1 - p_2)^2 \quad , \quad t' = (p_1 + p_3)^2 \quad . \tag{9.6}$$

One of the most basic and fruitful ideas to emerge in particle physics during the fifties was that the two scattering amplitudes $F(s,t)$ and $F'(s',t')$ are not only related, but are in fact the same analytic function [1]. That is to say, if one considered s' to be the analytic continuation of $s = (p_1 + p_2)^2$ to $p_2 \to -p_2$ and t' the analytic continuation of $t = (p_1 - p_3)^2$ to $p_3 \to -p_3$, then

$$F(s,t) = F'(s,t) \quad . \tag{9.7}$$

The process of Figure 9.1, for which $s > 0$, is called the s-channel and that of Figure 9.2, for which $t > 0$, the t-channel. The hypothesis (9.7) is based upon an analysis of Feynman diagrams and of axiomatic field theory [2]. It is related to the spectral condition, causality and the temperedness of the field-distributions.

Returning to the s-channel, an alternative pair of variables to (s,t) are $(s, \cos\theta)$,

$$F(s,t) = f(s, \cos\theta) \quad , \tag{9.8}$$

where θ is the angle between the <u>three</u>-momenta $\vec{p}_1$ and $\vec{p}_2$ in the center of mass frame of p_1 and p_2 (Figure 9.3). The relationship between t and $\cos\theta$ is

$$\cos\theta = \frac{s(t-u) + (m_3^2 - m_4^2)(m_1^2 - m_2^2)}{\{[s-(m_1-m_2)^2][s-(m_1+m_2)^2][s-(m_3-m_4)^2][s-(m_3+m_4)^2]\}^{\frac{1}{2}}} \qquad (9.9)$$

This looks complicated unless the masses are equal. However, the important point is that $\cos\theta$ <u>is linear in</u> t. In the analyses of scattering data it is usual [3] to make a "partial wave decomposition" of $f(s, \cos\theta)$ i.e., to expand $f(s, \cos\theta)$ in terms of Legendre functions

$$f(s, \cos\theta) = \sum_\ell (2\ell + 1)a_\ell(s)P_\ell(\cos\theta) \quad . \qquad (9.10)$$

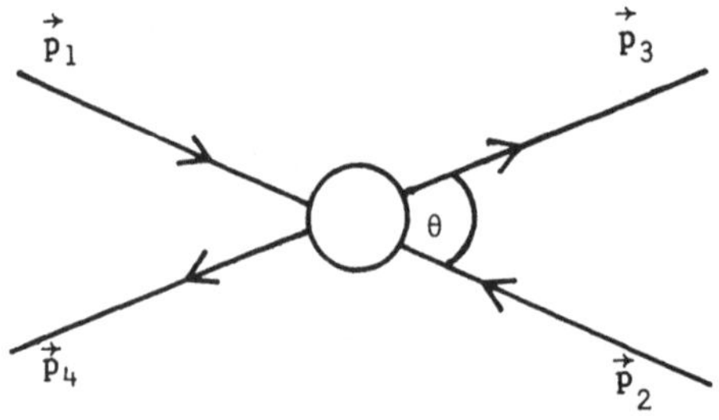

FIGURE 9.3. SCATTERING IN CM SYSTEM IN S-CHANNEL IN 3-SPACE

This is done for two reasons. (a) The <u>unitary</u> condition, which says that the total probability for scattering is unity, is diagonal in the P_ℓ basis. In fact, it reads

$$a_\ell(s) = \sin\delta_\ell(s)\exp i\delta_\ell(s) \quad ,$$

where $\delta_\ell(s)$ is real, and a scattering analysis is normally an analysis of the "phase-shifts" $\delta_\ell(s)$. (b) For low-energies, $s \sim (m_1 + m_2)^2$, the low values of $\ell(\ell = 0,1,2)$ dominate. (One can see this intuitively by noting that for low energy we have low relative angular momentum of the two particles, and as we shall see later, ℓ is the relative angular momentum.)

<u>Regge Theory</u>

One of the problems of scattering theory was how to combine the analyticity (9.7) with the expansion (9.10). As we go from the s channel, where $t < 0$ and $|\cos\theta| \leq 1$, to the t-channel, where $t > (m_1 + m_2)^2$ and $|\cos\theta| \geq 1$, the expansion (9.10) diverges. To overcome this difficulty, Regge [4] showed that, at least for a class of non-relativistic potential scattering theories, the way to continue $\cos\theta$ was to express the expansion (9.10) in integral form. First, one writes

$$f(s, \cos\theta) = \frac{1}{2\pi i}\int_C \frac{(2\ell + 1)d\ell}{\sin\pi\ell}\, a_\ell(s)P_\ell(\cos\theta) \quad , \qquad (9.11)$$

where C is the contour of Figure 9.4, then divides the integrand into + and - signature parts

$$f^{\pm}(s, \cos\theta) = \frac{1}{2\pi i}\int_C \frac{(2\ell + 1)d\ell}{\sin\pi\ell}\, a_\ell^{\pm}(s)[P_\ell(\cos\theta) \pm P_\ell(-\cos\theta)] \quad , \qquad (9.12)$$

which have independent physical properties, and then, because each converges separately on the circle at infinity, opens up the contour to L, which is the furthest

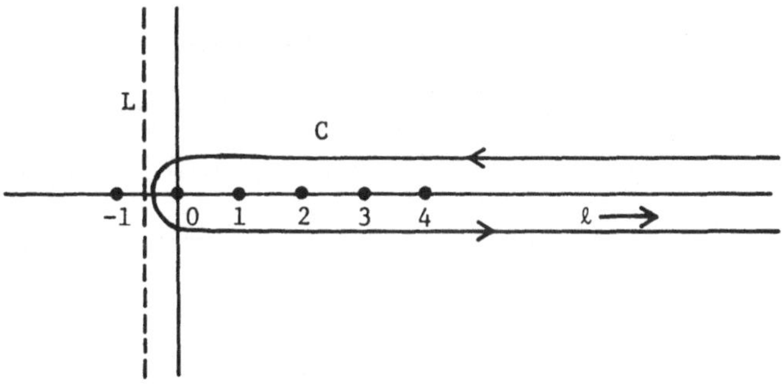

FIGURE 9.4. THE CONTOURS OF C AND L

line to the left allowed by the $P_\ell(\cos\theta)$. On the way, one picks up the poles of $a_\ell^{\pm}(s)$, which for the class of potentials considered is a meromorphic function of ℓ to the right of L, and obtains (simplifying for clarity to the case when $a_\ell^{\pm}(s)$ has only one pole to the right of L)

$$f^{\pm}(s, \cos\theta) = (2\alpha^{\pm}(s) + 1)\frac{\beta^{\pm}(s)}{\sin\pi\alpha^{\pm}(s)}[P_{\alpha^{\pm}(s)}(\cos\theta) \pm P_{\alpha^{\pm}(s)}(-\cos\theta)]$$
$$+ \frac{1}{2\pi i}\int_L \frac{(2\ell + 1)d\ell}{\sin\pi\ell}\, a_\ell^{\pm}(s)[P_\ell(\cos\theta)$$
$$\pm P_\ell(-\cos\theta)] \quad , \qquad (9.13)$$

where $\alpha^{\pm}(s)$ is the position of the pole, and $\beta^{\pm}(s)$ the residue of $a_\ell^{\pm}(s)$ at the pole. The expression (9.13) can now be continued in $\cos\theta$ into the t-channel, and indeed to $t \sim \cos\theta \to \infty$.

What is the relevance of all this to relativistic scattering? The point is that one now makes the hypothesis [5] that although relativistic scattering may be quite different from non-relativistic scattering, it retains at least one feature of it, namely, the fact that $a_\ell(s)$ is meromorphic to the right of L.

This is quite an assumption, and indeed, has had to be modified. But it is at least within the general philosophy that nature is simple if looked at the right way--and here the postulate is that the right way to look at $f(s, \cos\theta)$ is from the point of view of its properties in the ℓ-plane to the right of L! In any case, let us investigate [6] the physical implications of (9.13).

The physical implications of (9.13) are best seen by noting that the pole $\alpha(s)$ is not fixed, but varies with s, and drawing the path of its real part as a function of s (Figure 9.5). There is good reason to believe, as we shall see in

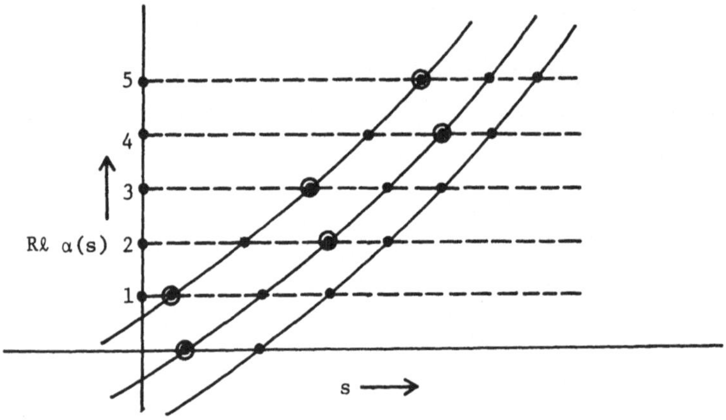

FIGURE 9.5. REGGE TRAJECTORIES

a moment, that its path is as in this figure. The physical implications are then two-fold:

(i) for $t \sim \cos\theta \to \infty$, $s < 0$, we have from (9.13)

$$f^{\pm}(s, \cos\theta) \to \frac{2\alpha^{\pm}(s) + 1}{\sin \pi\alpha^{\pm}(s)} \beta^{\pm}(s)(\cos\theta)^{\alpha^{\pm}(s)} \quad . \qquad (9.14)$$

This means that in the t-channel, as $t \to \infty$,

$$f^{\pm}(s,t) \to A^{\pm}(s)t^{\alpha^{\pm}(s)} \quad , \qquad (9.15)$$

i.e., we have an explicit statement about the behavior of the scattering amplitude as a function of the energy (t) for high energy. This is a result which could not be obtained experimentally and was not obtained theoretically before the advent of Regge theory. What was known theoretically before was that, because of the unitary condition for T, $f(s,t)$ was bounded, and probably decreased, as a function of t for $t \to \infty$. This is why $Rl\ \alpha^{\pm}(s)$ is assumed to be less than 1 for $s < 0$ in Figure 9.5. But the explicit t-dependence was first obtained in Regge theory, and is clearly controlled by the Regge-pole at $\ell = \alpha^{\pm}(s)$.

(ii) If $\text{Im}\ \alpha^{\pm}(s)$ is small, then when $\text{Rl}\ \alpha^{\pm}(s)$ = integer, $1/\sin \pi\alpha^{\pm}(s)$ is large. Hence, remembering the factor $P_{\alpha^{\pm}(s)}(\cos\theta) \pm P_{\alpha^{\pm}(s)}(-\cos\theta)$, which is small for $\text{Rl}\ \alpha^{\pm}(s)$ = even/odd integer, we see that $f^{\pm}(s, \cos\theta)$ is large for $\text{Rl}\ \alpha^{\pm}(s)$ = even/odd integer. Returning to the s-channel, $s > (m_1 + m_2)^2$, we see that the s-channel amplitude therefore becomes large, or resonates, whenever $\text{Rl}\ \alpha^{\pm}(s)$ = even/odd integer. Furthermore, a simple analysis of how the amplitude resonates near $\alpha^{\pm}(s)$ = even/odd integer, shows that it behaves as if it were the contribution to the s-channel scattering of an unstable bound state particle or resonance of mass = $\sqrt{s}$, spin = $\text{Rl}\ \alpha^{\pm}(s)$, and life time $\propto [\text{Im}\ \alpha^{\pm}(s)]^{-1}$, Figure 9.5. This result clearly suggests that the $\text{Rl}\ \alpha^{\pm}(s)$ = even/odd integer points on the Regge-trajectory of Figure 9.5 should be interpreted as unstable particles of increasing mass and spin. And indeed, if one examines Figure 5.1a, one sees that the baryons for which it can be checked do indeed lie on Regge trajectories. The mesons do not have sufficiently well-determined spins and parities for a direct check but other considerations support the conjecture that they also lie on Regge trajectories. A typical conjecture [11] is shown in Figure 9.6.

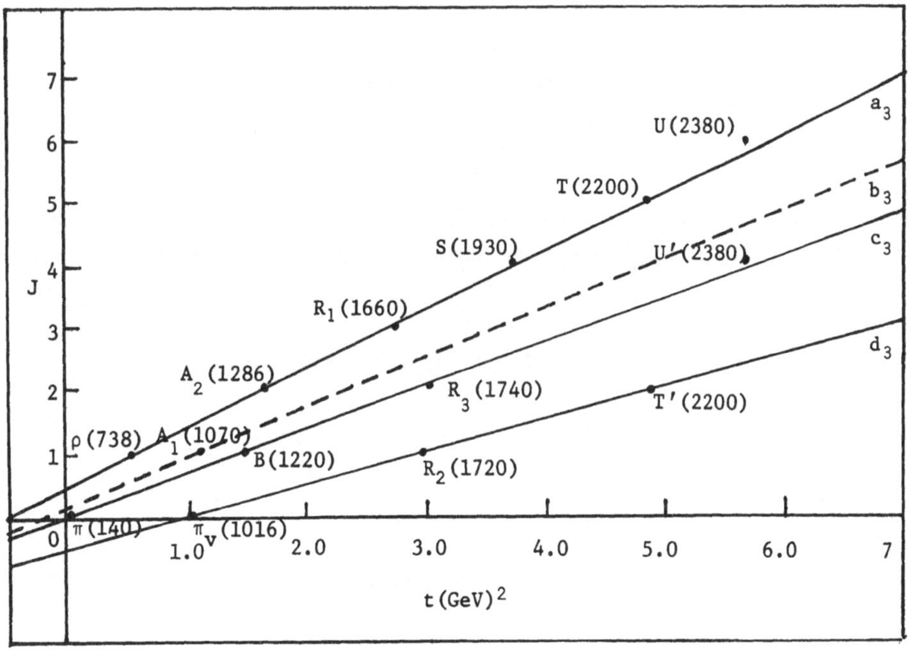

FIGURE 9.6.

The most beautiful part of the results (i) and (ii) lies in their combination. By combining them we see that the resonant states, or unstable particles, which are produced in the s-channel, dictate the high-energy behavior in the t-channel (and, of course, conversely). This unexpected relationship between these hitherto unconnected phenomena is a result that is almost certain to survive, no matter how the details of the Regge theory may have been modified.

A further beauty of the result is that it simultaneously solves a long-standing puzzle in scattering theory, namely, that if one were to continue the contribution to $F(s,t)$ of a particle with a fixed high spin (therefore high powers of $\cos\theta$, therefore high powers of t) from the s-channel to the t-channel, this contribution alone would violate the unitary condition for large t. The Regge result solves the problem by showing that the spin is really $\mathrm{Rl}\ \alpha^{\pm}(s)$, and hence is not fixed, but varies with s and becomes less than 1 for $s < 0$ in the t-channel.

After the above rather lengthy description of the background, let us turn at last to the little groups.

Consider first the two-particle state $|p_1p_2\rangle$ in the s-channel. This state can equally well be described by $|P,q\rangle$ where $P = p_1 + p_2$, $q = p_1 - p_2$. Since $p_1^2 = m_1^2$, $p_2^2 = m_2^2$, if we consider P as 4 independent variables there are two constraints on q. As a result, we can write $|P,q\rangle$ as $|P,R(\theta_1\phi_1)q_0\rangle = |s = P^2, \vec{P}, R(\theta_1\phi_1)q_0\rangle$, where q_0 is a fixed vector and $R(\theta_1\phi_1)$ is a rotation belonging to the little group SO(3) of P. (The angle $(\theta_1\phi_1)$ is the angle between a fixed z-axis and the $\vec{p}_1 - \vec{p}_2$ line in the 3-dimensional diagram of Figure 9.3.) In a similar way the state $|p_3p_4\rangle$ can be written as $|s' = P'^2, \vec{P}', R(\theta_2\phi_2)q_0\rangle$, where $P' = p_3 + p_4$, $R(\theta_2\phi_2)$ is an element of the little group SO(3) of P', and $(\theta_2\phi_2)$ is the angle between the fixed z-axis and $\vec{q}\,' = \vec{p}_3 - \vec{p}_4$. However, from energy momentum conservation we have

$$s = s' \quad , \quad \vec{P} = \vec{P}' \quad . \tag{9.16}$$

Hence $R(\theta_1\phi_1)$ and $R(\theta_2\phi_2)$ are elements of the <u>same</u> little group, namely that of $P = P'$. For the scattering amplitude, which is Poincaré, and therefore rotationally invariant, we then have

$$\begin{aligned}\langle p_3p_4|T|p_1p_2\rangle &= \langle s,\vec{P},R(\theta_2\phi_2)q_0|T|s,\vec{P},R(\theta_1\phi_1)q_0\rangle \\ &= \langle s,\vec{P},q_0|T|s,\vec{P},R^{-1}(\theta_2\phi_2)R(\theta_1\phi_1)q_0\rangle \\ &= \langle s,\vec{P},q_0|T|s,\vec{P},R(\theta_3,\phi_3)q_0\rangle \quad ,\end{aligned} \tag{9.17}$$

where $(\theta_3,\phi_3) = \theta$ is the angle between the lines $\vec{p}_1 - \vec{p}_2$ and $\vec{p}_2 - \vec{p}_4$ in Figure 9.3. Hence

$$f(s, \cos\theta) = \langle p_3p_4|T|p_1p_2\rangle = \langle s,\vec{P},q_0|T|s,\vec{P},R(\theta)q_0\rangle \quad . \tag{9.18}$$

But since $R(\theta)$ is an element of the little group SO(3) of P, this means that as a function of θ, $f(s, \cos\theta)$ is a function over the little group SO(3). Hence the expansion

$$f(s, \cos\theta) = \sum_{\ell} (2\ell + 1)a_{\ell}(s)p_{\ell}(\cos\theta) \quad , \qquad (9.19)$$

emerges as nothing but the expansion of the scattering amplitude as a function over the little group SO(3) in terms of the irreducible representations of SO(3) [7].

Once it is realized that the partial wave decomposition (9.19) which is restricted to the part of the $s - t$ plane which belongs to the s-channel, $s > 0$, is nothing but an expansion over the little group SO(3) of $P = p_1 + p_2$, a method for extending the expansion to other channels immediately suggests itself, namely, to make a little group expansion in the same variable in the other channel also. In the other channel, $s = P^2 < 0$, since $P = p_1 + p_2$ with $p_2^0 < 0$. Hence the little group is SO(2,1).

However, there is a snag. The snag is that whereas in the s-channel the unitary condition guarantees that $f(s, \cos\theta)$ will be square integrable over the little group SO(3), in the other channels there is no guarantee that it will be square-integrable over SO(2,1), and in general it is not.

At this point, however, one can return to Regge theory. Looking at the Regge expansion (9.13) one sees at once that the background integral is nothing but the expansion of the scattering amplitude in terms of the <u>principal</u> series of SO(2,1). Thus what Regge theory says is that, although the full scattering amplitude is not square-integrable over SO(2,1), when one removes the contribution of the Regge poles, the remainder <u>is</u> square-integrable. Thus equation (9.13) can be looked at from two points of view. From the Regge, or physical, point of view, the pole term is the important term and the integral just an incidental background term. From the group theoretical point of view, the integral is the interesting group expansion term, and the pole term just an incidental subtraction term to make the integral converge.

One may ask why only the principal series appears in the Regge formula. First there is a theorem due to Bargmann [8] which states that any square-integrable function over SO(2,1) can be expanded in terms of the principal series and the discrete series with $\ell > -\frac{1}{2}$. This theorem explains why the supplementary, trivial and $D^{\pm}(-\frac{1}{2})$ representations do not appear. Secondly, the discrete series with $\ell > -\frac{1}{2}$ do not appear in our case because we have left out the spin of the particles [9].

In conclusion, I should mention that the further generalization to an expansion of $f(s, \cos\theta)$ in terms of the Lorentz group SO(3,1) (to include SO(2,1) for $t \sim \cos\theta_s$, and SO(3) for s) has been considered. The SO(3,1) expansion becomes particularly interesting and illuminating at the point $s = 0$ in the continuation of s from the s to the t-channel, because we can choose $\vec{P} = 0$ as our reference frame, and then, for $s = 0$, $P = 0$, in which case SO(3,1) is itself the little group. The fact that at $s = 0$ the little group expands to SO(3,1) has physical consequences, notably that to every Regge trajectory crossing the line $s = 0$ in Figure 9.5 there is a family of trajectories with values $\alpha(0)$, $\alpha(0) - 1$, $\alpha(0) - 2$, ... corresponding to the representations $j = j_0, j_0 + 1, \ldots$ of SO(3) occurring in an irreducible representation of SO(3,1) [10].

10. INTERNAL SYMMETRIES

In the previous three chapters, we have considered the space time properties of relativistic Hilbert space in general, and of 1-particle states in particular. In the present lecture, we should like to consider some properties of the particles which are independent of space-time. Because they are independent of space-time, these properties, or symmetries, are called internal symmetries [1].

The first internal symmetries came to light when the structure of the atomic nucleus began to be investigated in the early thirties. The nucleus was found to consist of protons and neutrons (each about 2,000 times the mass of the electron, i.e., each about 10^{-25} grs.) and investigation of the forces that held them together (the nuclear forces) showed that

1) they were much stronger than the electromagnetic forces (they are the strong forces mentioned in Section 5) and
2) they were charge-independent. That is to say, apart from statistics, they did not distinguish between protons and neutrons--the force between two protons was the same as the force between two neutrons or the force between a proton and neutron. (This is in marked contrast to the electromagnetic forces which distinguish clearly between the proton and neutron, since the proton is charged and the neutron is not.)

To formulate charge-independence, it was convenient to introduce on the physical Hilbert space, an abstract invariance group. The group used was the SU(2) group (isotopic spin group) and the idea was to assign the proton and the neutron, respectively, to two orthogonal vectors $|p)$ and $|n)$ in the 2-dimensional representation of SU(2) and then to demand the invariance of the nuclear Hamiltonian H_N under the SU(2) group

$$[SU(2), H_N] = 0 \quad . \tag{10.1}$$

The generator I of SU(2) for which

$$I|p) = \frac{1}{2}|p) \quad , \quad I|n) = \frac{-1}{2}|n) \quad , \tag{10.2}$$

is called by convention I_3 or the third component of isospin. For many-nucleon states (nucleon = proton, neutron), one uses tensor products of the states $|p)$, $|n)$ in conjunction with (10.1).

The mathematics of the isospin group is the same as for the ordinary electron spin group SU(2), but the physics is quite different. First, every vector in the ordinary 2-dimensional spin space is realizable in nature, but only $|p)$ and $|n)$ in isotopic spin space are realizable. (Nobody has ever succeeded in constructing a state which is a linear superposition $a|p) + b|n) ab \neq 0$.) Secondly, the operators in spin space transform non-trivially under space rotations, whereas the operators in isospace are independent of space-time.

Later in the investigation of nuclear structure, it was found that the nuclear forces between protons and neutrons were mediated by π-mesons, of which there are three, π^0 and $\pi^{\pm}$, the index referring to the charge. It turned out that the π's could be incorporated into the isospin scheme by assigning them to the 3-dimensional representation of SU(2) with

$$I_3|\pi^{\pm}) = \pm|\pi^{\pm}) \qquad I_3|\pi^0) = 0 \quad . \tag{10.3}$$

Once higher (relativistic) energies became available, and the bombardment of nuclei with protons and neutrons of high energy resulted in the production of new particles, it was found that the forces producing the new particles were

a) of the same order of strength as the nuclear forces

b) still charge independent.

These forces are, therefore, called generically "strong" forces, and the associated Hamiltonian the strong Hamiltonian H_s.

From the charge independence of the strong interactions and the assignment of the creating particles (p,n) and $(\pi^{\pm 0})$ denoted by N and π, respectively, to irreducible representations of SU(2), it follows that the created particles should also belong to irreducible representations of SU(2). And this turns out to be the case. In fact, the metastable hadrons $K^{\overset{+}{0}}$, $\overline{K}^{\overset{+}{0}}$, η, Λ, $\Sigma^{\pm 0}$, $\Xi^{\overset{-}{0}}$, and Ω^- are found to belong to the 2,2,1,1,3,2 and 1-dimensional representations of SU(2) and, hence, are denoted by K, $\overline{K}$, η, Λ, Σ, Ξ, and Ω, respectively.

For the unstable hadrons, the same results are found. All can be assigned to irreducible representations of SU(2) with

$$[SU(2),H_s] = 0 \quad , \tag{10.4}$$

and, indeed, the number of unstable hadrons is now so large that one no longer refers to them individually but refers only to the isospin multiplets to which they belong. This method of referring to them has been anticipated in Table 5, in which

I (the total isospin) labels the representation of SU(2) to which the particles belong (dimension = 2I + 1). Note that all particles in the same SU(2) multiplet have the same mass (up to electromagnetic and weak interaction corrections) and the same spin, indeed have the same space-time properties in general. This is because the internal invariance group SU(2) and the space-time Poincaré group are independent, i.e.,

$$[SU(2),P_{+}^{\dagger}] = 0 \quad . \tag{10.5}$$

In the production of new particles from $N,\overline{N},\pi$, a new invariance law became evident, namely, that when new particles are produced from $N,\overline{N}$ and π they come only in certain combinations. The simplest description of the allowed combinations is obtained by introducing a new operator Y on $\mathcal{H}$, assigning the SU(2) multiplets to integer eigenvalues of Y and demanding that

$$[Y,H_s] = 0 \quad , \tag{10.6}$$

i.e., that Y be conserved in the strong interactions. Because of its analogy to the electric charge, Q, which takes integer values on the particle states and is conserved in interactions, Y is called the <u>hypercharge</u>. Note, however, that whereas Q is conserved in <u>all</u> interactions, Y (like SU(2)) is conserved only in the strong ones. The hypercharges of $\pi,K,\overline{K},\eta$ and $N,\Lambda,\Sigma,\Xi,\Omega$ are (0,1,-1,0) and (1,0,0,-1,-2), respectively. In general, we have also the relation

$$Q = I_3 + \frac{1}{2}Y \quad , \tag{10.7}$$

which was first discovered by Gell-Mann and Nishijima.

The strong interactions are found to be invariant, therefore, under SU(2) and Y, and hence under the group SU(2) × U(1) where U(1) is generated by Y. But, as Louis Michel has pointed out, the fact that Q in (10.7) is integer, means that only those representations of SU(2) × U(1) occur, for which Y = D (modulo 2), where D is the dimension of the SU(2) representation. Since such representations of SU(2) × U(1) are exactly those which are representations of U(2), it follows that we can replace the invariance under SU(2) × U(1) by invariance under U(2).

A glance at Figure 5.1 will show that U(2) is the <u>maximal</u> group with the property $[U(2),P_{+}^{\dagger}] = 0$, up to electromagnetic and weak corrections, since any other transformations among the particles will not commute with the mass and spin.

In spite of this, Figure 5.1 does suggest that one could go beyond U(2). The reason is that the different U(2) multiplets seem to fall into sets which, while they do not have the same mass, do have the same spin and have approximately the same mass. Examples of such sets are $(\pi, K, \overline{K}, \eta)$ and $(N, \Lambda, \Sigma, \Xi)$. For this reason, it has been proposed that U(2) be enlarged to a group SU(3). The original choice SU(3) rested primarily on two factors:

1) SU(3) allows two and only two commuting generators, and two and only two additive operators (I_3 and Y) are necessary to label the particles.
2) SU(3) has an irreducible representation (the 8-dimensional adjoint representation) which can accommodate the two sets of eight particles (π, K, $\overline{K}$, η) and (N, Λ, Σ, Ξ) with the correct I, I_3, and Y values. ($I(I+1) = I_1^2 + I_2^2 + I_3^2$) See Figure 5.1.

(Actually, the η was not known experimentally at the time SU(3) was proposed and was predicted by SU(3).) Having adopted SU(3), one tries to assign the other, unstable, particles to SU(3) multiplets. Success has already been achieved with the $J^P = \frac{3}{2}^+$ multiplet (N*, $\Sigma^* \equiv Y^*$, Ξ^*, Ω) which is assigned to the 10-dimensional representation (decimet) of SU(3) (Figure 5.1) and the $J^P = \frac{3}{2}^+, \frac{5}{2}^+, 1^-$, and 2^+ multiplets which are assigned to the 8-dimensional representation (octet). The higher J^P (spin-parity) multiplets look equally promising. The existence of Ω, like η, was actually predicted by the SU(3) assignment.

Thus, for particle assignments, SU(3) turns out to be as successful as the exact invariance group U(2) had been before it. Unlike U(2), however, SU(3) is not an invariance group, i.e.,

$$[X,H_s] \neq 0 \quad , \tag{10.8}$$

for those generators X of SU(3) which do not generate transformations within U(2). Indeed, if (10.8) were zero, SU(3) transformations would commute with the mass-operator and all the particles assigned to an SU(3) multiplet would have to have the same mass, which is manifestly not the case.

One might then ask: Of what use is SU(3) apart from classifying and predicting particles? The answer is that although SU(3) is not an invariance group, it is approximately an invariance group, and by stating how certain operators transform with respect to it, one can obtain physical predictions, correct to within the approximation (~20%).

The operators whose transformation properties are of most interest are

1) the Hamiltonian H_s
2) the mass operator M
3) the electric current $j_\mu(x)$
4) the weak current of the hadrons $j_\mu^\omega(x)$

The transformation properties of these operators are assigned on physical grounds and amount to a statement about the tensor character of the operators. The physical predictions are then extracted by using the WE theorem. Let us consider the operators 1) to 4) briefly:

1) The Hamiltonians: one usually assumes that

$$H_s = H_s^{(0)} + H_s^{(1)} \quad ,$$

where $H_s^{(0)}$ is SU(3) invariant and about five times as large as $H_s^{(1)}$. Hence, to within 20%, H_s can be regarded as SU(3) invariant. This allows us to obtain 20% estimates for the relations between strong decay processes such as

$$N^* \to N\pi \quad ,$$

$$\Sigma^* \to \Sigma\pi \quad ,$$

$$\to \Lambda\pi \quad ,$$

$$\Xi^* \to \quad \pi \quad ,$$

for example,

$$\frac{N^* \to N\pi}{\Sigma^* \to \Lambda\pi} = \frac{\rho}{\sigma}\frac{(N^*, H_s N\pi)}{(\Sigma^*, H_s \Lambda\pi)} \simeq \frac{\rho}{\sigma}\frac{(N^*, H_s^{(0)} N\pi)}{(\Sigma^*, H_s^{(0)} N\pi)}$$

$$= \frac{\rho}{\sigma}\frac{C^{10\ 8\ 8}_{N^*\ N\ \pi}\left(\frac{3}{2}^-, H_s^{(0)}\frac{1}{2}^-\ 0\right)}{C^{10\ 8\ 8}_{\Sigma^*\ \Lambda\ \pi}\left(\frac{3}{2}^-, H_s^{(0)}\frac{1}{2}^-\ 0\right)}$$

$$= \frac{\rho}{\sigma}\frac{C^{10\ 8\ 8}_{N^*\ N\ \pi}}{C^{10\ 8\ 8}_{\Sigma^*\ \Lambda\ \pi}} \quad , \tag{10.9}$$

where ρ, σ are kinematical phase-space factors, depending only on the masses (see Equation (10.15) below), the C's are Clebsch-Gordan coefficients, and $\left(\frac{3}{2}^-, H_s^{(0)}\frac{1}{2}^-\ 0\right)$ is a WE reduced matrix element. In a similar way, one can relate scattering processes such as

$$\pi N \to \pi N$$

$$\to K\Sigma$$

$$\to K\Lambda$$

$$\overline{K}N \to \pi\Sigma$$

$$K\Xi \quad ,$$

etc., for particles $(\pi, K, \overline{K})$ and $(N, \Sigma, \Lambda, \Xi)$ in the same SU(3) multiplets.

2) The Mass Operator M, like H_s, is assumed to be of the form

$$M = M^{(0)} + M^{(1)} \quad ,$$

where $M^{(0)}$ is SU(3)-invariant and $M^{(1)}$ is not. However, in the case of the mass, one goes further and also makes a positive statement about $M^{(1)}$, namely, $M^{(1)}$ transforms as the Y-component of the eight-dimensional representation of SU(3).

Using the WE theorem or general tensor techniques, it is then possible to show that operating on the space of an irreducible representation of SU(3),

$$M = \alpha + \beta Y + \gamma[I(I + 1) - \frac{1}{4} Y^2] \quad , \qquad (10.10)$$

where $I(I + 1) = I_1^2 + I_2^2 + I_3^2$ is the total isotopic spin and α, β and γ are SU(3) scalars. Thus, for any SU(3) multiplet, we have the mass-formula (10.10), where α is the mean value of M for the multiplet.

This formula agrees with experiment to within 4% for the metastable baryon octet, to within 0.5% for the $\frac{3^-}{2}$ baryon decimet (for which it makes two predictions, one of which is the prediction of the Ω at exactly the right mass value) and it agrees reasonably well (with some "representation mixing" modifications) for the remaining SU(3) multiplets.

At this point, we make a digression and use the SU(3) mass operator to illustrate the following point concerning tensor operators in finite-dimensional spaces. (There is a generalization to infinite dimensional spaces but it is more complicated in several respects.) Let $\mathcal{H}$ be a finite-dimensional Hilbert space, G a group implemented by unitary transformations on $\mathcal{H}$, and L(A) the space of linear operators on A. Then L(A) is itself a representation space for G. Hence any given operator $A \in L(A)$ can be expanded in terms of G. But this means that any operator is a tensor operator in the sense that it can be expanded as a series of irreducible tensor operators. Thus, the real content of the statement that an operator is a "tensor" is that in its expansion, one (or a few) irreducible tensor components dominate. For example, in the case of the mass operator M restricted to the octet space, M must be one of the 64 possible linear independent operators A on this space. With respect to SU(3), however, this 64-dimensional space (L(A)) splits into irreducible subspace of dimension $1,8,8,27,10,\overline{10}$ with respect to SU(3). Hence, a *priori*, M is limited to be

$$M = M^1 + M^8 + M^8 + M^{10} + M^{\overline{10}} + M^{27} \quad . \qquad (10.11)$$

In addition, the direction of M within each irreducible subspace is completely determined by the condition that M be a U(2) invariant

$$[U(2),M] = 0 \quad , \qquad (10.12)$$

and, in fact, this condition kills 10 and $\overline{10}$, and constrains M to be

$$M = M_0^1 + M_Y^8 + M_Y^8 + M_Y^{27} \quad , \qquad (10.13)$$

where Y is the direction of the hypercharge. Thus, the only assumption that goes into the mass formula (10.10) which is not already implied by general considerations is that in (10.13) the component of M in the 27 is suppressed. This assumption is made for other tensor operators also (see below), in which case it is called explicitly "octet dominance". Incidentally, it might be worth remarking that the analogue of (10.13) for the decimet is $M = M_0^1 + M_Y^8 + M_Y^{27} + M_Y^{64}$, so that the octet dominance assumption kills two parameters (27 and 64) in this case. This is why the mass-formula yields two predictions for the decimet.

The strong decay rates and the mass formulae provide stringent cross-checks on the SU(3) particle assignments. In fact, since the mass of a particle is usually known experimentally long before its spin-parity, in practice one assigns according to the mass formula first and checks with spin-parity afterwards.

An interesting feature emerges in the case of the mesons 0^-, 1^-, and 2^+. These come not in octets but in nonets and, to fit the mass formula, one must assume that the two U(2) scalar particles (η,η^*), (ω,φ) and (f,f^*), respectively, which occur in the nonets, do not have definite SU(3) properties; rather, the linear combinations

$$\begin{aligned} \eta_8 &= \cos\theta\ \eta + \sin\theta\ \eta^* \\ \eta_0 &= \cos\theta\ \eta^* - \sin\theta\ \eta \quad , \end{aligned} \qquad (10.14)$$

etc., belong to the SU(3) octet and scalar representations, respectively. This phenomenon is called "representation mixing" and it robs the mass formula of its direct predictions for these particles by adding the new unknown parameter θ. The best that the mass-formula can do is determine the various θ's. (They turn out to be $\theta_{0^-} \simeq 10^0$, $\theta_{1^-} \simeq 60^0$, $\theta_{2^+} \simeq 30^0$.) The interesting point, however, is that in spite of this, the mass formula is not empty. One can get indirect predictions, and indeed one of the most remarkable features of the decay rate analyses is that in two cases in which the experimental decay rates are in complete contradiction with phase space and SU(3) without mixing, the use of mixing accounts for the discrepancy. For example, experimentally

$$\frac{f \to 2\pi}{f \to K\bar{K}} \simeq 50 \quad , \quad \frac{f^* \to 2\pi}{f^* \to K\bar{K}} \leqslant \frac{1}{5} \quad , \tag{10.15}$$

whereas, on account of the much smaller mass of the pion, phase space would predict the ratio $\gg 1$ in both cases. With mixing, SU(3) predicts, using the WE theorem,

$$\frac{f \to 2\pi}{f \to K\bar{K}} \simeq \frac{\rho}{\sigma}\left[\frac{\cos\theta_2\left\langle\frac{8}{f_8}\middle|S^0\middle|\frac{8}{\pi}\right\rangle\left|\frac{8}{\pi}\right\rangle + \sin\theta_2\left\langle\frac{0}{f_0}\middle|S^0\middle|\frac{8}{\pi}\right\rangle\left|\frac{8}{\pi}\right\rangle}{\cos\theta_2\left\langle\frac{8}{f_8}\middle|S^0\middle|\frac{8}{K}\right\rangle\left|\frac{8}{\bar{K}}\right\rangle + \sin\theta_2\left\langle\frac{0}{f_0}\middle|S^0\middle|\frac{8}{K}\right\rangle\left|\frac{8}{\bar{K}}\right\rangle}\right]^2$$

$$\simeq \frac{\rho}{\sigma}\left[\frac{\cos\theta_2 C^{888_s}_{\pi\pi 8} + \alpha\sin\theta_2 C^{880}_{\pi\pi 0}}{\cos\theta_2 C^{888_s}_{K\bar{K}8} + \alpha\sin\theta_2 C^{880}_{K\bar{K}0}}\right]^2$$

$$= \frac{\rho}{\sigma}\,\frac{3(2\alpha\sin\theta_2 + \cos\theta_2)^2}{4(\alpha\sin\theta_2 - \cos\theta_2)^2} \quad , \tag{10.16}$$

where S^0 is the scalar approximation to the S-matrix S,

$$\alpha = \frac{\langle 0\|S^0\|88\rangle}{\langle 8\|S^0\|88\rangle} \quad ,$$

and the phase space factor ρ/σ is given by

$$\frac{\rho}{\sigma} = \left(\frac{p}{q}\right)^{2\ell+1} = \left[\frac{m_f(1 - 4m_\pi^2|m_f^2)^{1/2}}{m_f(1 - 4m_K^2|m_f^2)^{1/2}}\right]^5$$

$$\simeq [(1 - 4(500)^2|(1250)^2)^{-1/2}]^5 \simeq \left(\frac{5}{3}\right)^5 \simeq 15 \quad , \tag{10.17}$$

where m denotes mass, p and q are the final state π and K three-momenta, respectively, and $\ell = 2$ is the final state orbital angular momentum. Similarly,

$$\frac{f^* \to 2\pi}{f^* \to K\bar{K}} = \frac{\rho'}{\sigma'} \cdot \frac{3}{4}\left[\frac{2\alpha\cos\theta_2 - \sin\theta_2}{\alpha\cos\theta_2 + \sin\theta_2}\right]^2 \quad , \tag{10.18}$$

where

$$\frac{\rho'}{\sigma'} = \left[\frac{m_{f^*}(1 - 4m_\pi^2|m_{f^*}^2)^{1/2}}{m_{f^*}(1 - 4m_K^2|m_{f^*}^2)^{1/2}}\right]^5$$

$$\simeq [(1 - 4(500)^2|(1500)^2)^{-1/2}]^5 \simeq \left(\frac{3}{\sqrt{5}}\right)^5 \simeq 4.5 \quad .$$

The values of θ_2 and α can be calculated from the mass formula and from f and A_2 decays, respectively. The values calculated in this way yield

$$2\alpha \simeq \tan\theta_2 \quad ,$$

and from (10.18) and (10.16) we see at once that this is exactly what is required to explain (10.15). Note that if there were no mixing ($\sin\theta_2 = 0$), SU(3) would predict

$$\frac{f^* \to 2\pi}{f^* \to K\bar{K}} = 3\,\frac{\rho'}{\sigma'} \quad ,$$

which is even worse than phase space alone.

3) The Electromagnetic Current: Here one takes a lead from the charge. From the Gell-Mann-Nishijima result

$$Q = I_3 + \frac{1}{2}Y \quad , \tag{10.19}$$

we see that the charge is actually a generator of SU(3). Hence, it transforms like a component of an 8-tensor (octet). We call this the Q-component of the octet. Since the charge is constructed from the current according to

$$Q = \int j_0(x)d^3x \quad , \tag{10.20}$$

it is then natural to assume that the current $j_\mu(x)$ transforms in the same way, i.e., as the Q-component of an octet of currents. The assumption is not binding unless one uses other principles such as locality and minimal principle, but it is a good Ansatz. Algebraically, the Ansatz may be written

$$[J^\lambda, j^\mu_\mu(x)] = i f_{\lambda\mu\sigma} j^\sigma_\mu(x) \quad , \quad \lambda,\mu,\sigma = 1\ldots 8 \quad , \tag{10.21}$$

where J^λ are the generators of SU(3) and the EM current is the Q-component of the octet tensor $j^\lambda_\mu(x)$. As an application of the Ansatz, one can consider the magnetic moments of the stable baryons. The magnetic moment operator is a linear function of the current so that if the current transforms like the Q-component of an octet, so does the magnetic moment operator μ^8_Q. Now for any octet member α, we have from the WE theorem

$$\left\langle {8 \atop \alpha} \middle| \mu^8_Q \middle| {8 \atop \alpha} \right\rangle = C^{8\ 8\ 8}_{\alpha\ Q\ \alpha}(8,\overset{8}{\mu},8)_c + d^{8\ 8\ 8}_{\alpha\ Q\ 8}(8,\overset{8}{\mu},8)_d \quad , \tag{10.22}$$

where the c and d are Clebsch-Gordan coefficients and $(8,\overset{8}{\mu},8)_c$

and $(8,\overset{8}{\mu},8)_d$ the corresponding reduced matrix elements. (The appearance of <u>two</u> reduced matrix elements is due to the fact that the $8 \otimes 8$ representation of SU(3) happens to contain the 8 twice.) From (10.22), it follows that all eight of the magnetic moments of the octet can be predicted from two of them. The two used as input are $\mu(p)$ and $\mu(n)$, which are well-known. The only predicted one which has been measured with good accuracy so far is $\mu(\Sigma^+)$ and the result agrees quite well with the prediction.

Similar considerations can be applied to the electromagnetic mass differences of the $\frac{1+}{2}$ baryons, which are of order $(e/\hbar c)^2$, and the prediction obtained

$$m(\Xi^-) - m(\Xi^0) = m(\Sigma^-) - m(\Sigma^+) + m(p) - m(n) ,$$

agrees extremely well with experiment.

4) The Weak Current $j^{\omega}_{\mu}(x)$ of the Metastable Hadrons: $j^{\omega}_{\mu}(x)$ is assumed to determine their leptonic (e.g., $\Lambda \to p + e + \nu$) and non-leptonic (e.g., $\Lambda \to p + \pi$) decays through the Hamiltonians

$$H_{int} = \int d^3x j^{\omega}_{\mu}(x) j^{\ell}_{\mu}(x) \quad , \quad \int d^3x j^{\omega}_{\mu}(x) j^{\omega}_{\mu}(x) \quad , \tag{10.23}$$

respectively, where $j^{\ell}_{\mu}(x)$ is the leptonic current. (Note the analogy between these interactions and the interaction

$$\int d^3x j_{\mu}(x) A_{\mu}(x) \quad , \tag{10.24}$$

between the EM current $j_{\mu}(x)$ and the EM field $A_{\mu}(x)$.)

The weak current is actually a linear combination of a true vector current $v_{\mu}(x)$ and an axial (or pseudo) vector current $a_{\mu}(x)$

$$j^{\omega}_{\mu}(x) = v_{\mu}(x) + a_{\mu}(x) \quad , \tag{10.25}$$

and $v_{\mu}(x)$ and $a_{\mu}(x)$, in turn, consist of parts that change the hypercharge eigenvalues by 0 and 1 units, respectively,

$$j^{\omega}_{\mu}(x) = v^0_{\mu}(x) + v^1_{\mu}(x) + a^0_{\mu}(x) + a^1_{\mu}(x) \quad . \tag{10.26}$$

We have already seen that the EM current is assumed to be the $Q = I_3 + \frac{1}{2}Y$ component of an SU(3) octet. It is now assumed that $v^0_{\mu}(x)$ and $v^1_{\mu}(x)$ are the $I_{\pm}$ and $\Delta Y = \pm 1$ members of the <u>same</u> EM octet as $j_{\mu}(x)$. (The identity of the octet means that $v_{\mu}(x)$ and the EM $j_{\mu}(x)$ have the same reduced matrix elements.) The $a^0_{\mu}(x)$ and $a^1_{\mu}(x)$ are assumed to be the $I_{\pm}$ and $\Delta Y = \pm 1$ components of a new SU(3) octet. They cannot be components of the <u>same</u> octet as v and j since they have different space-time properties.

Using these transformation properties of $j_{\mu}^{\omega}(x)$ in the Hamiltonians (10.23), one can apply the WE theorem and obtain selection rules for the decays. One obtains the empirically observed $\Delta S \equiv \Delta(Y + B) = \Delta Q$ and $\Delta I = \frac{1}{2}$ rules for leptonic decays and (if one invokes also octet dominance, i.e., suppression of the 27-dimensional representation in $(8 \otimes 8)_{\text{symmetric}} = 1 + 8 + 27$), the empirically observed rules $\Delta S \neq 2$, $\Delta S = \Delta Q$, and $\Delta I = \frac{1}{2}$ rules for non-leptonic decays.

In sum, therefore, SU(3) is a group which is useful not only for classifying the elementary particles, but for predicting mass relationships between them and, because it is an approximate invariance group, it can be used for obtaining 20% estimates on the scattering, electromagnetic, and decay processes of the particles. The estimates are, of course, only on <u>relative</u> matrix elements. The dynamical content of the theory is hidden in the reduced matrix elements, which cancel out in the ratios.

At present, the origins of SU(3) symmetry and the 20% SU(3) symmetry-breaking are unexplained. Both are empirical discoveries which one has learned how to handle, but not to explain.

11. BEYOND SU(3)

It is natural to try to go beyond SU(3) and see if

1) the elementary particles have any further regularities

2) the SU(3) properties have any relation to space-time.

One regularity the particles certainly possess is the "Regge recurrence" mentioned in Section 9, namely, the property that each SU(3) multiplet of spin parity J^P reoccurs at higher masses with spin parity $(J + 2n)^P$ for $n = 1,2,3...$. Attempts to describe the Regge families $(J + 2n)^P$, $n = 1,2,3,...$, with infinite component wavefunctions do not seem successful, as we saw earlier.

Apart from the Regge recurrences and SU(3), the particles do not have any obvious regularities. However, the search for new regularities and the attempts to relate SU(3) to space-time have led to some interesting ideas. One of these is the use of new particles called <u>quarks</u> which are perhaps worth discussing.

The idea behind the quarks [1] is that the fundamental, 3-dimensional representation of SU(3) should, like the 8- and 10-dimensional representation, describe real particles (the quarks), and that since the 3-dimensional representation is fundamental, all the other particles should be bound states of the quarks. In particular, the 0^- and 1^- mesons should be bound states $q\bar{q}$ of

1 quark and 1 anti-quark, and the $\frac{1}{2}^+$ and $\frac{3}{2}^+$ baryons should be bound states qqq of 3 quarks. This would certainly be compatible with the SU(3) decompositions,

$$\bar{3} \times 3 = 1 + 8 \quad ,$$
$$3 \times 3 \times 3 = 1 + 8 + 8 + 10 \quad . \tag{11.1}$$

On the other hand, the charges and hypercharges of the particles are additive quantum numbers because the corresponding operators are generators of SU(3). Hence, the charges and hypercharges of the component quarks would have to add up to those of the composite mesons and baryons, and one can check that for this to be true the charges and hypercharges of the quarks would have to be $\frac{1}{3}$ integer (the charge of the proton having been normalized to 1). This means that physically the quarks would be rather unusual objects.

Much experimental effort has been devoted to finding quarks, but so far without success. Nevertheless, the quark idea is used extensively. The reason is that even if the quarks are only fictitious, they provide a basis for making educated guesses about the real particles. Their use also simplifies many mathematical calculations.

The existence of quarks would not explain SU(3) itself, since an SU(3) triplet of quarks is assumed from the outset. But their existence would go far to explain the existence and mass-spin values of the J^P multiplets which are observed with ever increasing mass and spin.

In particular, for the lower multiplets 0^-, 1^-, $\frac{1}{2}^+$, and $\frac{3}{2}^+$ one has even been able to go beyond SU(3) to a larger group SU(6) [2] by using the quark model. The procedure is the following: The quarks are assumed to be spin $\frac{1}{2}$ particles. Hence, in their rest frames their wavefunctions $f^i_\alpha(x)$ are labeled by two sets of indices, $i = 1,2,3$ referring to SU(3) and $\alpha = 1,2$ referring to ordinary spin. One can now consider the group SU(6) of all unitary unimodular <u>x-independent</u> transformations on the 6-dimensional space $f^i_\alpha(x)$. This group contains SU(3) and the spin group SU(2) as subgroups in direct product form. If we now make the physical assumption that when the quarks bind together to form the 0^-, 1^-, $\frac{1}{2}^+$, and $\frac{3}{2}^+$ particles, the binding is in some sense SU(6) invariant, then we see that the bound state particles should belong to the $\bar{6} \times 6$ and $6 \times 6 \times 6$ representations of SU(6), respectively. To see whether this prediction agrees with experiment, one makes the SU(6) decompositions:

$$\bar{6} \times 6 = 1 + 35 \quad ,$$
$$6 \times 6 \times 6 = 70 + 56 + 20 \quad , \tag{11.2}$$

and asks whether any of the irreducible representations obtained have the correct SU(3) and spin content to accommodate the observed 0^-, 1^-, $\frac{1}{2}^+$, $\frac{3}{2}^+$ multiplets. The answer is yes. Indeed, if one makes the SU(3) × SU(2) decompositions of the

SU(6) 35 and 56, one finds

$$35 = (1,1) + (8,1) + (8,0) \quad ,$$
$$56 = (10,\tfrac{3}{2}) + (8,\tfrac{1}{2}) \quad , \qquad (11.3)$$

where the first figure refers to the dimension of the SU(3) representation and the second to the spin. This means that the 35 and 56 of SU(6) can accommodate the mesons 0^- and 1^- and the baryons $\frac{3}{2}^+$ and $\frac{1}{2}^+$, respectively. Thus, for the lower lying SU(3) multiplets, the assumption of SU(6) invariant binding for the quarks leads to the correct SU(3)-spin relationships. Attempts to extend the hypothesis of SU(6) invariant binding to the higher multiplets does not seem to work (presumably because the orbital angular momentum as well as the spin must be taken into account). For these, one falls back on more explicit dynamical quark models.

Having discovered that the lower lying SU(3) multiplets are predicted by SU(6) it is of interest to see whether SU(6) could be exploited farther. The investigation takes two forms, practical and principle. The practical investigations ask whether, following SU(3), we can make postulates about the SU(6) transformation character of H_s, M, $j_\mu(x)$ and $j_\mu^\omega(x)$ and obtain useful predictions.

For H_s, the answer is no. Although the quark-binding appears to be SU(6) invariant, the scattering matrix certainly is not. With the mass operator, one does a little better. By assuming that the mass breaking is additive with respect to SU(3) and spin, one can predict with 10% accuracy the mass spacing within the two higher SU(3) multiplets ($\frac{3}{2}^+$ and 1^-) in terms of the mass spacing within the two lower ones ($\frac{1}{2}^+$ and 0^-), respectively.

For the electric current $j_\mu(x)$, only the magnetic moment is considered, and this is assumed to transform like a member of an SU(6) 35. The matrix elements to be calculated are then of the form $\left\langle {}^{56}_{\alpha} \middle| \mu^{35}_{\beta} \middle| {}^{56}_{\gamma} \right\rangle$, where α and γ refer to members of the 56, i.e., $\frac{3}{2}^+$ and $\frac{1}{2}^+$ particles, and β refers to the magnetic moment member of the 35. But since for SU(6), $56 \times 35 (= 56 + 70 + 300 + 1134)$ contains the 56 only <u>once</u>, there is only <u>one</u> reduced matrix element

$$\left\langle {}^{56}_{\alpha} \middle| \mu^{35}_{\beta} \middle| {}^{56}_{\gamma} \right\rangle = C^{56\ 35\ 56}_{\alpha\ \ \beta\ \ \gamma} \left(56, \mu^{35}\ 56 \right) \quad ,$$

to be inserted in the WE theorem, and we obtain very strong predictions. In particular, using the proton magnetic moment $\mu(p)$ as input for the reduced matrix element, one obtains the SU(3) independent predictions

$$\mu(n) = -\frac{2}{3}\mu(p) \quad ,$$
$$\mu(10) = q\mu(p) \quad , \qquad (11.4)$$
$$\mu_{N^* \to N+\gamma} = \frac{2\sqrt{2}}{3}\mu(p) \quad ,$$

where $\mu(n)$ is the magnetic moment of the neutron, $\mu(10)$ the magnetic moment, and q the charge, of any member of the $\frac{3}{2}^{+}$ decimet, and $\mu_{N^*\to N+\gamma}$ the magnetic moment contribution (which is the largest contribution) to the EM decay $N^* \to N\gamma$ The first equation is in unbelievably good agreement with experiment (~3%), the $\mu(10)$ have not yet been measured, and the third equation is in reasonable agreement with experiment (~30%). Thus, the assignment of SU(6) properties to the magnetic moment seems to be quite successful.

Finally, for the weak current, SU(6) has the advantage of being able to carry the vector and axial vector currents in the same representation. This is because in the SU(6) limit, which is assumed to be non-relativistic, $v_\mu(x) \to v_0(x)$ which is spin 0, $a_\mu(x) \to \vec{a}(x)$ which is spin 1, and we already know from the meson classification that the 35 of SU(6) has exactly the right content to carry spin 0 and 1 SU(3) octets. Assigning v_0 and $\vec{a}$ to the 35, one can make some interesting predictions, and they agree reasonably well with experiment.

In brief, therefore, the attempt to push SU(6) beyond a mere classification group for the particles, while not spectacularly successful, is not unsuccessful. It is only when SU(6) invariance is demanded for the scattering matrix that we get a complete breakdown.

The other kind of investigation into SU(6) is more a question of principle. The relationship between the space-time symmetries of particles and the internal SU(2),SU(3) symmetries has never been properly understood, and with the advent of SU(6), in which SU(3) and the spin group are simultaneously embedded, it looked as if one might have a handle on this problem. The question also arises as to how SU(6), whose formulation is completely non-relativistic, should be made relativistic. These two questions are related and hinge on the question as to how the spin group SU(2) in SU(6) is to be interpreted. Three possibilities suggest themselves

a) as the little group of $p = \alpha$ for $p^2 > 0$

b) as a subgroup of SL(2,C) in the manifestly covariant Lorentz transformations

$$\psi(x) \to S(\Lambda)\psi(\Lambda^{-1}x) \quad , \quad \Lambda \in SL(2,C)$$

c) as a subgroup of $P_{+}^{\uparrow}$.

Each of these possibilities suggest a way of making SU(6) relativistic.

In Case a), it is simply a question of expressing the SU(6) theory in a manifestly covariant formalism and this has been done explicitly in Ref. [3]. There are no new predictions.

In Case b), one takes the quark wavefunctions

$$f_\alpha^i(x) \overset{\Lambda,a}{\to} S_{\alpha\beta}(\Lambda) f_\beta^i(\Lambda^{-1}(x - a)) \quad , \tag{11.5}$$

where α is now a Dirac index and $S_{\alpha\beta}(\Lambda)$ the Dirac representation of SL(2,C), and considers the pseudo-unitary unimodular x-independent group SU(6,6) on the index space (β,i) [4]. This group contains SU(3) and SL(2,C) as subgroups in direct product form, and thus replaces SU(6) directly. Using SU(6,6), one can proceed exactly as with SU(6). But one obtains very few new good predictions, and encounters a lot of trouble [5].

The difficulty stems from the fact that to relate the manifestly covariant wavefunctions to the physical particles, one must eliminate the auxiliary parts of the wavefunctions. This is done by means of the manifestly invariant projection operators

$$\frac{1}{2m}(\gamma^{\mu}p_{\mu} - m) \quad , \tag{11.6}$$

etc., discussed in Section 7. But while the operators (11.6) are manifestly Poincaré invariant, they are not SL(2,C) invariant (p_{μ} is an SL(2,C) scalar) and, hence, certainly not SU(6,6) invariant. Hence, the auxiliary components of the wavefunction cannot be eliminated in a way which is simultaneously SU(6,6) and $P_{+}^{\dagger}$ invariant.

The problem becomes particularly acute in connection with probability conservation in scattering theory. Probability conservation is expressed through the unitarity condition

$$S^{\dagger}S = SS^{\dagger} = 1 \quad ,$$

for the scattering matrix. Now consider this equation in matrix notation,

$$\sum_{n} (i,S^{\dagger}n)(n,Sj) = S_{ij} \quad .$$

If in the sum $\sum_{n}$ we put in all the SU(6,6) states, then we have SU(6,6) invariance, but we do not have true probability conservation since the sum is not over the physical states. If, on the other hand, we include in the $\sum_{n}$ only the physical states, then we have true probability conservation, but we do not have SU(6,6) invariance since the projections on the physical states, as we have just seen, are not SU(6,6) invariant. Thus, for the scattering matrix, physical unitary and SU(6,6) invariance are mutually incompatible.

Of course, one might legitimately ask: Why should the S-matrix be SU(6,6) invariant? After all, it is not SU(6) invariant. The point is that by making SU(6) relativistic one had hoped to overcome the defect that SU(6) was not an invariance group. The failure to overcome that defect is a serious setback for SU(6,6), and together with the failure of SU(6,6) to provide useful new predictions, it has led to its abandonment.

The third attempt (c)) to make SU(6) relativistic is, in a sense, more ambitious than b). It rests on the observation that SU(3) cannot be completely independent of the space-time coordinates x since it does not commute with the space-time mass operator. So the attempt is to combine SU(3) and the full

Poincaré group $P_+^\uparrow$ in a larger group G. Of course, for the combination to be useful, some restrictions must be placed on G. (The group of all possible unitary transformations on Fock space obviously contains both SU(3) and $P_+^\uparrow$, but this observation contains no useful information.) The two main restrictions which have been suggested for G are:

1) In the limit of SU(3)-symmetry, G is an invariance group for the S-matrix.

2) Whether or not it is an invariance group, G is a Lie group.

Unfortunately, both suggestions run into trouble. In Case 1), one can show [6] that under very general conditions either $S = 1$ (no scattering) or $G = P_+^\uparrow \otimes G_0$, where $\otimes$ denotes direct product and G_0 contains SU(3), i.e., either $S = 1$ or the combination is trivial. In Case 2), one can show that in any irreducible representation of G the mass spectrum of $P_+^\uparrow$ has no gaps [7]. Hence, G would be unsuitable for classifying the hadrons. Even apart from this kind of trouble, the difficulties of combining SU(3) and $P_+^\uparrow$ in a larger group G can be seen by considering the action of G on $P_+^\uparrow$ and SU(3) space respectively. One can see that the action cannot make much physical sense unless the combination is trivial [8].

The failure of attempts b) and c) to make SU(6) relativistic make it appear that if SU(6) is to be regarded as anything other than a nonrelativistic accident, one must look elsewhere for a framework in which to embed it. Such a framework is provided by current algebra, which we shall discuss in the next chapter.

12. CURRENT ALGEBRA

In the last two sections, we saw that the elementary particles exhibit regularities or symmetries other than those demanded by Poincaré invariance. However, none of the symmetries is exact. U(2) symmetry becomes exact only in the limit that weak and electromagnetic interactions are neglected, SU(3) symmetry is broken to within about 20% by even the strong interactions, and SU(6) symmetry works at best in a haphazard and empirical way. The question is: Could one find a framework within which the U(2), SU(3), and SU(6) results could be understood in a coherent fashion? We have already seen that the idea of putting SU(3) and SU(6) into larger groups is rather unsuccessful. In the present lecture, we wish to discuss a more successful approach, namely, current algebra [1].

The starting point for the introduction of current algebra is the idea that the fundamental objects for strong interaction physics are not the fields $\psi(x)$ but the currents

$$j_\mu^X(x) \qquad (12.1)$$

(which in a field theory would be constructed out of the fields). The role of the currents is to mediate the interactions. For example, the electromagnetic interactions of all the particles (strongly interacting or not) are assumed to take place via an interaction Hamiltonian of the form

$$H_e = e \int d^3x j_\mu(x) A_\mu(x) \quad , \tag{12.2}$$

where e is the electric charge, $j_\mu(x)$ the electric current, and $A_\mu(x)$ the electromagnetic potential. Similarly, the leptonic and non-leptonic weak decays of the (otherwise) strongly interacting particles are assumed to take place via interaction Hamiltonians of the form

$$H_{n\ell} = G \int d^3x j_\mu^\omega(x) j_\mu^\omega(x) \quad , \quad H_\ell = G \int d^3x j_\mu^\omega(x) j_\mu^\ell(x) \quad , \tag{12.3}$$

where G is the weak coupling constant, $j_\mu^\omega(x)$ is the weak current of the strongly interacting particles, and $j_\mu^\ell(x)$ is the weak current of the leptons.

From the <u>currents</u> $j_\mu^X(x)$, we can define charges

$$X(t) = \int d^3x j_0^X(x) \quad . \tag{12.4}$$

What current algebra assumes is that independent of the form or even the existence of an underlying field theory, the charges and currents satisfy simple algebraic relations among themselves (analogous to $[X,P] = i\hbar$ in quantum mechanics). The postulated relations are

charge-charge algebra $\quad [X,Y] = iZ \quad ,$ (12.5a)

charge-current algebra $\quad [X,j_\mu^Y(x)] = ij_\mu^Z(x) \quad ,$ (12.5b)

current-current algebra $\quad [j_0^X(x),j_0^Y(x')] = ij_0^Z(x)\delta(x - x') \quad ,$ (12.5c)

$\quad [j_0^X(x),\vec{j}^Y(x')] = i\vec{j}^Z(x)\delta(x - x') + S(x,x')$ (12.5d)

(12.5a)–(12.5d): At equal times

where the structure constants of the algebra in question are in practice those of SU(2), SU(3), SU(2) × SU(2), SU(3) × SU(3) (and, with a modification to be discussed later, SU(6) × SU(6)). "At equal times" means that the time variable in X, Y and Z, etc., has the same value. The term $S(x,x')$ in the last equation is called a Schwinger term [2]. It is inserted because it can be shown that without it the equation would not be consistent. $S(x,x')$ is unknown, but it is usually assumed to be purely symmetric in X and Y, so that at least the antisymmetric part of the last equation is not empty.

Note particularly that since the algebraic relations (12.5) are nonlinear, they normalize the currents and hence make it meaningful to say that the coupling constants in (12.2), (12.3) are small, large, universal, etc. In fact, the need to normalize the weak currents was one of the motivations for current-algebra [3].

In general, it is not assumed that the charges are time independent. However, we have the equivalence relations

$$\frac{dX(t)}{dt} = 0 \Leftrightarrow [H,X(t)] \Leftrightarrow \partial_\mu j_\mu^X(x) = 0 \Leftrightarrow X(t)|0\rangle = 0 \quad , \qquad (12.6)$$

where H is the Hamiltonian under consideration and $|0\rangle$ is the vacuum state. The first two relations are fairly obvious. The last follows from a theorem due to Colemen [4].

The question now is: How are physical consequences to be extracted from this formal algebra?

Let us first consider the exact symmetry limit, e.g., SU(2) with weak and electromagnetic interactions neglected or SU(3) with the 20% SU(3) breaking interaction neglected. In that limit (12.6) holds for all the charges and the charge × charge algebra becomes the usual SU(2) or SU(3) symmetry algebra, with the charges as generators. In particular, if the physical Hilbert $\mathcal{H}$ is decomposed with respect to the charge algebra (12.5a), the mass degenerate particles can be, and are, assigned to irreducible subspaces of the algebra. We then obtain the usual SU(2) or SU(3) theory. In particular, the charge × current algebra (12.5b) then becomes the assignment of tensor properties to the current as described in the last two chapters.

The real advantage of the current algebra appears when the symmetry is not exact. In that case, it is assumed that the current algebra relations (12.5) are exact, but that (12.6) does not hold and, hence, that the assignment of particles to SU(3) subspaces of $\mathcal{H}$ is incorrect. However, it is assumed that there is at least a subalgebra of the charge algebra which is exact and is large enough to locate the particles in $\mathcal{H}$ relative to the algebra. (The subalgebra is that of U(2) for SU(3) and that of U(1) for SU(2).)

Having placed the particles relative to the algebra, the physical information is then extracted as follows: Consider the charge × charge relation $[X,Y] = iZ$. The presently measurable matrix elements of X, Y, Z are their values between 1-particle states. It is, therefore, suggestive to sandwich the equation $[X,Y] = iZ$ between 1-particle states. Let us denote 1-particle states by (n) and 2-or-more-particle states by (c). We obtain

$$\sum_{n'} (n, X\, n')(n', Y\, n'') + \sum_{c} (n, X\, c)(c, Y\, n'') - X \Leftrightarrow Y = i(n, Z\, n'') , \quad (12.7)$$

where the sum $\sum_{n'}$ runs over all the 1-particle states and $\sum_c$ over all the many-particle states. Now if

$$\sum_c (n, X\, c)(c, Y\, n'') = 0 \quad , \qquad (12.8)$$

we would be in a strong position with regard to experiment, since we would have a direct algebraic statement about the measurable quantities $(n, X\, n')$. However, in general, (12.8) is not true. Indeed, (12.8) is true essentially only in the exact symmetry limit since (12.8) implies that at least one of X and Y leaves the 1-particle states invariant and, in general, this can only happen if they leave the vacuum invariant as well.

Thus, in general one cannot omit the c-summation in (12.7), and one must proceed otherwise. How one proceeds depends on the matrix elements to be calculated. We shall mention here only two well-known examples:

1) Adler-Weisberger calculation [5]: One uses $SU(2) \times SU(2)$ algebra, namely one assumes that the isospin charges T_i and the charges A_i belonging to the axial vector current $j^A_\mu(x)$ satisfy the relations

$$[T_i,T_j] = i\varepsilon_{ijk}T_k \quad ,$$

$$[T_i,A_j] = i\varepsilon_{ijk}A_k \quad , \qquad i = 1,\ 2,\ 3 \qquad (12.5a)'$$

$$[A_i,A_j] = i\varepsilon_{ijk}T_k \quad .$$

Then one chooses $n = n'' =$ proton, and $X, Y = A_\pm = (A_1 \pm iA_2)/2$. It follows that $Z = I_3$ and $n' =$ neutron. If

$$g_A = (n\ A^+\ p) \quad ,$$

denotes the weak coupling constant between the neutron and proton, (12.7) reduces to

$$|g_A|^2 + \sum_c (pA^+c)(cA^-p) - (pA^-c)(cA^+p) = 1 \quad . \qquad (12.10)$$

Thus, we would have a prediction for $|g_A|^2$ if we could evaluate $\sum_c$. (In particular, $|g_A|^2$ would be 1 if $\sum_c$ were zero.) To evaluate $\sum_c$, one makes the so-called PCAC (partially conserved axial current) hypothesis, namely, that

$$\partial_\mu A^\mu_\pm(x) = \kappa g_A \pi_\pm(x) \quad , \qquad (12.11)$$

where $\pi_\pm(x)$ is the field of the $\pi^\pm$-meson, and the constant κ is determined from the decays $n \to p +$ leptons and $\pi \to$ leptons. Integrating (12.11) to

$$\frac{1}{i}\frac{d}{dt} A^\pm(t) = \kappa g_A \int d^3x\ \pi_\pm(x) \quad , \qquad (12.12)$$

and inserting the result into $\sum_c$, one obtains

$$|g_A|^2 \left\{1 + \kappa^2 \sum_c \frac{(p\pi^+c)(c\pi^-p) - (p\pi^-c)(c\pi^+p)}{(E_c - E_p)^2}\right\} = 1 \quad . \qquad (12.13)$$

The point now is that the Σ_c-term can be directly related to the cross-sections $\sigma_\pm(c)$ for $\pi_\pm p$ scattering. Inserting the observed values for $\sigma_\pm(c)$, one obtains $|g_A|^2 \simeq 1.18$, in excellent agreement with experiment. Note that the entire departure of $|g_A|^2$ from unity comes from the non-conservation of $A_\pm(t)$ (Equation (12.12)).

2) The second example [6] uses $SU(6) \times SU(6)$ algebra, or at least that part of it in which

$$[\vec{A},\vec{A}] = T + \vec{A} \quad , \tag{12.14}$$

where T is the isospin charge (generator of the isospin group) and $\vec{A}$ is the spatial charge

$$\vec{A}(t) = \int d^3x \, \vec{a}(x) \quad , \tag{12.15}$$

where $\vec{a}(x)$ is the space-part of the $SU(3)$ axial vector current. The use of the spatial charge is peculiar to $SU(6) \times SU(6)$. Only the time-component charges (12.4) are used for $SU(2)$, $SU(3)$, $SU(2) \times SU(2)$, and $SU(3) \times SU(3)$. Inserting Equation (12.14) between $\frac{1}{2}^+$ and $\frac{3}{2}^+$ states, denoted by N, we obtain

$$(N \, \vec{A}\left[\sum_n n)(n + \sum_c c)(c\right]\vec{A}' \, N) - \vec{A} \Leftrightarrow \vec{A}' = (N, T + \vec{A}'' \, N) \quad . \tag{12.16}$$

If one now makes the approximation of replacing the sum over n and c by a sum over N only, i.e.,

$$\sum_n n)(n + \sum_c c)(c \to \sum_N N)(N \quad , \tag{12.17}$$

in (12.16), one obtains

$$\sum_{N'} (N,\vec{A}N')(N',\vec{A}'N) - \vec{A} \Longleftrightarrow \vec{A}' = (N,T + \vec{A}''N) \quad , \tag{12.18}$$

and by choosing appropriate members of N and $\vec{A}$, one can derive from (12.18) practically all the interesting $SU(6)$ results. Thus, $SU(6)$ can be simply understood as a combination of the charge-algebra (12.14) and the saturation assumption (12.17). It should be emphasized that the masses of the particles N are not assumed to be the same and the charges $\vec{A}(t)$ are not assumed to be time-independent.

These examples and other applications of the charge algebra support the view that the correct way to understand $SU(3)$, $SU(3) \otimes SU(3)$, etc., is not as exact symmetry groups, but as exact charge algebras. Any approximations to be made are made in the saturation of the algebra (the sum over intermediate states).

So far, we have discussed only the charge $\times$ charge algebras (when the symmetry is not exact). However, the charge $\times$ current algebras can be similarly handled, and in recent years most of what is called current algebra theory has been devoted to systematically (and very successfully) exploiting the charge $\times$ current algebras.

Let us sketch very briefly the kind of idea involved for one of the most important applications [1] of current algebra, namely, the derivation of what are called low energy theorems. For an interaction involving an external π-meson, the matrix element of interest can be written as

$$M = (a, T\int e^{ipx+qy\ldots}\pi(x)\varphi(y)\ldots d^4(xy\ldots), b) \quad , \tag{12.19}$$

where a and b are initial and final states, p_μ is the meson 4-momentum, $\pi(x)$ is the meson field, $\varphi(y)$ any other typical field or current (possibly another π-meson field), and T is the time ordering operator $(T(\pi(x)\varphi(y)) = \varphi(y)\pi(x)$, $\pi(x)\varphi(y)$ for $x_0 < y_0$, $x_0 > y_0)$.

Replacing $\pi(x)$ by $\partial_\mu A_\mu(x)$ according to (12.11), we obtain

$$\begin{aligned} M &= (a, T\int \ldots\partial_\mu A_\mu(x)\ldots b) \\ &= p_\mu(a, T\int \ldots A_\mu(x) b) - \sum (a, T'\int \ldots [A_0(x), \varphi(y)]_{ET}\ldots b) \quad , \end{aligned} \tag{12.20}$$

where the second term comes from the fact that the time derivative does not commute with the time-ordering T. The non-commutativity of ∂_0 and T can be expressed in the form $\frac{d}{dt}\theta(t) = \delta(t)$ and hence leads to <u>equal-time</u> commutators such as the commutator $[A_0(x), \varphi(y)]_{ET}$ exhibited, together with a residual time ordering T' for the remaining unequal times.

Now because the mass of the pion is small, for processes for which the pion 3-momentum is small, it is legitimate to let $p_\mu \to 0$. Then the first term in (12.20) vanishes and in the second term $\int e^{ipx} d^3x\, A_0(x) \to \int d^3x\, A_0(x) = A_0$ where A_0 is the axial charge. Hence, in the "soft-pion limit" $p_\mu \to 0$, M is dominated by the second term in (12.20), and the second term, in turn, is determined by the equal time charge × current commutator $[A_0, \varphi(y)]$ of the charge × current algebra. In this way the charge × current algebra determines the low energy or soft pion limit of π-meson processes. The argument generalizes, of course, to processes with more than one π, e.g., $\pi - p$ scattering (Figure 12.1).

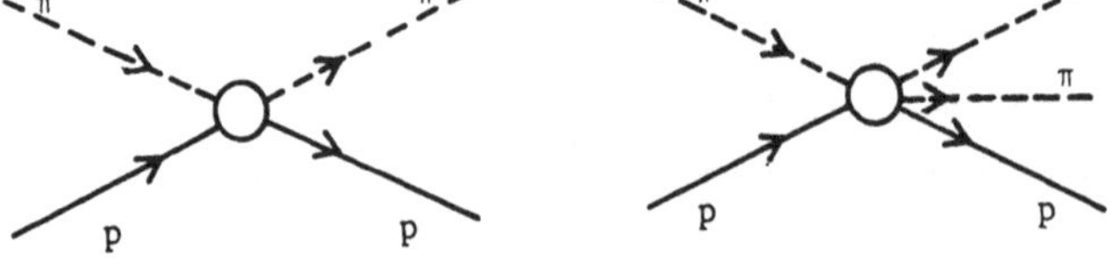

FIGURE 12.1. ELASTIC AND π-PRODUCING π-p SCATTERING

The success achieved with charge × charge and charge × current algebra tempts one to go farther and assume the current × current algebra. The current

× current algebra has not yet been severely tested experimentally, but its simplicity is appealing, as is the fact that it yields the charge × current and charge × charge algebras on integration. Note that neither the charge × current nor the current × current algebra is a Lie algebra, and a mathematical problem of some interest at present is to find all the unitary irreducible representations of an algebra of this form, i.e., an algebra of the form

$$[X_\alpha(x),X_\beta(y)] = f_{\alpha\beta\gamma}X_\gamma(x)\delta(x - y) \quad , \tag{12.21}$$

where the $f_{\alpha\beta\gamma}$ are the structure constants of a simple Lie group and $x \in R^3$.

An algebra of the form (12.21) would be particularly useful if the sum over all intermediate states to be inserted between the operators on the left hand side of (12.21) could be approximated (saturated) by a sum over a number of 1 particle states (not necessarily a finite number). This is because a saturation with 1-particle states would clearly yield algebraic relations for quantities of the form

$$\langle \text{1-particle, } X_\alpha(x) \ \text{ 1-particle}\rangle \quad , \tag{12.22}$$

and such quantities have the property that their Fourier transforms with respect to x are the form-factors for the particles and so are within reach of experiment. Unfortunately, the saturation with 1-particle states raises some difficulties of principle. One can show, for example, that unless the current $j_\mu(x)$ is trivial, the current × current algebra (12.21) cannot be even approximately saturated with 1-particle states (even if an infinite number of 1-particle states are used) unless the masses are degenerate. However, it has been conjectured [7] that in the limit that P_z, the third component of the total momentum of all the states, becomes infinite, the saturation with 1-particle states may become exact and lead to predictions for the mass-spectrum and the form factors, or at least to correlations between the two. This conjecture, which is based on experience with the free-Dirac equation and the charge × current algebra, is still open. Preliminary investigations, using, for simplicity, the special case of a <u>factored</u> current

$$\begin{aligned} j^0_\alpha(x) &= \lambda_\alpha j^0(x) \quad , \\ [\lambda_\alpha,\lambda_\beta] &= i\varepsilon_{\alpha\beta\gamma}\lambda_\gamma \quad , \\ j^0(x)j^0(y) &= j^0(x)\delta(x - y) \quad , \end{aligned} \tag{12.23}$$

show that in the factored case the solutions can be written as infinite component wave equations. This result furnishes another link between conventional physics and infinite component wave equations, but since, as we have seen in Lecture 8, infinite component equations have some undesirable physical properties, the result may only be an indication that the factorization hypothesis (12.23) is too strong.

13. REFERENCES

Chapter 1

[1] E. Whittaker, *Analytical Dynamics*, Cambridge (1937). C. Lanczos, *Variational Principles of Mechanics*, Toronto Univ. Press (1962).

[2] P. Dirac, *Rev. Mod. Phys.*, 21, 392 (1949). T. Jordan, E. C. G. Sudarshan, *ibid.*, 33, 515 (1961). M. Pauri, G. Prosperi, *J. Math. Phys.*, 7, 366 (1966). M. Hamermesh, *Group Theory*, Addison-Wesley, Cambridge, Mass. (1962).

Chapter 2

[1] N. Bohr, *Constitution of Atoms and Molecules*, Munksgaard, Copenhagen and Benjamin, New York (1963).

[2] E. Wigner, *Phys. Rev.*, 77, 711 (1950).

[3] F. Rellich, *Nachr. Akad. Wiss. Göttingen*, 11A, 107 (1946). J. Dixmier, *Comp. Math.*, 13, 263 (1958).

[4] H. Weyl, *Theory of Group Representations and Quantum Mechanics*, Methuen, London (1931).

[5] J. von Neumann, *Math. Ann.*, 104, 570 (1931).

[6] F. Riesz, B. Nagy, *Functional Analysis*, Blackie, London (1956).

[7] A. Wightman, *Proceedings Fifth Eastern Theoretical Conference*, edited by D. Feldman, Benjamin, New York (1967).

[8] M. Hamermesh, *Ann. Phys.*, 9, 518 (1960). J. Levy-Leblond, *J. Math. Phys.*, 4, 776 (1963). J. Voisin, *ibid.*, 6, 1822 (1965).

[9] E. Inönü, E. Wigner, *Nuovo Cimento*, 9, 705 (1952).

Chapter 3

[1] E. Wigner, *Group Theory*, Academic Press, New York (1959), p. 233.

[2] A list of papers is given in [3] below. See also J. Jauch, in *Group Theory and its Applications*, edited by E. Loebl, Academic Press, New York (1968), G. Ludwig, *Grundlagen der Quanten Mechanik*, Springer, Göttingen (1954), A. Messiah, *Quantum Mechanics*, Vol. II, North-Holland, Amsterdam (1962), L. O'Raifeartaigh and G. Rasche, *Ann. Phys.*, 25, 155 (1963).

[3] V. Bargmann, *J. Math. Phys.*, 5, 862 (1964).

[4] E. Condon, C. Shortley, *Theory of Atomic Spectra*, Cambridge (1935).

[5] P. Kramer and M. Moshinsky, in *Group Theory and its Applications*, edited by E. Loebl, Academic Press, New York (1968).

[6] The SO(4) symmetry of the H-atom was first analyzed by W. Pauli, *Z. Physik*, 36 (1926), V. Fock, *ibid.*, 98, 145 (1935), V. Bargmann, *ibid.*, 99, 576 (1936).

[7] C. Fronsdal, *Phys. Rev.*, 156, 1665 (1967). A. Barut, H. Kleinert, *ibid.*, 160, 1149 (1967).

Chapter 4

[1] C. Eckart, *Rev. Mod. Phys.*, 2, 302 (1930). E. Wigner, *Group Theory*, Academic Press, New York (1959).

[2] N. Akhiezer, I. Glazman, *Theory of Linear Operators in Hilbert Space II*, Ungar, New York (1963).

[3] L. Gårding, *Proc. Nat. Acad. Sci. U. S.*, 33, 331 (1947).

[4] I. Segal, *Duke Math. J.*, 18, 221 (1951).

[5] E. Nelson, *Lecture Notes*, ETH, Zürich (1963).

[6] P. Cartier, J. Dixmier, *Amer. J. Math.*, 80, 131 (1958).

[7] E. Nelson, *Ann. of Math.*, 70, 572 (1959).

[8] L. Gårding, *Bull. Soc. Math. France*, 88, 73 (1960).

[9] Harish-Chandra, *Proc. Nat. Acad. Sci. U. S.*, 37, 170 (1951).

[10] R. Goodman, *J. Functional Analysis*, 3, 246 (1969).

[11] R. Goodman, *Trans. Amer. Math. Soc.*, (to appear).

Chapter 5

[1] For more detailed information, see for example, E. Segré, *Nuclei and Particles*, Benjamin, New York (1964).

[2] P. Dirac, *Proc. Roy. Soc.*, 114A, 243 (1927). W. Heisenberg, W. Pauli, *Z. Physik*, 56, 1 (1929), 59, 160 (1930).

[3] For precise mathematical formulation, see R. Streater and A. Wightman, *PCT, Statistics and All That*, Benjamin, New York (1964) and R. Jost, *The General Theory of Quantized Fields*, Amer. Math. Soc., Providence, R. I. (1965).

Chapter 6

[1] The lifting problem has been analyzed for general topological groups by V. Bargmann, *Ann. of Math.*, 59, 1 (1954), D. Simms, *Lecture Notes in Mathematics*, Springer, Heidelberg (1968), K. Parthasarathy, *ibid.* (1969).

[2] H. Bacry, J. Levy-Leblond, *J. Math. Phys.*, 9, 1605 (1968).

[3] E. Wigner, *Ann. of Math.*, 40, 39 (1939). See also Y. Shirokov, *JETP*, 6, 919 (1958), H. Joos, *Fortschr. Physik*, 10, 65 (1962), A. Chakrabarti, *J. Math. Phys.*, 7, 949 (1966). J. Lomont, H. Moses, *ibid.*, 3, 405 (1962). N. Mukunda, *ibid.*, 9, 532 (1968). H. Moses, *ibid.*, 9, 16 (1968) and references therein.

[4] G. Mackey, *Induced Representations of Groups and Quantum Mechanics*, Benjamin, New York (1968).

[5] R. Newton, E. Wigner, *Rev. Mod. Phys.*, 21, 400 (1949). A. Wightman, *ibid.*, 34, 845 (1962).

[6] B. Schroer, *Fortschr. Physik*, 2, 1 (1963). I. Segal, *Proc. Nat. Acad. Sci. U. S. A.*, 57, 194 (1967).

Chapter 7

[1] The ideas of the present chapter are implicit in the work of Wigner *(Ann. of Math.* (1939)), V. Bargmann and E. Wigner, *Proc. Nat. Acad. Sci. U. S. A.*, 34 211 (1946), and are treated explicitly by Pursey, *Ann. Phys.*, 32, 157 (1965), to which the reader is referred for many other references.

[2] L. Foldy, *Phys. Rev.*, 102, 568 (1956).

[3] M. Jacob, G. Wick, *Ann. Phys.*, 7, 404 (1959).

[4] A. Wightman, *Symmetry Principles at High Energy*, edited by A. Perlmutter et al. Benjamin, New York (1968). S. Weinberg, *Phys. Rev.*, 133B, 1318 (1964).

[5] K. Johnson, E. Sudarshan, *Ann. Phys.*, 13, 126 (1961).

[6] I. Gelfand, R. Minlos, Z. Shapiro, *Representations of the Rotation and Lorentz Groups*, Pergamon, New York (1963).

[7] M. Fierz, *Helv. Phys. Acta*, 12, 3 (1939).

[8] W. Rarita, J. Schwinger, *Phys. Rev.*, 60, 61 (1941).

[9] V. Bargmann, E. Wigner, *Proc. Nat. Acad. Sci. U. S. A.*, 34, 211 (1946).

Chapter 8

[1] E. Majorana, *Nuovo Cimento*, 9, 335 (1932).

[2] See reference [6] of Chapter 7.

[3] E. Abers, I. Grodsky, R. Norton, *Phys. Rev.*, 159, 1222 (1967).

[4] G. Feldman, P. Mathews, *Phys. Rev.*, 154, 1241 (1967). C. Fronsdal, *ibid.*, 156, 1665 (1967). Further references can be found in L. O'Raifeartaigh, *Symmetry Principles at High Energy*, edited by A. Perlmutter et al., Benjamin, New York (1968) and I. Todorov, *Proceedings of Theoretical High Energy Conference*, Rochester (1967).

[5] I. Grodsky, R. Streater, *Phys. Rev. Lett.*, 20, 695 (1968).

[6] N. Bogoliubov, V. Vladimirov, *Nauch. Dokl. Vysshei*, Shkoly (1958), 3, p. 26. R. F. Streater, *Ph.D. Thesis*, London (1959), p. 41. J. Bros. H. Epstein, V. Glaser, *Comm. Math. Phys.*, 6, 77 (1967).

[7] A. Oksak, I. Todorov, *Degeneracy of the Mass-Spectrum for Infinite-Component Fields*, Princeton Institute for Advanced Study, Preprint (1970).

Chapter 9

[1] J. Hamilton, *Theory of Elementary Particles*, Oxford (1959). G. Chew, *S-matrix Theory of Strong Interactions*, Benjamin (1961). G. Chew, M. Jacob, *Strong Interaction Physics*, Benjamin (1964).

[2] R. Eden, P. Landshoff, D. Olive, J. Polkinghorne, *The Analytic S-matrix*, Cambridge (1966). R. Eden, *High Energy Collisions of Elementary Particles*, Cambridge (1967). G. Källen, *Elementary Particle Physics*, Addison-Wesley, New York (1964). A. Wightman, *Dispersion Relations and Elementary Particles*, edited by C. de Witt & R. Omnes, Wiley, New York (1960).

[3] L. Schiff, *Quantum Mechanics*, McGraw-Hill, New York (1949).

[4] T. Regge, *Nuovo Cimento*, 18, 947 (1960).

[5] G. Chew, S. Frautschi, *Phys. Rev. Lett.*, 8, 41 (1962). M. Gell-Mann, S. Frautschi, F. Zachariasen, *Phys. Rev.*, 126, 2204 (1962).

[62 E. J. Squires, *Complex Angular Momentum and Particle Physics*, Benjamin, New York (1963). M. Froissart, R. Omnes, *Mandelstam Theory and Regge Poles*, Benjamin, New York (1963). R. Newton, *The Complex J-plane*, Benjamin, New York (1964). S. Frautschi, *Regge-Poles and S-matrix Theory*, Benjamin, New York (1963).

[7] M. Toller, *Nuovo Cimento*, 37, 631 (1965). H. Joos, in *Lectures in Theoretical Physics*, University of Colorado, Boulder (1964), *Fortschr. Physik*, 10, 65 (1962).

[8] V. Bargmann, *Ann. of Math.*, 48, 586 (1947).

[9] J. Boyce, *J. Math. Phys.*, 8, 675 (1967).

[10] D. Freedman, J. Wang, *Phys. Rev.*, 153, 1596 (1967). G. Domokos, G. Tindle, *ibid.*, 165, 1906 (1968). M. Toller, *Nuovo Cimento*, 54, 295 (1968).

[11] A. Ahmadzadeh, R. Jacob, *Phys. Rev.*, 176, 1719 (1968).

Chapter 10

[1] A few references for isotopic spin are: J. Blatt, V. Weisskopf, *Theoretical Nuclear Physics*, Wiley, New York (1952); P. Roman, *Theory of Elementary Particles*, North-Holland, Amsterdam (1960); S. Schweber, *Relativistic Quantum Field Theory*, Row-Peterson, New York (1961).

Some references for SU(3) are: M. Gell-Mann, Y. Ne'eman, *The Eightfold Way*, Benjamin, New York (1964); M. Gourdin, *Unitary Symmetries*, North-Holland, Amsterdam (1967); P. Carruthers, *Introduction to Unitary Symmetry*, Wiley, New York (1966); E. Loebl, *Group Theory and its Applications*, Academic Press, New York (1968).

[2] See also: F. Lurçat, L. Michel, *Nuovo Cimento*, 21, 575 (1961); L. Michel in *Group Theoretical Concepts and Methods in Elementary Particle Physics - Istanbul Summer School 1962*, edited by F. Gürsey, Gordon & Breach, New York (1964).

Chapter 11

[1] M. Gell-Mann, *Phys. Lett.*, 8, 214 (1964). G. Zweig, *CERN Reports nos 8/82/TH.401 and 8419/TH.412* (1964).

[2] Although it is convenient to describe SU(6) in terms of quarks, they were not used explicitly in the original introduction, F. Gürsey, *Phys. Rev. Lett.*, 13, 173 (1964), A. Pais, L. Radicati, *ibid.*, 13, 175 (1964), F. Gürsey, A. Pais, L. Radicati, *ibid.*, 13, 299 (1964), B. Sakita, *Phys. Rev.*, 136, B1756 (1964). For a review article on SU(6), containing an extensive list of references, see A. Pais, *Rev. Mod. Phys.*, 38, 215 (1966).

[3] K. Bitar, F. Gürsey, *Phys. Rev.*, 164, 1805 (1964).

[4] B. Sakita, K. Wali, *Phys. Rev.*, 139, B1355 (1965). A. Salam, R. Delbourgo, J. Strathdee, *Proc. Roy. Soc.*, 284A, 146 (1965). M. Beg, A. Pais, *Phys. Rev. Lett.*, 14, 267 (1965).

[5] M. Beg, A. Pais, *Phys. Rev. Lett.*, 14, 509 (1965).

[6] S. Coleman, *Phys. Rev.*, 138, B1262 (1965). S. Coleman, J. Mandula, *Phys. Rev.*, 159, 1251 (1967).

[7] For a review of the mathematical aspects of the mass-spectrum problem and other mathematical aspects of the difficulty of combining SU(3) and $P_+^\uparrow$ in G see G. Hegerfeldt, J. Henning, *Fortschr. Physik*, 16, 491 (1968), 17, 463 (1969).

[8] L. Michel, *Phys. Rev.*, 137, B405 (1965). H. Lipkin, in *Symmetry Principles at High Energy*, edited by A. Perlmutter et al., Benjamin, New York (1968). See also: W. McGlinn, *Phys. Rev. Lett.*, 12, 467 (1964), E. C. G. Sudarshan, *J. Math. Phys.*, 6, 1329 (1965) and reference [6].

Chapter 12

[1] Current Algebra was originally proposed by M. Gell-Mann, *Physics*, 1, 63 (1964); *Phys. Rev.*, 125, 1067 (1962). The two standard books on current algebra are: S. Adler and R. Dashen, *Current Algebras*, Benjamin, New York, (1968); B. Renner, *Current Algebras and their Applications*, Permagon Press, Oxford (1968). See also A. Völkel, U. Völkel, *Nuovo Cimento*, 63A, 203 (1969).

[2] J. Schwinger, *Phys. Rev. Lett.* 3, 296 (1959).

[3] M. Gell-Mann, Proceedings Conference High Energy Physics held in Rochester, 1960, p. 508 in *The Eightfold Way*, Benjamin, New York (1964).

[4] S. Coleman, *J. Math. Phys.*, 7, 787 (1966).

[5] S. Adler, *Phys. Rev. Lett.*, 25, 1051 (1965). W. Weisberger, *ibid.*, 25, 1047 (1965).

[6] B. Lee, *Phys. Rev. Lett.*, 14, 676 (1965).

[7] R. Dashen, M. Gell-Mann, *Phys. Rev. Lett.*, 17, 340 (1966). S. Fubini, *Proceedings Fourth Coral Gables Conference 1967*, W. H. Freeman & Co., San Francisco (1967).

[8] S.-J. Chang, R. Dashen, L. O'Raifeartaigh, *Phys. Rev. Lett.*, 21, 1026 (1968). B. Hamprecht, H. Kleinert, *Phys. Rev.*, 180, 1410 (1969). M. Gell-Mann, D. Horn, J. Weyer, *Proceedings Heidelberg International Conference*, North-Holland, Amsterdam (1968). H. Leutwyler, *Phys. Rev. Lett.*, 20, 561 (1968). H. Bebié, F. Ghielmetti, V. Gargé, H. Leutwyler, *Phys. Rev.*, 177, 2196 (1969).

ON CERTAIN UNITARY REPRESENTATIONS
WHICH ARISE FROM A QUANTIZATION THEORY

by

Bertram Kostant*

In this paper we are concerned with certain explicit constructions of unitary representations which arise from a general theory relating quantization and unitary representations. We shall not go into the general theory here but we can refer the reader to a forthcoming publication entitled "Quantization and Unitary Representations, Part I - Prequantization" which will appear as part of the series "Lectures in Modern Analysis and Applications" edited by C. T. Taam, in *Lecture Notes in Mathematics* published by Springer-Verlag. Those considerations here for solvable groups are part of a joint work of L. Auslander and myself.

1. THE REPRESENTATION $\mathrm{ind}_G(\eta_g, h)$

Let G be a Lie group, not necessarily connected, and let $\mathfrak{g}$ be its Lie algebra.

Now let $g \in \mathfrak{g}'$ be a linear functional on $\mathfrak{g}$ and let $\mathfrak{g}_g$ be the Lie algebra of the isotropy subgroup $G_g \subseteq G$ with respect to the coadjoint representation of G on $\mathfrak{g}'$. Thus if B_g is the alternating bilinear form on $\mathfrak{g}$ given by $B_g(x,y) = \langle g,[y,x]\rangle$ then

$$\mathfrak{g}_g = \{x \in \mathfrak{g} \mid B_g(x,y) = 0 \quad \text{for all} \quad y \in \mathfrak{g}\} \quad .$$

That is $\mathfrak{g}_g$ is the radical of B_g.

We may regard g as a complex valued linear functional on $\mathfrak{g}_{\mathbb{C}} = \mathfrak{g} + i\mathfrak{g}$. A polarization at g is a complex subalgebra $h \subseteq \mathfrak{g}_{\mathbb{C}}$ such that

(1) $\mathfrak{g}_g \subseteq h$ and $\mathfrak{g}_g$ is stable under $\mathrm{Ad}\, G_g$ (note that G_g is not necessarily connected even if G is connected)

(2) $\dim_{\mathbb{C}} \mathfrak{g}_{\mathbb{C}}/h = 1/2 \dim_{\mathbb{R}} \mathfrak{g}/\mathfrak{g}_g$ (recall $\dim_{\mathbb{R}} \mathfrak{g}/\mathfrak{g}_g$ is even since $\mathfrak{g}_g$ is the radical of B_g)

* Department of Mathematics, Massachusetts Institute of Technology, Cambridge, Massachusetts. Currently at Tata Institute, Department of Mathematics, Bombay, India.

(3) $g|[h,h] = 0$, i.e., $g|h$ is a homomorphism

(4) $h + \bar{h}$ is a Lie algebra of $g_{\mathbb{C}}$.

Now let $d = h \cap g$ so that if $d_{\mathbb{C}} = d + id$ one has

$$d_{\mathbb{C}} = h \cap \bar{h} \quad .$$

Also let $e = (h + \bar{h}) \cap g$ so that if $e_{\mathbb{C}} = e + ie$ one has

$$e_{\mathbb{C}} = h + \bar{h} \quad .$$

Now clearly h is equal to its own orthogonal subspace relative to the extension of B_g to $g_{\mathbb{C}}$. It follows easily then that d is the orthogonal subspace to e relative to B_g and hence if $\hat{x} \in e/d$ denotes the image of $x \in e$ under the quotient map $e \to e/d$ one defines a non-singular alternating bilinear form $\hat{B}_g$ on e/d by the relation

$$(\hat{x},\hat{y}) = \langle g,[y,x]\rangle$$

for $x,y \in e$. Next note that we may identify $(e/d)_{\mathbb{C}}$ with $e_{\mathbb{C}}/d_{\mathbb{C}}$ so that

$$(e/d)_{\mathbb{C}} = h/d_{\mathbb{C}} \oplus \bar{h}/d_{\mathbb{C}}$$

is a linear direct sum. Since $\bar{h}/d_{\mathbb{C}} = (\overline{h/d_{\mathbb{C}}})$ relative to conjugation over the real form e/d of $(e/d)_{\mathbb{C}}$ one defines a non-singular operator $j \in \operatorname{End} e/d$ where $j^2 = -I$ and (upon complexification) $j = -i$ on $h/d_{\mathbb{C}}$ and $j = i$ on $\bar{h}/d_{\mathbb{C}}$.

Remark 1. Note that if $u \in e/d$ one has

$$u + iju \in h/d_{\mathbb{C}} \quad \text{and} \quad u - iju \in \bar{h}/d_{\mathbb{C}} \quad .$$

Let S_g be the bilinear form on e/d given by

$$\{u,v\} = (ju,v) \quad .$$

Proposition 1

S_g is a non-singular symmetric bilinear form on e/d. Moreover, j is orthogonal relative to both S_g and $\hat{B}_g$. That is, if $u,v \in e/d$ one has

$$\{ju,jv\} = \{u,v\} \quad \text{and} \quad (ju,jv) = (u,v) \quad .$$

Proof. It is clear that by definition $h/d_{\mathbb{C}}$ is orthogonal to itself relative to the extension of $\hat{B}_g$ to $(e/d)_{\mathbb{C}}$. Thus by Remark 1, one has for $u,v \in e/d$

$$0 = (u + iju, v + ijv) = [(u,v) - (ju,jv)] + i[(ju,v) + (u,jv)] \quad .$$

Since the imaginary part is zero this implies that

$$(ju,v) = -(u,jv) = (jv,u) \quad . \tag{1.1}$$

That is $\{u,v\} = \{v,u\}$ and hence S_g is symmetric. It is clearly non-singular since j is non-singular. The relation (1.1) together with $j^2 = -I$ clearly implies j is orthogonal relative to both S_g and $\hat{B}_g$.

We will say that the polarization h is positive in case S_g is a positive definite bilinear form. (This includes the case where $e/d = 0$, that is where $h = \overline{h}$.)

Remark 2. A simple criterion for the positivity of the polarization h without going to the quotient e/d is as follows: We assert that h is a positive polarization if and only if

$$-i(z,\overline{z}) \geq 0$$

for all $z \in h$. Indeed if $z \in h$ write $z = x + iy$ where $x,y \in e$. Thus $\hat{y} = j\hat{x}$ and hence $-i(z,\overline{z}) = -i(x + iy, x - iy) = 2(y,x) = 2(\hat{y},\hat{x}) = 2(j\hat{x},\hat{x}) = 2\{\hat{x},\hat{x}\}$. The relation then follows since the correspondence $z \mapsto x$ maps h onto e/d.

Now let $b = \{x \in d \mid \langle g,x \rangle\} = 0$. It follows that b has codimension 1 in d if and only if $g|d \neq 0$.

Remark 3. If g is nilpotent one knows that $g|g_g \neq 0$ and hence $g|d \neq 0$ if and only if $g \neq 0$.

Now let D_0 and E_0 be the connected Lie subgroups of G corresponding to $d = h \cap g$ and $e = (h + \overline{h}) \cap g$. Since h is stable under $\mathrm{Ad}\, G_g$ it follows that D_0 and E_0 are normalized by G_g and $D = G_g D_0$ and $E = G_g E_0$ are subgroups of G.

Proposition 2

The groups D *and* D_0 *are closed in* G. *Also* D_0 *is the identity component of* D *so that* d *is the Lie algebra of* D.

Proof. Since d and e are each other's orthogonal subspaces relative to B_g, one has that if $x \in g$. then $\langle x \cdot g, y \rangle = 0$ for all $y \in e$ if and only if $x \in d$. Thus

$$\langle a \cdot g - g, y \rangle = 0$$

for all $a \in D_0$ and hence for all $a \in \overline{D}_0$. But if x lies in the Lie algebra of $\overline{D}_0$ then clearly $\langle x \cdot g, y \rangle = 0$ for all $y \in e$ so that $x \in d$. Thus D_0 and $\overline{D}_0$ have the same Lie algebras and hence $D_0 = \overline{D}_0$.

Now let D_1 be the identity component of $\overline{D} = \overline{D_0 G_g}$. Then if $a \in D_1$ one has $\langle a \cdot g - g, y \rangle = 0$ for all $a \in D_1$, and $y \in e$. Then if d_1 is the Lie algebra of D_1 one has $d_1 \subseteq d$. But of course $d \subseteq d_1$ since $D_0 \subseteq D_1$. Thus

$d = d_1$ so that $D_0 = D_1$ is the identity component of $\overline{D}$. But $D_0 \subseteq D \subseteq \overline{D}$. Hence D is also closed and D_0 is the identity component of D.

QED

Now consider the D-orbit $D \cdot g \subseteq g'$. For any subspace $a \subseteq g$ let $\tilde{a}$ be its orthogonal subspace in g'.

Proposition 3

$D \cdot g$ *is an open set of the affine plane* $g + \tilde{e}$ *in* g'. *Also* $D \cdot g = D_0 \cdot g$.

Proof. We first observe that $g + \tilde{e}$ is stable under the action of D. Indeed since e is stable under Ad D clearly $\tilde{e}$ is stable under D. However, since $D = D_0 G_g$ one has $D \cdot g = D_0 \cdot g$ and hence if $b \in D$ and $f \in \tilde{e}$ one has $b \cdot (g + f) - g = a \cdot g - g + b \cdot f$ for some $a \in D_0$. But then $b \cdot (g + f) - g \in \tilde{e}$ (as above) so that $g + \tilde{e}$ is stable under D.

But now clearly $d \cdot g \subseteq \tilde{e}$. On the other hand one has a natural isomorphism $d \cdot g \cong d/g_g$. But then $\dim d \cdot g = \dim d/g_g = \dim \tilde{e}$. Hence $d \cdot g = \tilde{e}$. But $d \cdot g$ is the tangent space at g to the orbit $D_0 \cdot g \subseteq g + \tilde{e}$. Thus $D \cdot g$ is open in $g + \tilde{e}$.

QED

We will say that the polarization h satisfies the Pukansky condition (see [4]) if $E \cdot g$ is closed; in which case E is closed and

$$D \cdot g = g + \tilde{e} \quad . \tag{1.2}$$

Lemma 1

If h *satisfies the Pukansky condition then* $D_0 \cap G_g = (G_g)_0$, *the identity component of* G_g. *Furthermore, if* D_1 *is the simply connected covering group to* D_0 *and* $\tau: D_1 \to D_0$ *is the covering map then* $\tau^{-1}((G_g)_0) = (G_g)_1$ *is connected.*

Proof. As a D_0 homogeneous space one has $D \cdot g = D_0 \cdot g \cong D_0/D_0 \cap G_g$. But since $(G_g)_0 \subseteq D_0$ one has that $(G_g)_0$ is the identity component of $D_0 \cap G_g$. However by (1.2) one has that $D_0 \cdot g$ is simply connected so that $D_0 \cap G_g$ is connected. Thus $D_0 \cap G_g = (G_g)_0$. But now also since $D_1/(G_g)_1 \cong D_0/(G_g)_0$ the simple connectivity of $D_0 \cdot g$ implies that $(G_g)_1 = \tau^{-1}((G_g)_0)$ is also connected.

QED

Now g vanishes on $[g_g, g]$ so that in particular g vanishes on $[g_g, g_g]$ or $g|g_g$ is a homomorphism $g_g \to \mathbb{R}$ of Lie algebras. We will say that g is integral if there exists a character $\eta_g: G_g \to \mathbb{T}$ whose differential is

$2\pi i g|\mathcal{g}_g$. That is if for all $x \in \mathcal{g}_g$

$$\frac{d}{dt}\eta_g(\exp tx)\Big|_{t=0} = 2\pi i\langle g,x\rangle \ .$$

When this is satisfied we will say that η_g corresponds to g.

Remark 5. If G is connected and simply connected one knows that the existence of η_g is equivalent to the integrality of the de Rham class of the canonical symplectic 2-form on the orbit $G \cdot g \subseteq \mathcal{g}'$ (see Kostant, Quantization and Unitary Representations, Part I).

Now since $\langle g,[d,e]\rangle = 0$ then $g|d$ also defines a Lie algebra homomorphism $d \to \mathbb{R}$.

Until otherwise stated we will assume g is integral and η_g is a character on G_g corresponding to g.

Proposition 4

If the Pukansky condition is satisfied then η_g *extends to a unique character*

$$\chi_g: D \to \mathbb{T}$$

whose differential is $2\pi i g|d$.

Proof. Now let the notation be as in Lemma 1 so that D_1 is the simply-connected covering group to D_0. Now since $\langle g,[d,d]\rangle = 0$ there exists a unique character $\chi_g^1: D_1 \to \mathbb{T}$ whose differential is $2\pi i g|d$. Now if the Pukansky condition is satisfied, then by Lemma 1 $(G_g)_1$ is connected and clearly $\chi_g^1|(G_g)_1 = \eta_g|(G_g)_0 \circ \tau$. But then if Z is the kernel of the covering map $\tau: D_1 \to D_0$ one has $Z \subseteq (G_g)_1 = \tau^{-1}((G_g)_0)$ and $\chi_g^1|Z$ is trivial. Hence there exists a unique character $\chi_g^0: D_0 \to \mathbb{T}$ such that $\chi_g^1 = \chi_g^0 \circ \tau$. Clearly $2\pi i g|d$ is the differential of χ_g^0.

Now G_g normalizes D_0 and hence G_g operates on the character group of D_0. However, χ_g^0 is invariant under this action since $G_g \cdot g = g$ and hence $G_g \cdot g|d = g|d$ (of course a character on a connected Lie group is determined by its differential). It follows then that if we form the semi-direct product $G_g \times D_0$ then (η_g,χ_g^0) defines a character on this group. However by Lemma 1 $G_g \cap D_0 = (G_g)_0$ and $\eta_g = \chi_g^0$ on $(G_g)_0$ so that (η_g,χ_g^0) is trivial on the kernel K of the surjection $\sigma: G_g \times D_0 \to D$ given by $(a,b) \to ab$. Thus (η_g,χ_g^0) is of the form $\chi_g \circ \gamma$ where χ_g is a character on D satisfying the conditions of the proposition. As such it is unique since $D = D_0G_g$ and χ_g is obviously uniquely determined on G_g and D_0. QED

Assume that h is a polarization satisfying the Pukansky condition.

Now let $X = E/D$. Since $E_0D = E$ it is clear that X is connected. On the other hand since $\hat{B}_g$ is a non-singular alternating bilinear form on e/d which is invariant under the action of D it is clear that X has a measure μ_X invariant under the action of E.

Now consider the space $M(E,\chi_g)$ of all measurable functions ϕ on E such that $\phi(ab) = \chi_g(b)^{-1}\phi(a)$ for all $a \in E$, $b \in D$. Then $M(E,\chi_g)$ is an E-module where if $a \in E$, $\phi \in M(E,\chi_g)$ then $a \cdot \phi \in M(E,\chi_g)$ is given by $(a \cdot \phi)(b) = \phi(a^{-1}b)$. Then if $\mathcal{H}(E,\chi_g)$ is the space of equivalence classes (defined by sets of measure zero) of $\phi \in M(E,\chi_g)$ such that $\|\phi\|^2 = \int|\phi|^2 d\mu_X$ is finite then $\mathcal{H}(E,\chi_g)$ is the Hilbert space associated with the unitary representation $\text{ind}_E\chi$. Since μ_X is an E-invariant measure one has $((\text{ind}_E\chi)(a))\phi = a \cdot \phi$ for $a \in E$, $\phi \in \mathcal{H}(E,\chi_g)$ (conforming to the usual abuse of language).

Now recall $h \cap \bar{h} = d_{\mathbb{C}}$ and $h + \bar{h} = e_{\mathbb{C}}$.

If $C^\infty(E)$ is the space of all C^∞ functions on E we note that $C^\infty(E)$ is a right $e_{\mathbb{C}}$ module where if $z = x + iy \in e_{\mathbb{C}}$ with $x,y \in e$ then if $\phi \in C^\infty(E)$ one puts $\phi \cdot z = \phi \cdot x + i\phi \cdot y$ and if $a \in E$

$$(\phi \cdot x)(a) = \frac{d}{dt}\phi(a \exp - tx)\Big|_{t = 0} \quad .$$

Clearly if $\phi \in C^\infty(E)$, $a \in E$, $z \in e_{\mathbb{C}}$ then

$$(a \cdot \phi) \cdot z = a \cdot (\phi \cdot z) \quad . \tag{1.3}$$

Now if $o \in X = E/D$ is the coset D then the tangent space $T_o(X)$ at o may be identified with e/d. Hence upon complexification

$$(T_o(X))_{\mathbb{C}} = e_{\mathbb{C}}/d_{\mathbb{C}} = h/d_{\mathbb{C}} \oplus \bar{h}/d_{\mathbb{C}} \quad .$$

<u>Proposition 5</u>

There is an E*-invariant complex structure on* X *such that* $h/d_{\mathbb{C}}$ *is the space of anti-holomorphic vectors at* o.

<u>Proof</u>. We define a complex distribution F on X such that for any $p \in X$ one has

$$(T_p(X))_{\mathbb{C}} = F_p \oplus \bar{F}_p$$

by putting $F_p = a_*(h/d_{\mathbb{C}})$ where $a \cdot o = p$, $a \in E$. This depends only on p and not on $a \in E$ since $h/d_{\mathbb{C}}$ is invariant under Ad D. Clearly F is E-invariant. By Nirenberg-Newlander, to prove that F_p is the space of anti-holomorphic tangent vectors at p, we have only to prove that F is involutory. That is, if ξ,η are two complex vector fields on X such that $\xi_p,\eta_p \in F_p$ for all X then $\zeta_p \in F_p$ for all $p \in X$ where $\zeta = [\xi,\eta]$. But this condition is purely local. If

$p \in X$ let $U \subseteq X$ be a neighborhood of p with the property that

$$\sigma: U \to E$$

is a smooth section of the projection $\pi: E \to E/D = X$. Then there exists an open neighborhood V of the identity on D such that the map

$$\tilde{\sigma}: U \times V \to W \in E$$

is a diffeomorphism onto an open set $W \subseteq E$ where $\tilde{\sigma}(a,b) = \sigma(a)b$. But let $\tilde{\xi},\tilde{\eta}$ be the complex vector fields on W defined by $\tilde{\xi} = (\tilde{\sigma})_*(\xi,0)$, $\tilde{\eta} = (\tilde{\sigma})_*(\eta,0)$. Clearly $\pi_*\tilde{\xi} = \xi, \pi_*\tilde{\eta} = \eta$. But then if F_h is the left invariant complex distribution on E defined by h, then F_h is involutory since h is a subalgebra (we are in the group case). However, $\tilde{\xi}_a, \tilde{\eta}_a \in (F_h)_a$ for any $a \in W$ since $h = \pi_*^{-1}(h/d_{\mathbb{C}})$. Then $[\tilde{\xi},\tilde{\eta}]_a \in (F_h)_a$ for any $a \in W$. However, $\zeta = \pi_*[\tilde{\xi},\tilde{\eta}]$ since $\tilde{\xi}$ is π-related to ξ, and $\tilde{\eta}$ is π-related to η. Thus $(\zeta)_p \in F_p$ for all $p \in U$. Hence F is involutory. QED

We can now speak of holomorphic functions on any open set $V \subseteq X = E/D$. In fact if

$$\pi: E \to X$$

is the quotient map then these are just the elements of $\phi \in C^\infty(V)$ such that, for all $z \in h$,

$$(\phi \circ \pi) \cdot z = 0 \tag{1.4}$$

in $\pi^{-1}(V)$.

Now let $C(E,\chi_g,h)$ be the set of all C^∞ functions ψ in $M(E,\chi_g)$ such that

$$\psi \cdot z = 2\pi i\langle g,z\rangle\psi$$

for all $z \in h$. By (1.3) it is clear that $C(E,\chi_g,h)$ is stable under the action of E and hence if

$$\mathcal{H}(E,\eta_g,h) = C(E,\chi_g,h) \cap \mathcal{H}(E,\chi_g)$$

(abuse of language) then $\mathcal{H}(E,\eta_g,h)$ is stable under $\mathrm{ind}_E\, \chi_g$.

<u>Remark 6</u>. Since χ_g is determined by η_g and h we use η_g in the notation rather than χ_g.

<u>Proposition 6</u>

$\mathcal{H}(E,\eta_g,h)$ is a closed subspace of the Hilbert space $\mathcal{H}(E,\chi_g)$.

<u>Proof</u>. We may assume $\mathcal{H}(E,\eta_g,h) \neq 0$. Let $a \in E$ and $p = \pi a \in X$. Since $\mathcal{H}(E,\eta_g,h) \neq 0$ there exists (by translation if necessary) an element $\psi \in \mathcal{H}(E,\eta_g,h)$ such that $\psi(a) \neq 0$. Let U be an open neighborhood of a with compact closure

such that $A > |\psi| > \epsilon > 0$ in U. Let $V = \pi(U) \subseteq X$.

Now if $\beta \in M(E,\chi_g)$ then clearly one has that $\beta = (\phi \circ \pi)\psi$ in U where ϕ is a measurable function on V. Also $\phi \in C^\infty(V)$ if and only if $\beta|U \in C^\infty(U)$. But now $\beta \in \mathcal{H}(E,\eta_g,h)$ so that for $z \in h$ one has $2\pi i\langle g,z\rangle\beta = \beta \cdot z$ $= ((\phi \circ \pi) \cdot z)\psi + (\phi \circ \pi)(\psi \cdot z)$. But also $\psi \cdot z = 2\pi i\langle g,z\rangle\psi$ so that one has $((\phi \circ \pi) \cdot z)\psi = 0$ which implies $(\phi \cdot \pi) \cdot z = 0$. Thus by (1.4) one has ϕ is holomorphic and hence $\beta \mapsto \phi$ defines a map

$$\mathcal{H}(E,\eta_g,h) \to B_0(V)$$

where $(B_0(V))$ is the space of all bounded holomorphic functions in V.

On the other hand (taking U small enough) if $z^1,\cdots,z^m$ are the holomorphic coordinates in V then the measure $i^{m^2} dz_1\wedge\cdots\wedge dz_m\wedge d\bar{z}_1\wedge\cdots\wedge d\bar{z}_m$ is absolutely continuous with bounded (from above and below) Radon-Nikodyn derivative with respect to $\mu_X|V$. But now if β_n is Cauchy in $\mathcal{H}(E,\eta_g,h)$ and $\beta_n = (\phi_n \circ \pi)\psi$ in U where $\phi_n \in B_0(V)$ then clearly $\phi_n dz^1\wedge\cdots\wedge dz^m$ is Cauchy in $B(V)$ using the notation of (Weil, [5], p. 59). Since $B(V)$ is complete (see again Weil, p. 59) it follows that $\phi_n dz^1\wedge\cdots\wedge dz^m \to \rho dz^1\wedge\cdots\wedge dz^n$ in $B(V)$ where ρ is holomorphic in V. But ϕ_n converges to ρ uniformly on compact subsets of V by Proposition 5 in Weil. On the other hand if $\beta_n \to \beta$ in $\mathcal{H}(E,\chi_g)$ where $\beta = (\phi \circ \pi)\psi$ in U for ϕ a measurable function on V one has $\phi_n \to \phi$ almost everywhere. Thus $\phi = \rho$ almost everywhere. But clearly $((\rho \circ \pi)\psi) \cdot z = 2\pi i\langle g,z\rangle(\rho \circ \pi)\psi$ on U for $z \in h$. Thus the equivalence class of β contains an element in $\mathcal{H}(E,\eta_g,h)$ proving that $\mathcal{H}(E,\eta_g,h)$ is complete. QED

Now since $\mathcal{H}(E,\eta_g,h)$ is stable under $\mathrm{ind}_E\,\chi_g$ it defines a subrepresentation $\mathrm{ind}_E(\eta_g,h)$ of $\mathrm{ind}_E\,\chi_g$. But since

$$\mathrm{ind}_G(\mathrm{ind}_E\,\chi_g) = \mathrm{ind}_G\,\chi_g$$

it follows that if

$$\mathrm{ind}_G(\eta_g,h) = \mathrm{ind}_G\,\mathrm{ind}_E(\eta_g,h)$$

then $\mathrm{ind}_G(\eta_g,h)$ is a subrepresentation of $\mathrm{ind}_E\,\chi_g$. We denote the corresponding Hilbert space by $\mathcal{H}(G,\eta_g,h)$.

<u>Remark 7</u>. It is clear that if μ_Z is a G-quasi invariant measure on G/E then $\mathcal{H}(G,\eta_g,h)$ can be taken to be the set of all equivalence classes of measurable functions ϕ on G such that $\phi_a \in \mathcal{H}(E,\eta_g,h)$ for all $a \in G$, and such that

$$\int_Z \|\phi_a\|^2 d\mu_Z(\bar{a}) < \infty$$

where $\phi_a(b) = \phi(ab)$ for $b \in E$ and $\bar{a} \in Z$ is the image of a in Z.

Remark 8. We recall for emphasis that $\mathrm{ind}_G(\eta_g,h)$ is defined when (1) $g \in \mathfrak{g}'$ is integral and (2) h is a polarization satisfying the Pukansky condition. However it may reduce to the zero representation if $\mathcal{H}(E,\eta_g,h)$ reduces to zero. From the point of view of the general quantization theory $\mathrm{ind}_G(\eta_g,h)$ is a "zero cohomology" representation.

2. THE SOLVABLE CASE, EXISTENCE OF ADMISSIBLE POLARIZATIONS

Although one is forced into considering higher cohomology representations in the case where G is semi-simple, L. Auslander and I have shown that the representations of the form $\mathrm{ind}_G(\eta_g,h)$ for a solvable Lie group G of type I are sufficient to give $\hat{G}$, the set of equivalence classes of irreducible unitary representations of G.

More precisely assume G is a solvable simply connected Lie group. Then for one thing we have shown that G is of type I if and only if (1) $g \in \mathfrak{g}'$ are integrable and (2) all orbits $G \cdot g = \mathcal{O} \subseteq \mathfrak{g}'$ are the intersections of a closed and open set. Furthermore in such a case we may explicitly give $\hat{G}$.

To do this consider first the maximal nilpotent ideal $n \subseteq \mathfrak{g}$. Let $g \in \mathfrak{g}$ and let $f = g|n \in n'$. Since n is stable under Ad G one may consider contragrediently the representation of G on n'. Let G_f be the isotropy subgroup of G at f. Obviously $G_g \subseteq G_f$ and $\mathfrak{g}_g \subseteq \mathfrak{g}_f$ where $\mathfrak{g}_f$ is the Lie algebra of G_f.

A polarization h at g is called admissible in case (1) it is positive (i.e., the bilinear form S_g on e/d is positive definite) and (2) $h \cap n_{\mathbb{C}}$ is stable under G_f and is a polarization at f.

Then the following is proved in [1].

Theorem 1

For any $g \in \mathfrak{g}'$ *whether or not* G *is of type* I *there exists an admissible polarization at* g. *Moreover, any admissible polarization* h *satisfies the Pukansky condition so that if* g *is integrable,* $\mathrm{ind}_G(\eta_g,h)$ *is defined. Furthermore, assuming* g *is integrable then* $\mathrm{ind}_G(\eta_g,h)$ *is independent of the choice of polarizations* h *and if* G *is of type* I *then* $\mathrm{ind}_G(\eta_g,h)$ *is irreducible and every irreducible unitary representation is equivalent to a representation of this form. Finally if* G *is type* I *then* $\mathrm{ind}_G(\eta_g,h)$ *and* $\mathrm{ind}_G(\eta^1_{g_1},h_1)$ *are equivalent if and only if* $G \cdot g = G \cdot g_1$ *and* η_g *corresponds to* $\eta^1_{g_1}$ *under the action of an element* $a \in G$ *such that* $a \cdot g = g_1$.

We cannot go into the proof of this theorem here but we will prove two relevant facts which are needed in the proof. The first of these asserts the independence of the polarization in the nilpotent case. This generalizes a result of

Kirillov who proved a similar theorem for the case of real polarizations, i.e., where $h = \bar{h}$ or $e = d$. One is forced into non-real polarizations by the second fact to be proved. To begin with we need

Theorem 2

Assume that $\mathfrak{g}$ *is nilpotent,* $0 \neq g \in \mathfrak{g}'$ *and the polarization* h *at* g *is positive. Let* $b = \text{Ker}\,(g|d)$. *Then* b *is an ideal in* e *and* e/b *is a Heisenberg Lie algebra with* d/b *as the 1-dimensional center.*

In particular d *is an ideal in* e *and* e/d *is commutative.*

Proof. If $x \in d$ let $\pi(x) \in \text{End}\; e/d$ be the operator on e/d induced by ad x. Since ad x is nilpotent so is $\pi(x)$. On the other hand the relation $\langle g,[d,e]\rangle = 0$ implies $\langle g,[d[e,e]]\rangle = 0$ since e is an algebra, it follows that $\pi(x)$ is skew-symmetric relative to $\hat{B}_g$. However, $\pi(x)$ obviously commutes with j so that it is skew-symmetric relative to S_g. Thus $\pi(x)$ is both nilpotent and skew-symmetric relative to a positive definite bilinear form. Hence $\pi(x) = 0$ so that d is an ideal in e.

But the relation $\langle g,[d,e]\rangle = 0$ then implies $[d,e] \subseteq b$ so that in particular $[b,e] \subseteq b$. Hence b is also an ideal in e. Furthermore d/b is obviously central in e/b. Also d/b is 1-dimensional since $g \neq 0$ (see Remark 3).

Now to prove that e/b is a Heisenberg Lie algebra with d/b as center, it suffices to show that e/d is abelian and d/b is the center of e/b. But for this it suffices only to show that e/d is abelian. Indeed if this were the case then for $x \in e-d$ one has $[x + b,y + b] \subseteq d/b$ for all $y \in e-d$. But from the non-singularity of $\hat{B}_g$ we can choose y so that $\langle g,[y,x]\rangle \neq 0$. This however implies $[x + b,y + b] = d/b$. Hence d/b is exactly the center of e/b.

We assert that to prove the theorem it suffices only to prove

Lemma 2

The center of e/d *is stable under* j.

Indeed assume Lemma 2 is true and let a be the center of e/d. Now S_g is non-singular on a since S_g is positive definite. But since a is stable under j it follows that $\hat{B}_g$ is also non-singular on a. Let v be the orthogonal complement to a in e/d relative to $\hat{B}_g$. We assert that v is a subalgebra. Indeed if $\hat{y},\hat{z} \in v$ and $\hat{x} \in a$ where $x,y,z \in e$ we must show

$$(\hat{x},[\hat{y},\hat{z}]) = 0 \quad . \tag{2.1}$$

But $(\hat{x},[\hat{y},\hat{z}]) = \langle g,[x[y,z]]\rangle = \langle g,[[x,y]z]\rangle + \langle g,[y,[x,z]]\rangle$. But $[x,y],[x,z] \in d$ since a is central in e/d. But then $[[x,y]z]$ and $[y,[x,z]]$ lie in b since $[d,e] \subseteq b$. This proves (2.1) so that v is a subalgebra. But it is obviously nilpotent so that if $\mathcal{V} \neq 0$ then center $\mathcal{V} \neq 0$. However, clearly center $\mathcal{V} \subseteq$ cent $e/d = a$ which is a contradiction. Thus $\mathcal{V} = 0$ so that $a = e/d$ is abelian. We proceed now to the

<u>Proof of Lemma 2</u>. Let $u \in$ center e/d. We must prove ju is central in e/d. Let $v \in e/d$. We first observe that

$$j[ju,v] = [ju,jv] \quad . \tag{2.2}$$

That is j commutes with ad ju. Indeed $u + iju$ and $v + ijv$ lie in $h/d_{\mathbb{C}}$ and since u is central

$$[u + iju, v + ijv] = -[ju,jv] + i[ju,v] \quad .$$

However since $h/d_{\mathbb{C}}$ is an algebra it follows that $[ju,v] = -j[ju,jv]$. Applying j to both sides yields (2.2). Now let $B = \text{ad } ju$ so the problem is to show that $B = 0$. Let $A = B + B^t$ where superscript t denotes the transpose relative to S_g. Hence $A = A^t$ is a symmetric operator. We next establish the relation

$$\{Av,w\} = \{[jw,v],u\} \tag{2.3}$$

for any $v,w \in e/d$. Indeed we first observe that for any $z_i \in e/d$, $i = 1,2,3$ one has

$$([z_1,z_2],z_3) + ([z_2,z_3],z_1) + ([z_3,z_1],z_2) = 0 \quad . \tag{2.4}$$

This of course follows from the relation $([z_1,z_2],z_3) = \langle f,[y_3,[y_1,y_2]]\rangle$ where $y_i \in e$ and $\hat{y}_i = z_i$.

Now $\{Bv,w\} = \{[ju,v],w\} = (j[ju,v],w) = -([ju,v],jw)$ by (1.1). On the other hand $\{B^t v,w\} = \{v,Bw\} = (jv,[ju,w]) = -(v,j[ju,w])$ again by (1.1). But $j[ju,w] = [ju,jw]$ by (2.2) so that $\{B^t v,w\} = -([jw,ju],v)$ since $\tilde{B}_f$ is alternating. Thus

$$\{Av,w\} = -(([ju,v],jw) + (([jw,ju],v)) \quad .$$

Hence $\{Av,w\} = ([v,jw],ju)$ by (2.4). But then $\{Av,w\} = (j[jw,v],u) = \{[jw,v],u\}$ by (1.1) establishing (2.3).

As a consequence of (2.3) note that $Au = 0$ and since A is symmetric one therefore has, by (2.3),

$$0 = (Av,u) = \{[ju,v],u\} \tag{2.5}$$

for all $v \in e/d$. We now assert that AB is skew-symmetric or that $AB + (AB)^t = 0$. That is since A is symmetric we assert

$$\{ABv,w\} + \{Av,Bw\} = 0 \tag{2.6}$$

for all $v,w \in e/d$.

Indeed $\{ABv,w\} = \{A[ju,v],w\} = \{[jw,[ju,v]],u\}$ by (2.3) where $[ju,v]$ replaces v. On the other hand $\{Av,Bw\} = \{Av,[ju,w]\} = \{[j[ju,w],v],u\}$ by (2.3) where $[ju,w]$ replaces w. But $j[ju,w] = [ju,jw]$ by (2.2) so that

$$\{(AB + (AB)^t)v,w\} = \{([jw,[ju,v]] + [[ju,jw],v]),u\}$$
$$= \{[ju,[jw,v]],u\} \qquad (2.7)$$

by Jacobi. However, (2.7) vanishes by (2.5) where $[jw,v]$ replaces v. This proves AB is skew-symmetric.

Now $AB = (B + B^t)B = B^2 + B^tB$. But $AB = -(AB)^t = -B^tA = -((B^t)^2 + B^tB)$. Thus $B^2 + B^tB = -(B^t)^2 - B^tB$ or $B^2 + (B^t)^2 = -2B^tB$. Therefore, $A^2 = (B + B^t)^2 = B^2 + (B^t)^2 + BB^t + B^tB = BB^t - B^tB$. But then $\mathrm{tr}\, A^2 = 0$ since $\mathrm{tr}\, BB^t = \mathrm{tr}\, B^tB$. However, since A is symmetric A^2 is positive semi-definite so that $\mathrm{tr}\, A^2 = 0$ implies $A = 0$. Thus B is skew-symmetric. But B is clearly nilpotent. Hence $B = 0$. QED

One now deduces the following generalization of a result of Kirillov. (See [3]).

Theorem 3

Let G *be any simply connected nilpotent Lie group and let* $\mathfrak{g}$ *be its Lie algebra. Let* $g \in \mathfrak{n}'$ *and let* $\mathfrak{h}$ *be any positive polarization at* g. *Then* $\mathrm{ind}_G(\eta_g,\mathfrak{h})$ *is irreducible and up to equivalence is independent of* $\mathfrak{h}$.

Proof. (Sketched). It follows from Theorem 2 that $\mathrm{ind}_E(\eta_g,\mathfrak{h})$ is just the Bargmann-Segal (see e.g., [2]) holomorphic construction of an irreducible unitary representation of the Heisenberg group E/B. ($B \subseteq E$ is the subgroup corresponding to $\mathfrak{b} = \mathrm{Ker}\, g|\mathfrak{d}$.) One knows therefore that $\mathrm{ind}_E(\eta_g,\mathfrak{h})$ is equivalent to $\mathrm{ind}_E\, \beta_g$ where $B \subseteq K \subseteq E$, K/B is a maximal commutative subgroup of E/B and β_g is the character on K whose differential is $2\pi i g|\mathfrak{k}$. Here $\mathfrak{k}$ is the Lie algebra of K. But then $\mathrm{ind}_G(\eta_g,\mathfrak{h})$ is equivalent to $\mathrm{ind}_G\, \beta_g$. However, since K is "half-way" between D and E it is also "half-way" between $\mathfrak{g}_g$ and $\mathfrak{g}$. One thus has that $\mathfrak{k}$ defines a real polarization at g. By Kirillov's result one knows that $\mathrm{ind}_G\, \beta_g$ is irreducible and that any real polarization gives rise to an equivalent representation. QED

Now returning to previous notation where $\mathfrak{g}$ is solvable one is forced into considering complex polarizations of the nil-radical $\mathfrak{n}$ of $\mathfrak{g}$ since, in general, there exists no real polarization at $f = g|\mathfrak{n}$ which is stable under G_f. However, by the next lemma there exists complex polarizations and in fact positive polarizations stable under G_f. Since the commutator group $G' \subseteq N$ where $N \subseteq G$ corresponds to $\mathfrak{n}$ it follows that $G_f' \subseteq N$ so that the hypothesis of the following lemma is satisfied where $F = G_f$.

Lemma 3

Let N *be a simply connected nilpotent Lie group and let* n *be its Lie algebra. Let* Aut n *be the group of all Lie algebra automorphisms of* n *so that* Ad N *is a subgroup of* Aut n.

Regard Aut n *as operating by contragredience on the dual* n'. *Let* $f \subseteq n'$. *Assume* F *is a group and a homomorphism* $F \to \text{Aut } n$ *(so that* F *operates on* n *and* n'*) such that* (1) *the commutator subgroup* F' *maps into* Ad N *and* (2) $F \cdot f = f$. *Then there exists a positive polarization* h_1 *at* f *which is stable under* F.

Proof. We assume inductively that the result is true for all simply connected nilpotent Lie groups of dimension smaller than dim n.

Let m = Ker f|center n. Assume this space has positive dimension. Clearly m is an ideal in n which is stable under F. Thus F operates on n/m inducing a map $F \to \text{Aut } n/m$ where $F' \to \text{Ad } N/M$ if M is the subgroup corresponding to m. Moreover, if $f_0 \in (n/m)'$ is induced by f then f_0 is fixed by F_0. Now by induction there exists $h_0 \subseteq (n/m)_{\mathbb{C}}$, a positive polarization at f_0 stable under F_0. But then $\pi^{-1}h_0 = h$ is clearly a positive polarization at f stable under F, where $\pi: n \to n/m$ is the quotient map (indeed $e = \pi^{-1}e_0$, $d = \pi^{-1}d_0$ and $e/d \cong e_0/d_0$). Thus we are done in this case so that we may assume dim m = 0 and hence center n is one-dimensional, spanned by an element z where $\langle f,z\rangle = 1$. Since f is fixed by F clearly z is also fixed under the action of F.

Now consider k = center $n/(z)$ so that $k = k_1/(z)$ where $k_1 \subseteq n$ is an ideal. Clearly Aut n operates on $n/(z)$ and k is clearly stable under the action of this group. However Ad N operates trivially on k since $[n,k_1] \subseteq \mathbb{R}z$. Thus the abelian group F/F' operates on k. Let $p \subseteq k$ be an irreducible subspace under the action of F/F' so that dim p is either 1 or 2. Now since $\langle f,z\rangle = 1$ we may write $k_1 = k_0 \oplus \mathbb{R}z$ where $k_0 = \text{Ker } f|k_1$. Since f is fixed under F and k_1 is stable under F it follows that k_0 is stable under F and that if $\pi: n \to n/(z)$ is the quotient map then π induces an F-isomorphism $k_0 \to k$. Let $p_0 \subseteq k_0$ be the F-irreducible subspace corresponding to $p \subseteq k$. Note then that F' must operate trivially on k_0.

Case 1. Assume dim $p_0 = 1$ so that $p_0 = \mathbb{R}w$. In this case we proceed along the lines used by Kirillov. That is, let $g \in n'$ be the linear functional defined by the relation $[y,w] = \langle g,y\rangle z$. One has $g \neq 0$ since otherwise w would be central in n contradicting the fact that center $n = \mathbb{R}z$. Thus there exists $x \in g$ such that $[x,w] = z$ and hence

$$n = \mathbb{R}x \oplus n_0$$

where $n_0 = \text{Ker } g$. But then n_0 is the centralizer of w and hence n_0 is a subalgebra stable under F. However, since n_0 has codimension 1 in n and n is nilpotent, n_0 is an ideal in n. In particular $N = XN_0$ where X and N_0 are the subgroups corresponding to $\mathbb{R}x$ and n_0.

Now the action of F on n_0 induces an epimorphism $F \to F_0 \subseteq \text{Aut } n_0$ where $F' \to F_0'$. However, $F' \to \text{Ad}_n N = \text{Ad}_n X \text{Ad}_n N_0$. But $\text{Ad}_n N_0$ operates trivially on $\mathbb{R}w$ since clearly $w \in \text{center } n_0$. On the other hand F' operates trivially on $w \in P_0$ as observed above. But since $[x,w] = z$ no non-trivial element of $\text{Ad}_n X$ operates trivially on w so we must have $F' \to \text{Ad}_n N_0$ which implies $F_0' \subseteq \text{Ad}_{n_0} N_0$.

Now clearly $f_0 = f|n_0$ is invariant under F_0. Furthermore, we assert that

$$(n_0)_{f_0} = n_f \oplus \mathbb{R}w \quad . \tag{2.8}$$

Indeed $w \in (n_0)_{f_0}$ since $w_0 \in \text{center } n_0$. To see that $n_f \subseteq (n_0)_{f_0}$ we have only to observe that $n_f \subseteq n_0$. But this is clear since otherwise there exists $y \in n_f$ such that $[y,w] = z$. But then

$$1 = \langle f,[y,w]\rangle = -\langle y \cdot f,w\rangle$$

contradicting the fact that $y \cdot f = 0$. Also one has $n_f \cap \mathbb{R}w = 0$ since $\langle w \cdot f,x\rangle = \langle f,[x,w]\rangle = \langle f,z\rangle = 1$. Finally if $y \in (n_0)_{f_0}$ let $c = \langle y \cdot f,x\rangle = \langle f,[x,y]\rangle$. But $\langle cw \cdot f,x\rangle = \langle f,cz\rangle = c$. Thus $\langle (y - cw) \cdot f,x\rangle = 0$. But $(y - cw) \cdot f|n_0 = (y - cw) \cdot f_0 = 0$ since $w \in (n_0)_{f_0}$. But then $y - cw = y_1 \in n_f$ so that $y \in n_f + \mathbb{R}w$. This establishes (2.8).

Now by induction there exists a positive polarization $h_0 \subseteq (n_0)_{\mathbb{C}}$ at f_0 which is stable under F_0. Clearly then one has

$$(n_f)_{\mathbb{C}} \subseteq ((n_0)_{f_0})_{\mathbb{C}} \subseteq h_0 \subseteq (n_0)_{\mathbb{C}} \subseteq n_{\mathbb{C}} \quad .$$

But since h_0 is "half-way" between $((n_0)_{f_0})_{\mathbb{C}}$ and $(n_0)_{\mathbb{C}}$ it is also "half-way" between $(n_f)_{\mathbb{C}}$ and $n_{\mathbb{C}}$ because n_f has codimension 1 in $(n_0)_{f_0}$ and n_0 has codimension 1 in n. Thus, if $h = h_0$ it follows that h is a positive polarization at f which is stable under the action of F.

Now if $\dim p_0 = 2$ we may write $p_0 = \mathbb{R}w_1 \oplus \mathbb{R}w_2$. If we define $g_j \in n'$, $j = 1,2$ by the relation $[y,w_j] = \langle g_j,y\rangle z$ then g_1 and g_2 are linearly independent since otherwise $p_0 \cap \text{center } n \neq 0$. But of course $p_0 \cap \text{center } n = 0$ since $\text{center } n = \mathbb{R}z$.

But then we may find elements $x_1, x_2 \in n$ such that

$$[x_i,w_j] = \delta_{ij} z \quad . \tag{2.9}$$

Clearly then

$$n = \mathbb{R}x_1 \oplus \mathbb{R}x_2 \oplus n_0 \tag{2.10}$$

where $n_0 = \text{Ker } g_1 \cap \text{Ker } g_2$ is the centralizer of the subspace p_0. Since p_0 is stable under F it follows that n_0 is a subalgebra stable under F. In fact since $[n,n]$ annihilates $k_1 \supseteq k_0 \supseteq p_0$, it follows that

$$[n,n] \subseteq n_0 \tag{2.11}$$

and hence n_0 is an ideal in n. The action of F on n_0 induces an epimorphism $F \to F_0 \subseteq \text{Aut } n_0$ where F' maps into F_0'. But the map $X_1 \times X_2 \times N_0 \to N$ is bijective where $(a_1,a_2,b) \to a_1a_2b$ and where $N_0 \subseteq N$ is the subgroup corresponding to n and X_j is the subgroup corresponding to $\mathbb{R}x_j$, $j = 1,2$. But now N_0 operates trivially on $p_0 \subseteq k_0$. But since no non-trivial element of X_1X_2 operates trivially on p_0 by the relations (2.9) it follows that $F' \to \text{Ad}_n N_0$ and hence $F_0' \subseteq \text{Ad}_{n_0} N_0$.

Now let $f_0 = f|n_0$. By induction there exists a positive polarization h_0 at f_0 which is stable under the action of F_0.

As in the case where $\dim p_0 = 1$ one has $[n_f, p_0] = 0$ so that $n_f \subseteq n_0$ and hence

$$n_f \subseteq (n_0)_{f_0} \tag{2.12}$$

Next observe that

$$(n_0)_{f_0} \subseteq n_f + p_0 = n_f \oplus p_0 \tag{2.13}$$

Indeed if $y \in (n_0)_{f_0}$ and c_j, $j = 1,2$ are defined by $c_j = \langle y \cdot f, x_j \rangle$ then $g = (y - c_1w_1 - c_2w_2) \cdot f$ is orthogonal to $\mathbb{R}x_1 + \mathbb{R}x_2$ by the relations (2.9). However, clearly g is orthogonal to n_0 so that $g = 0$ which implies $y - c_1w_1 - c_2w_2 \in n_f$ and hence $y \in n_f + p_0$. Now $n_f \cap p_0 = 0$ since by the relation (2.9) any non-zero element $w \in p_0$ is such that $z \in \text{Im ad } w$. But since $\langle f,z \rangle \neq 0$ this implies $w \in n_f$. Hence (2.13) is established.

Case 2. Assume $[w_1,w_2] = 0$. Then $p_0 \subseteq n_0$ and hence $p_0 \subseteq \text{center } n_0$ which implies $p_0 \subseteq (n_0)_{f_0}$. Thus by (2.12) and (2.13) one has $(n_0)_{f_0} = n_f \oplus p_0$ so that n_f has codimension 2 in $(n_0)_{f_0}$. Since n_0 has codimension 2 in n this implies that h_0 is "half-way" between $(n_f)_{\mathbb{C}}$ and $n_{\mathbb{C}}$ and hence $h = h_0$ defines a positive polarization at f which is stable under F.

Case 3. Assume $[w_1,w_2] \neq 0$. Now since F' operates trivially on p_0 it follows that F operates, irreducibly, as an abelian group on the 2-dimensional space. The commuting ring in $\text{End } p_0$ is therefore isomorphic to $\mathbb{C}$ and hence w_1 and w_2 may be chosen in p_0 so that $\mathbb{C}u, \mathbb{C}\bar{u} \subseteq (p_0)_{\mathbb{C}}$ are stable under the action of F where $u = w_1 + \sqrt{-1}\, w_2$ and $\bar{u} = w_1 - \sqrt{-1}\, w_2$. Furthermore, it is clear that since they are necessarily independent we may choose w_1, w_2 so that $[w_1,w_2] = z$. But then we may choose x_1 and x_2 so that $x_1 = w_1$, $x_2 = -w_2$ and hence (2.10) becomes

$$\mathbb{R}w_1 \oplus \mathbb{R}w_2 \oplus n_0 = n \ .$$

But then $p_0 \cap n_0 = 0$ so that, since $n_f \subseteq (n_0)_{f_0} \subseteq n_f + p_0$ by (2.12) and (2.13) one has $n_f = (n_0)_{f_0}$. But then since n_0 has codimension 2 in n, it follows that h_0 fails by one dimension of being a maximum isotropic subspace (m.i.s.) of $n_{\mathbb{C}}$ relative to B_f.

Now put

$$h = h_0 + \mathbb{C}u \quad .$$

Since $h_0 \subseteq (n_0)_{\mathbb{C}}$ and $u \in (p_0)_{\mathbb{C}}$ it follows that $[u,h_0] = 0$ so that not only h is a m.i.s. of $n_{\mathbb{C}}$ but h is a subalgebra stable under the action of F. Also since $n_f \subseteq h$ it follows that h is stable under Ad N_f. But now $h + \bar{h} = h_0 + \bar{h}_0 + \mathbb{C}u + \mathbb{C}\bar{u} = (h_0 + \bar{h}_0) + (p_0)_{\mathbb{C}}$. However, $h_0 + \bar{h}_0$ is a subalgebra since h_0 is a polarization at f_0. But $h_0 + \bar{h}_0 \subseteq (n_0)_{\mathbb{C}}$ and since $[p_0, n_0] = 0$ it follows that $h + \bar{h}$ is a subalgebra since $[(p_0)_{\mathbb{C}}, (p_0)_{\mathbb{C}}] = \mathbb{C}z$ and $z \in n_f = (n_0)_{f_0} \subseteq h$. Thus h is a polarization at f. We have only to show that h is positive.

But now since $p_0 \cap n_0 = 0$ one has $d = h \cap n = h_0 \cap n = h_0 \cap n_0 = d_0$. But if $e = (h + \bar{h}) \cap n$ and $e_0 = (h_0 + \bar{h}_0) \cap n = (h_0 + \bar{h}_0) \cap n_0$ then one has

$$e/d = e_0/d_0 \oplus (d_0 \oplus p_0)/d_0 \quad .$$

But this is an orthogonal direct sum relative to both $\hat{B}_f$ and S_f. Indeed this is clear since e_0 and d_0 are orthogonal relative to B_{f_0} and hence relative to B_f. But also $[p_0, e_0] = 0$. Furthermore $(d_0 + p_0)/d_0$ is stable under j since $(p_0)_{\mathbb{C}} = (p_0)_{\mathbb{C}} \cap h \oplus (p_0)_{\mathbb{C}} \cap \bar{h} = \mathbb{C}u \oplus \mathbb{C}\bar{u}$. But by assumption S_f is positive definite on e_0/d_0. However, it is positive definite on $(d_0 + p_0)/d_0$ since if $[w_i] = w_i + d_0, i = 1,2$ one has $j[w_1] = [w_2]$ and $j[w_2] = -[w_1]$. Thus, $\{[w_1],[w_2]\} = 0$ and $\{[w_1],[w_1]\} = (j[w_1],[w_1]) = ([w_2],[w_1]) = \langle f,[w_1,w_2]\rangle = \langle f,z\rangle = 1$. Similarly $\{[w_2],[w_2]\} = 1$. Hence S_f is positive definite. QED

Lemma 3 shows that there exists a positive polarization h_1 at f which is stable under G_f. Now let $e = g|g_f$. We assert there exists a positive polarization h_2 at e (for the identity component $(G_f)_0$ of G_f) which is stable under G_g. To see this one cannot directly apply Lemma 3 since g_f is not necessarily nilpotent. However, if $n_f = g_f \cap n$ and $a = \text{Ker } f|n_f$ then a is ideal in g_f and g_f/a is indeed nilpotent. Furthermore e induces a linear functional on g_f/a and since $G_g' \subseteq N_f$, the subgroup corresponding to n_f, it follows from Lemma 3 that h_2 exists by passing to the quotient g_f/a. But now if we put $h = h_1 + h_2$ then it follows easily that h is an admissible polarization at g. But then we may form $\text{ind}_G(n_g, h)$ giving the most general irreducible unitary representation of a simply connected solvable Lie group of type I.

REFERENCES

[1] Auslander, L. and Kostant, B., "Quantization and Representations of Solvable Lie Groups", to appear (see announcement in *Bull. Amer. Math. Soc.*, 73, 692-695 (1967).

[2] Bargmann, V., "On A Hilbert Space of Analytic Functions and An Associated Integral Transform", *Comm. Pure Appl. Math.*, 14, 187-214 (1961).

[3] Kirillov, A. A., "Unitary Representations of Nilpotent Lie Groups", *Uspehi. Mat. Nauk.*, 17, 57-110 (1962).

[4] Pukansky, L., "On The Theory of Exponential Groups", *Trans. Amer. Math. Soc.*, 126, 487-507 (1967).

[5] Weil, A., *Variétés Kählériennes*, Hermann, Paris (1958).

DERIVATION AND SOLUTION OF AN INFINITE-COMPONENT WAVE EQUATION FOR THE RELATIVISTIC COULOMB PROBLEM

by

I. T. Todorov*

SUMMARY

The aim of these notes is to give a self-contained exposition of the derivation and solution of an infinite-component wave equation. They cover some of the results of recent work by C. Itzykson, V. Kadyshevsky, and the author [1,2,3].

First we sketch the derivation of a three-dimensional quasi-potential equation in momentum space involving integration over the mass-shell hyperboloid $p^2 = m^2$. We show that for the relativistic Coulomb potential $V(p,q) = \frac{\alpha}{(p - q)^2}$ this equation can be written in an equivalent algebraic form in terms of rational functions of the generators of a degenerate ("metaplectic") representation of $SO(4,2)$. The solution of the bound-state eigenvalue problem is carried out by reducing the representation of $SO(4,2)$ with respect to the irreducible representations of its subgroup $SO(3) \otimes SO(2,1)$ and by an extensive use of the Bargmann realization of the discrete series of unitary representations of $SO(2,1)$.

---

* Institute for Advanced Study, Princeton, New Jersey. On leave from Joint Institute for Nuclear Research, Dubna, USSR and from Physical Institute of the Bulgarian Academy of Sciences, Sofia, Bulgaria.

TABLE OF CONTENTS

INTRODUCTION

This paper consists of three parts. First, I will try to persuade you that the equation we are going to solve has something to do with physics. We will consider a class of relativistic quasi-potential equations for the two-body problem and will single out a simple equation of this class corresponding to the scalar Coulomb interaction. Second, we shall show that our simple equation is equivalent to an infinite-component wave equation written in terms of the generators of a unitary representation of the conformal group $SO(4,2)$. Finally, we shall solve the arising eigenvalue problem by applying some known tools of the theory of representations of the pseudo-unitary group.

In Section 1 we will have to use, without much explanation, some of the physicists' jargon (which is introduced in the first few chapters of any textbook on quantum field theory). The rest of my talk (Sections 2,3) is practically self-contained and does not require any special knowledge of physics.

1. QUASI-POTENTIAL EQUATION FOR THE RELATIVISTIC TWO-BODY PROBLEM [1,2,3]

1.1 Old-fashioned Perturbation Theory and Feynman-Dyson Rules

We will be concerned in what follows with the scattering and bound-states problems of two relativistic particles.

Let us have two equal-mass particles of initial (4)-momenta q_1, q_2 and final momenta p_1, p_2. Taking into account the energy-momentum conservation $(p_1 + p_2 = q_1 + q_2)$, we can express p_i and q_i in terms of three 4-vectors: the center-of-mass momentum

$$P = p_1 + p_2 = q_1 + q_2 \ , \tag{1.1}$$

and the relative momenta

$$p = \frac{1}{2}(p_1 - p_2), \quad q = \frac{1}{2}(q_1 - q_2) \ . \tag{1.2}$$

On the mass-shell, i.e., for $p_1^2 = p_2^2 = q_1^2 = q_2^2 = m^2$ we have the identities

$$pP = qP = 0, \ \frac{1}{4} P^2 + p^2 = \frac{1}{4} P^2 + q^2 = m^2.$$

(We use the system of units for which $c = \hbar = 1$ throughout these notes.) In the framework of quantum field theory, to each particle one usually makes correspond a local field operator. So, we associate with particles 1 and 2 the complex scalar fields $\psi_1(x)$ and $\psi_2(x)$, of mass m and assume that their interaction is given by the local Hamiltonian density

$$\mathcal{H}(x) = -g(:\psi_1^*(x)\psi_1(x):+:\psi_2^*(x)\psi_2(x):)\varphi(x) \ , \tag{1.3}$$

where : : is the sign for the Wick "normal" product

$$:\psi^*(x)\psi(x): = \lim_{y\to 0} [\psi^*(x + y)\psi(x - y) - \langle 0|\psi^*(x + y)\psi(x - y)|0\rangle] ,$$

($|0\rangle$ is the "free vacuum") and $\varphi(x)$ is a hermitian field of mass μ. Then, the scattering amplitude can be written as a (formal) power series in the coupling constant g. There have been two different presentations of this formal expansion: the old-fashioned (non-covariant) perturbation theory and the modern Feynman-Dyson covariant technique. The second one is much more familiar nowadays. Each term of the series is represented in this approach as a sum of multiple integrals corresponding to the so-called Feynman diagrams (see Figure 1).

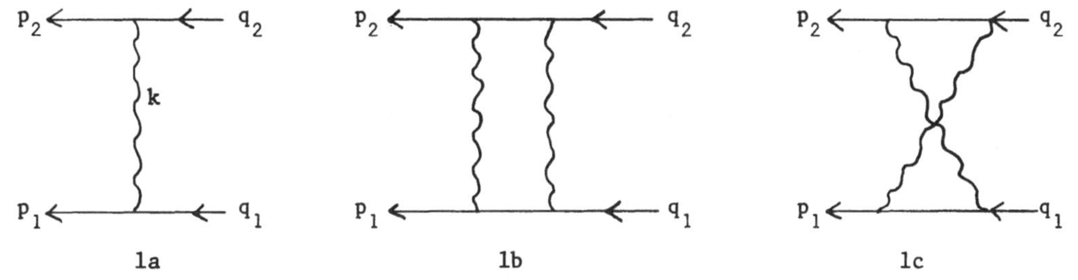

FIGURE 1

An important property of the Feynman rules is that they involve 4-momentum conservation in each vertex of the graph (a factor $g\delta(p + k - q)$ corresponding to a vertex with momentum q on the incoming line and momenta p and k on the outgoing lines). This tempts the physicists to interpret individual Feynman graphs as multiple emission and absorption amplitudes (although, strictly speaking, only the sum of all graphs for a given process has a well-defined physical meaning). Such an interpretation, however, only makes sense for off-mass shell intermediate particles, since, according to the Feynman rules, to an internal (say wavy) line with mass μ and momentum k corresponds a factor $\frac{1}{\mu^2 - k^2 - i0}$ (integration being carried out subsequently over all 4-dimensional internal momenta k), and this factor becomes infinite on the mass shell (i.e., for $k^2 = \mu^2$).

More recently [4] a graphic picture was also given for the old-fashioned perturbation expansion. To describe it, we associate with any Feynman graph with N vertices N! new graphs constructed in the following way. We start with the set of all oriented graphs with the same picture as the original one and with all possible enumerations of the vertices 1, ..., N. Every internal line is oriented toward the vertex with smaller number. Further, we let a spurion (dotted) line enter vertex 1, connect 1 with 2, 2 with 3 and so on (always oriented toward the vertex with larger number), and finally go out of the vertex N. For instance, to the second order Feynman graph of Figure 1a correspond the two diagrams of Figure 2.

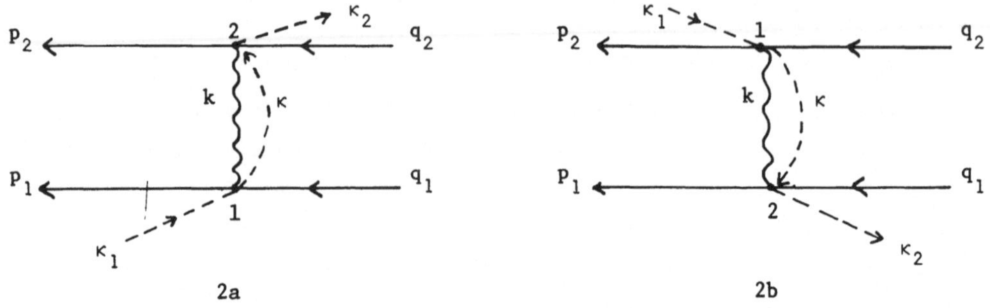

FIGURE 2

Here to the oriented wavy line with mass μ and momentum k corresponds the "on-mass-shell propagator"

$$\delta_\mu^+(k) = \theta(k_0)\delta(k^2 - \mu^2), \text{ where } \theta(k_0) = \begin{cases} 1 & \text{for } k_0 > 0 \\ 0 & \text{for } k_0 < 0 \end{cases} . \qquad (1.4)$$

However, the energy of the particles (represented by solid lines) is not conserved, the conservation law in each vertex taking into account the energies of the dotted lines. For instance, to vertex 1 of the diagram in Figure 2a corresponds a factor

$$-\frac{g}{\sqrt{2\pi}}\,\delta(q_1 + k - p_1 + (\kappa_1 - \kappa)n) ,$$

where n is a 4-dimensional unit vector in the direction of the time axis. Finally, to an internal dotted line of "energy" κ we make correspond the propagator

$$\frac{1}{2\pi}\,\frac{1}{\kappa - i0} . \qquad (1.5)$$

Integration is carried out over κ from $-\infty$ to $+\infty$ (along with the integration over the internal momenta k).

Remark. For those familiar with the formalism of quantum field theory we mention that the splitting of a Feynman graph of N vertices into $N!$ non-covariant graphs (containing dotted lines) corresponds to the decomposition of a time-ordered product of N local operators $H(x_1) \ldots H(x_N)$ into $N!$ ordinary products (with appropriate θ-functions). On the energy shell, i.e., for $\kappa_1 = \kappa_2 = 0$, the sum of the contributions of these $N!$ graphs coincides with the (on-mass-shell) contribution of the original Feynman graph.

Example. The contribution from the two diagrams of Figure 2 is

$$\frac{1}{(2\pi)^2}\,\delta(p_1 + p_2 - q_1 - q_2 + (\kappa_2 - \kappa_1)n)T^{(2)} , \qquad (1.6)$$

where

$$T^{(2)} = \frac{g^2}{2}\left(\frac{1}{\omega_{p_1-q_1}} \frac{1}{\kappa_1 + q_1^0 - p_1^0 + \omega_{p_1-q_1} - i0} + \frac{1}{\omega_{p_2-q_2}} \frac{1}{\kappa_1 + q_2^0 - p_2^0 + \omega_{p_2-q_2} - i0}\right), \tag{1.7}$$

where $\omega_k = \sqrt{\mu^2 + \underline{k}^2}$. On the energy shell, for $\kappa_1 = \kappa_2 = 0$, $q_1 - p_1 = p_2 - q_2$ the right-hand side of Equation (1.7) reduces to the covariant Feynman rule for the on-shell amplitude $\underline{T}$:

$$\underline{T}^{(2)} = \frac{g^2}{\omega^2_{p_1-q_1} - (p_1^0 - q_1^0)^2 - i0} = \frac{g^2}{\mu^2 - (p_1 - q_1)^2 - i0}.$$

1.2. Off-mass-shell Bethe-Salpeter Equation and Off-energy-shell Quasi-potential Equation for the Scattering Amplitude

Two types of linear equations for the scattering amplitude have been considered corresponding to the two types of expansions discussed in the previous section. Historically, the first one is the Bethe-Salpeter (B-S) equation which was, actually, first proposed by Nambu (1950) (for a complete bibliography on the B-S equation see the recent review article [5]). It is an off-mass-shell equation which originates from the Feynman-Dyson rules. In order to write it down we need the notions of the "complete Feynman propagator" $\Delta'_F(p)$ and of the sum of all $\psi_1 + \psi_2$ irreducible graphs $I_P(p,q)$.

The complete (sometimes also called modified) Feynman propagator $\Delta'_F(p)$ is defined as the sum of the contributions of all Feynman graphs to the two point Green's function (see Figure 3).

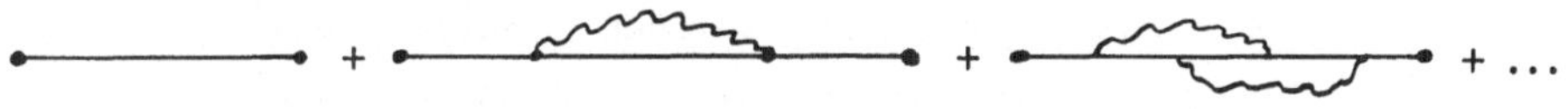

FIGURE 3

$$\Delta'_F(p) = \frac{1}{m^2 - p^2 - i0} + \frac{g^2}{(2\pi)^4} \int_{(m+\mu)^2}^{\infty} \frac{f(x)dx}{(x - p^2)^2(x - p^2 - i0)} + O(g^4), \tag{1.8}$$

where f is defined by the phase-space integral

$$\frac{1}{\pi}\int \delta_m^+(p - k)\delta_\mu^+(k)d^4k = \theta(p_0)\theta(p^2 - (m + \mu)^2)f(p^2),$$

$$f(x) = \frac{1}{8x}[x^2 - 2(m^2 + \mu^2)x + (m^2 - \mu^2)^2]^{1/2}. \tag{1.9}$$

Remark. The graphs in Figure 3 correspond in general to divergent integrals (this is for instance the case with the second order term whose contribution is written explicitly in (1.8)). We choose the renormalization in such a way that the regularized integrals vanish for $p^2 = m^2$ together with their first derivatives. This permits cancellation of the pole terms $\frac{1}{(m^2 - p^2 - i0)^2}$ coming from the two external lines in all graphs of Figure 3 except the first one. Hence, according to our definition, only the first term in the expansion (1.8) has a pole-type singularity for $p^2 = m^2$.

A connected diagram D of the $\psi_1 + \psi_2$ (elastic) scattering process is called reducible (or more specifically $\psi_1 + \psi_2$-reducible) if it can be decomposed into two graphs D' and D" of the same process connected by one ψ_1 and one ψ_2 lines such that D' contains both incoming lines of D (with momenta q_1, q_2) and D" contains both outgoing lines of D (with momenta p_1, p_2) (see Figure 4).

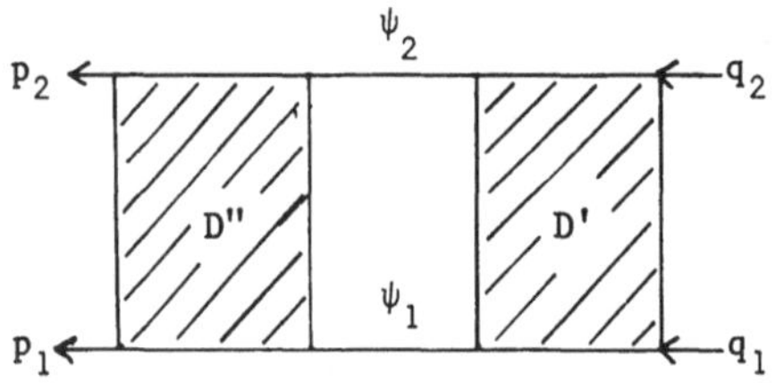

FIGURE 4

Otherwise, if this is not possible, the diagram is called $\psi_1 + \psi_2$-irreducible. According to this definition the graph shown in Figure 1b is reducible while the graphs of Figures 1a and 1c are irreducible. We denote the sum of the contributions of all irreducible graphs by $I_P(p,q)\delta(p_1 + p_2 - q_1 - q_2)$.

Let $T_P(p,q)$ be the off-mass-shell $\psi_1 + \psi_2$-scattering amplitude (in other words let $T_P(p,q)\delta(p_1 + p_2 - q_1 - q_2)$ be the sum of all connected Feynman graphs of the $\psi_1 + \psi_2$-elastic scattering without radiative corrections on the external lines). Then the B-S equation can be written in the form

$$T_P(p,q) = I_P(p,q) - \frac{i}{(2\pi)^2} \int I_P(p,k)\Delta_F'(\tfrac{1}{2} P + k)\Delta_F'(\tfrac{1}{2} P - k)T_P(k,q)d^4k \; . \qquad (1.10)$$

It can be checked directly that the iterative solution of Equation (1.10) coincides with the sum of all Feynman graphs for T_P. Equation (1.10) is a source of non-trivial approximations for T_P. Even if we restrict ourselves to the first terms of the expansions in g^2 for Δ_F' and T_P we find that the solution of (1.10) has g dependent poles as a function of P^2 which never occurs in any finite order in perturbation theory. These poles are interpreted as the squares of the masses of the two-particle bound states.* They coincide with the eigenvalues of P^2 for which

* It should be realized that such an interpretation is not a consequence of the principles of quantum field theory. We shall discuss below the advantages of an alternative definition of the bound-state energy eigenvalues.

the homogeneous equation

$$[\Delta_F'(\frac{1}{2}P + p)\Delta_F'(\frac{1}{2}P - p)]^{-1}\Phi_P(p) = \frac{-i}{(2\pi)^2}\int I_P(p,k)\Phi_P(k)d^4k \ , \tag{1.11}$$

(corresponding to (1.10)) has a non-trivial solution satisfying certain boundary conditions.

Equation (1.11) has a number of undesirable features as compared to the non-relativistic Schrödinger equation (for a concise discussion of the diseases of the B-S equation see the elegant paper by Wick [6]). First of all, it involves a fourth coordinate--the relative energy $p_0(k_0)$ (or the relative time in the original B-S formulation), which does not have a clear physical meaning. Its presence makes obscure the non-relativistic limit of the B-S equation and leads to extra (unphysical) solutions, the energy eigenvalues $(W^2 = P^2)$ being labeled by one more quantum number than in the Schrödinger equation. This point is clarified by the Wick-Cutkosky model [6]--the only exactly solvable example of the B-S equation we know. (In this example $[\Delta_F'(k)]^{-1}$ is replaced by $(\Delta_F(k))^{-1} = m^2 - k^2$ and $I_P(p,q)$ is given by (minus) the scalar Coulomb potential $I_P(p,q) = \frac{g^2}{-(p-q)^2}$.) If g^2 belongs to a certain interval it has been shown that some extra energy eigenvalues do in fact appear (for more details see Reference [3]). In the lowest order approximation with respect to the coupling constant g (which has only been considered in practice) the operator on the left-hand side of (1.11) is a fourth-order polynomial in p (i.e., a fourth-order differential operator in coordinate space). This is another source of extra solutions of the B-S equation. No probabilistic interpretation is possible for the wave-function Φ, since it is not normalizable.

The three-dimensional "quasi-potential" approach to the two-particle bound state problem, based on the off-energy shell old-fashioned perturbation theory (see [7,1]), seems free of all these difficulties of the B-S equation and our further discussion will be based on it.

First of all, we choose the unit vector n, of the time axis (which appeared in the formula of the old-fashioned perturbation theory) along the center of mass momentum P. In this frame, taking into account the conservation law

$$p_1 + p_2 - \kappa_1 n = q_1 + q_2 - \kappa_2 n \ , \tag{1.12}$$

(see (1.6)) and the mass-shell condition

$$p_1^2 = p_2^2 = q_1^2 = q_2^2 = m^2 \ , \tag{1.13}$$

we can write

$$\underline{p}_1 = -\underline{p}_2 = \underline{p}, \quad \underline{q}_1 = -\underline{q}_2 = \underline{q}, \quad p_1^0 = p_2^0 = p^0, \quad q_1^0 = q_2^0 = q^0 \ ;$$

$$|p_0| = E_{\underline{p}} = \sqrt{m^2 + \underline{p}^2} \ , \quad p_0 - 1/2\ \kappa_1 = q_0 - 1/2\ \kappa_2 \equiv E \ . \tag{1.14}$$

Further, we introduce the notion of an irreducible graph in the Kadyshevsky diagram technique. We call a graph D, corresponding to the old-fashioned perturbation expansion of the $\psi_1\psi_2$-elastic scattering amplitude, irreducible if it cannot be split into two solid-line connected diagrams D_1 and D_2 in the way shown on Figure 5.

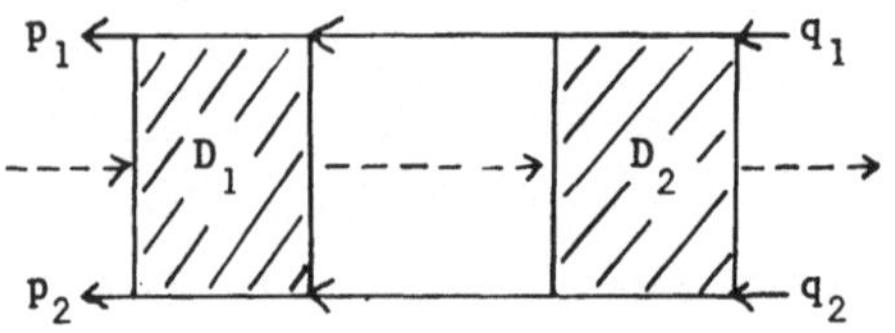

FIGURE 5. Reducible Graph

We denote the sum of all irreducible graphs (which do not contain radiative corrections on external lines) by

$$-V_E(p,q)\delta(p_1 + p_2 - \kappa_1 n - q_1 - q_2 + \kappa_2 n) .$$

Finally we define the total Green's function

$$2E_k G_E(k_0)\delta(\kappa_1 - \kappa_2)\delta(\underline{k} - \underline{k}') ,$$

as the sum of all solid line disconnected self-energy diagrams of the $(\psi_1\psi_2)$-scattering amplitude with the following property: the line ψ_1 with (incoming) momentum $k = (k_0,\underline{k})$ (and with all possible radiative corrections) may be connected with the line ψ_2 with (incoming) momentum $(k_0,-\underline{k})$ only by a dotted line. The first two terms in the expansion of $G_E(k_0)$ (with respect to g^2) are given by

$$2E_k G_E^{(2)}(k_0) = \frac{1}{4\pi}\left\{\frac{1}{k_0 - E - i0} + \left(\frac{g}{2\pi}\right)^2 \int_{x_0(k_0)}^{\infty} f(x^2 + m^2 - k_0^2) \frac{(x + 2k_0 - 2E)(k_0^2 + x) + xk_0^2}{(x^2 - k_0^2)^2[(\frac{1}{2}x + k_0 - E - i0)^2 - k_0^2} dx\right\} , \tag{1.15}$$

where $x_0(k_0) = [(m + \mu)^2 + \underline{k}^2]^{1/2} = (2m\mu + \mu^2 + k_0^2)^{1/2}$ and f is defined by (1.9). The solid-line connected off-energy-shell scattering amplitude $T_E(p,q)$ (without radiative corrections on the external lines) satisfies the "quasi-potential" equation

$$T_E(p,q) + V_E(p,q) + \int V_E(p,k)G_E(k_0)T_E(k,q)\delta_m^+(k)d^4k = 0 . \tag{1.16}$$

In order to obtain the corresponding homogeneous equation we assume that there exists an r-fold degenerate $(r \geq 1)$ bound state of mass $2B < 2m$ in the $\psi_1\psi_2$-system. Furthermore, in analogy with the Bethe-Salpeter equation we postulate that the scattering amplitude $T_E(p,q)$ has a simple pole for $E = B$. In the neighborhood of this pole we put

$$G_E(p_0)T_E(p,q)G_E(q_0) = \frac{1}{4\pi}\sum_{a=1}^{r}\frac{\phi_{Ba}(p)\overline{\phi}_{Ba}(q)}{B - E - i0} + \text{regular terms for } E \rightarrow B , \tag{1.17}$$

where $\phi_{Ba}(p)$ will be interpreted as the wave function of the bound state of mass $2B$ and other quantum numbers specified by a. Inserting (1.17) in Equation (1.16) and comparing the residues for the pole $E = B$, we obtain

$$\sum_{a=1}^{r} [\phi_{Ba}(p) + G_B(p_0)\int V_B(p,k)\phi_{Ba}(k)\delta_m^+(k)d^4k]\overline{\phi}_{Ba}(q) = 0 .$$

Taking into account that $\overline{\phi}_{Ba}(q)$ are linearly independent we find the following homogeneous equation for each of the wave functions $\phi_B(p)$:

$$[G_B(p_0)]^{-1}\phi_B(p) + \int V_B(p,k)\phi_B(k)\delta_m^+(k)d^4k = 0 . \tag{1.18}$$

The normalization condition for ϕ_B may be also obtained from Equation (1.16) by first applying to both sides the integral operator

$$(KF)(p) = \int T_E(p,p')G_E(p_0')F(p')\delta_m^+(p')d^4p'$$

and then inserting (1.17) and comparing the residues for $E = B$. The result is [8]:

$$\iint \overline{\phi}_{Ba}(k_1)\left\{-\frac{\partial}{\partial B}\left[\frac{\partial}{\partial B}(G_B(k_{10})^{-1}2E_{k1}\delta(\underline{k}_1 - \underline{k}_2) + V_B(\underline{k}_1,\underline{k}_2)\right]\right\}$$
$$\phi_{Bb}(k_2)\delta_m^+(k_1)\delta_m^+(k_2)d^4k_1d^4k_2 = \delta_{ab} . \tag{1.19}$$

Equation (1.18) does not have the defects of the Bethe-Salpeter equation discussed above. In particular, it has a straightforward (and transparent) non-relativistic limit.

1.3. A Simple Model: The Scalar Coulomb Problem

In the lowest order in g the bound-state Equation (1.18) has the form

$$p_0(E - p_0)\phi_E(p) = \frac{1}{8\pi}\int V_E^{(2)}(p,k)\phi_E(k)\delta_m^+(k)d^4k , \tag{1.20}$$

where according to (1.7),

$$V_E^{(2)}(p,q) = \frac{g^2}{\omega_{p-q}(2E - p_0 - q_0 - \omega_{p-q} + i0)} , \tag{1.21}$$

$$\omega_k = \sqrt{\mu^2 + \underline{k}^2} .$$

The "potential" (1.21) is quite complicated so that Equation (1.20) does not allow an exact solution even in the limit of zero-mass exchange $(\mu = 0)$. In what follows we shall study the model equation in which $V_E^{(2)}$ is replaced by the relativistic scalar Coulomb potential

$$V(p,q) = \frac{g^2}{(p - q)^2} \tag{1.22}$$

and the integration is carried over the two-sheeted hyperboloid $k^2 = m^2$ ($\theta(k_0)$ being replaced by $\varepsilon(k_0) = \theta(k_0) - \theta(-k_0)$ in the right-hand side of Equation (1.19)).

Let us make a few remarks about the place of this model in the study of the relativistic two-body problem.

Originally, back in 1963, Logunov and Tavkhelidze [9] have postulated the following three dimensional quasi-potential equation

$$T_{E_q}(\underline{p},\underline{q}) + V_{E_q}(\underline{p},\underline{q}) + \frac{1}{4\pi}\int V_{E_q}(\underline{p},\underline{k}) \frac{T_{E_k}(\underline{k},\underline{q})}{E_k^2 - (E_q + i0)^2} \frac{d^3k}{2E_k} = 0 \tag{1.23}$$

(we have changed the sign convention for V adopted in Reference [9] in order to be consistent with the non-relativistic limit for the potential). This equation differs from our Equation (1.16) both in the Green's function and in the potential (the second order off-shell amplitude and potential being defined by

$$T_{E_q}^{(2)}(\underline{p},\underline{q}) = -V_{E_q}^{(2)}(\underline{p},\underline{q}) = \frac{g^{(2)}}{\mu^2 + (\underline{p} - \underline{q})^2} \tag{1.24}$$

in [9]). However, the perturbative solutions of both Equations (1.16) and (1.23) coincide on the energy shell provided that we put the exact expressions for G_E and V_E (i.e., the sum of all irreducible graphs in our case), reproducing in both cases the on-mass-shell Feynman rules. The non-uniqueness of the quasi-potential equation originates in the non-uniqueness of the off-energy-shell extrapolation of the scattering amplitude. There exists in fact an infinite family of three dimensional equations of the type

$$T + V + VGT = 0 \tag{1.25}$$

which give the same on-shell amplitude and which ensure the elastic unitarity condition

$$T - T^* = T(G - G^*)T^* \tag{1.26}$$

for Hermitian potentials V. It is easy to see that our model equation with Green's function $G_E^{(0)} = [8\pi E_k(k_0 - E - i0)]^{-1}$ and potential (1.22) can be obtained in second order from an equation of this family (it is sufficient to check that on the energy shell, i.e., for $p_0 = q_0 = E_p = E$, the "relativistic Coulomb potential" (1.22) coincides with (1.21) and (1.24) for $\mu = 0$, and that the Green's functions of Equations (1.17) and (1.23) have the same discontinuity $G_E - G_E^*$). At the same time (1.22) provides a natural generalization of the non-relativistic Coulomb potential. The main approximation to the real electromagnetic interaction of two charged particles consists in the replacement of the vector potential (which gives rise to an angular momentum dependence of the energy eigenvalues) with a scalar potential (this is known to lead to an error of the order of 10^{-4}). Another model equation of the same class (with E_k replaced by E in $G_E^{(0)}$) is considered in [1].

2. ALGEBRAIZATION OF THE RELATIVISTIC COULOMB PROBLEM

2.1. Introductory Remarks

We shall deal from now on with Equations (1.20), (1.22). Noting that the coupling constant g has the dimension of mass and that m is the only mass in the Coulomb problem we introduce dimensionless variables by

$$\frac{1}{m^2} g^2 = \frac{2}{\pi}\alpha,\ \frac{1}{m} p \to p,\ \frac{1}{m} k \to k,\ \frac{1}{m} E \to E\ . \tag{2.1}$$

In these variables our quasi-potential equation assumes the form

$$p_0(E - p_0)\phi_E(P) = \frac{\alpha}{(2\pi)^2}\int \frac{\phi_E(k)}{(p - k)^2}\,\varepsilon(k_0)\delta(k^2 - 1)d^4k\ . \tag{2.2}$$

We are looking for the eigenvalues of E, for which Equation (2.2) has a non-trivial solution. Our first step to the solution of this problem will be its "algebraization". We will show that the free-particle energy operator p_0 and the integral operator on the right-hand side of Equation (2.2) can be expressed as simple rational functions of the generators of certain unitary representation of the conformal group SO(4,2). A similar algebraization has been carried out for the Bethe-Salpeter equation (in terms of the generators of SO(5,2)) in Reference [10]. Before going into the technical details we would like to make a comment about the meaning of this step.

The advantage of the algebraic form of an equation is in its independence of the realization of the algebra under consideration. The representation of a given Lie algebra is specified by a set of identities in its enveloping algebra. It may have many different (though unitarily equivalent) realizations. The choice of the most appropriate realization for the given equation is suggested by the symmetry of the problem which is most easily seen in its algebraic, i.e., realization-independent formulation. A famous example of an algebraic presentation of a physical theory is the Dirac formulation of non-relativistic quantum mechanics which is given in terms of the generators p and q of the Heisenberg algebra. Some special problems of high symmetry such as the harmonic oscillator can be solved directly in the invariant formulation. For many others the algebraic picture, being the most flexible one, suggests a convenient choice of coordinates.

We will start with a brief description of the conformal group and of the peculiar degenerate unitary representation we are going to use.

2.2. A Remarkable Representation of the Conformal Group

The conformal group SO(4,2) can be defined as the set of pseudo-orthogonal transformation in six dimensions which preserves a non-degenerate real symmetric quadratic form, xgx, with signature (2,4). For an appropriate choice of the basis we can write

$$xgx = x_A g^{AB} x_B = x_0^2 - x_1^2 - x_2^2 - x_3^2 - x_5^2 + x_6^2 \tag{2.3}$$

(in order to be consistent with traditional notation in physics (where often $x_4 = ix_0$ is used) we omit the index 4 in labeling x_A and g^{AB}). We will be interested actually in the restricted conformal group which consists of the connected component of the identity element of SO(4,2) and is denoted by $SO_0(4,2)$.

The Lie algebra of $SO_0(4,2)$ is generated by the infinitesimal rotations $i\Gamma_{AB}$ (in the AB plane). They form an antisymmetric tensor ($\Gamma_{AB} = -\Gamma_{BA}$) with 15 independent components satisfying the commutation relations

$$[\Gamma_{AB},\Gamma_{CD}] = i(g_{AD}\Gamma_{BC} + g_{BC}\Gamma_{AD} - g_{AC}\Gamma_{BD} - g_{BD}\Gamma_{AC}) \; . \tag{2.4}$$

The lowest faithful representation of this Lie algebra is 4-dimensional and is given by the set of Dirac γ-matrices:

$$\Gamma_{a6} \to \gamma_{a6} = \frac{1}{2}\gamma_a, \; \Gamma_{ab} \to \gamma_{ab} = \frac{i}{4}[\gamma_a,\gamma_b], \; a, b = 0, 1, 2, 3, 5 \tag{2.5}$$

where γ_a satisfy the identity

$$\{\gamma_a,\gamma_b\} \equiv \gamma_a\gamma_b + \gamma_b\gamma_a = 2g_{ab} \cdot \mathbb{1} \; . \tag{2.6}$$

The γ's are in fact the generators of the defining representation of the pseudo unitary group SU(2,2) which is a two-fold covering group of $SO_0(4,2)$. In other words there exists a hermitian matrix β with two positive and two negative eigenvalues such that

$$\beta\gamma_\mu = \gamma_\mu^*\beta \; . \tag{2.7}$$

We will also use the notation Γ_a for Γ_{a6}.

Now we are going to describe the particular irreducible unitary representation R_0 of $SO_0(4,2)$ which we will use for the algebraization of Equation (2.2). This representation has been used for many years by physicists but has been usually omitted in the mathematical classification of the unitary representations of the pseudo unitary (or of the pseudo orthogonal) group (see, however, References [11,12] where the place of the "ladder" representations of SU(2,2) is indicated). The representation R_0 is characterized by the following properties: (i) it remains irreducible when restricted to any of the five-dimensional rotation subgroups $SO_0(3,2)$ and $SO_0(4,1)$ of $SO_0(4,2)$ as well as to its Poincaré subgroup; (ii) when restricted to the subgroup SO(4) the representation R_0 splits into the direct sum of tensor representations* $\bigoplus_{n=1}^{\infty}(n,n)$ each (n,n) appearing with multiplicity one; (iii) the n^2-dimensional subspace $\mathcal{H}(n,n)$ (in which acts the representation (n,n) of SO(4) is an eigen subspace for the generator $\Gamma_0(=\Gamma_{06})$ of the subgroup SO(2) which commutes with SO(4): $f_n \in \mathcal{H}(n,n) \Rightarrow \Gamma_0 f_n = n f_n$.

We will describe here a particular realization of the representation R_0 on the space $\mathcal{H}$ of functions $\phi(p)$ defined on the double sheeted hyperboloid $V_1 = \{p: p^2 = 1\}$

$$(\phi,\phi) \equiv \frac{1}{\pi^4}\iint \frac{\overline{\phi(p)}\phi(q)}{-(p-q)^2}\,\delta(p^2-1)\delta(q^2-1)d^4p\,d^4q < \infty \tag{2.8}$$

(cf. [13]).

First of all we introduce homogeneous coordinates on V_1:

$$p_\mu = \frac{u_\mu}{u_5}, \quad q_\mu = \frac{v_\mu}{v_5} \tag{2.9}$$

* We recall that the group SO(4) is locally isomorphic to the direct product SU(2) ⊗ SU(2). Accordingly, each (unitary, irreducible) representation of SO(4) can be characterized by two integers (k,l) equal to the dimensions of the corresponding representations of the two groups SU(2).

and consider $\phi(p)$ as a restriction to the manifold $\{u = (|p_0|, \underline{p} \in (p_0)), p \in V_1\}$ of a homogeneous function $F(u)$ of degree of homogeneity-2 defined on the light-cone $C^+_{1,4}$

$$F(\lambda u) = \lambda^{-2}F(u) \quad \text{for} \quad \lambda > 0,\ u \in C^+_{1,4} \tag{2.10}$$

$$C^+_{1,4} = \{u:\ u_0 = |\vec{u}| \equiv \sqrt{u_1^2 + u_2^2 + u_3^2 + u_5^2}\}\ . \tag{2.11}$$

Taking into account that

$$-(p - q)^2 = 2\,\frac{uv}{u_5 v_5}\ (uv \equiv u_0 v_0 - \vec{u}\vec{v})$$

we find that the scalar product (2.8) assumes the form

$$\begin{gathered}(F,F)_{-2} = \frac{1}{2\pi^4} \iint \overline{F}(u)\, \frac{1}{uv}\, F(v)\delta(u_0 - 1)\delta(v_0 - 1)\delta(u^2)\delta(v^2)d^5u d^5v \\ [\text{for} \quad \phi(p) = F(|p_0|, \underline{p}\ \varepsilon(p_0)) \quad \text{or} \quad F(u) = u_5^{-2}\phi(\frac{u_0}{u_5}, \frac{\underline{u}}{u_5})]\ .\end{gathered} \tag{2.12}$$

The restriction of the representation R_0 on the $SO_0(4,1)$ subgroup of $SO_0(4,2)$ is defined as a set of argument transformation

$$SO_0(4,1) \ni \Lambda \to [U(\Lambda)F](u) = F(\Lambda^{-1}u)\ . \tag{2.13}$$

That is the Majorana representation of the complementary series of unitary representations of $SO_0(4,1)$, i.e., the only representation of the complementary series which can be extended to a representation of $SO_0(4,2)$. To see this we first remark that the representation (2.13) in the space $\mathcal{H}_{-2}$ with scalar product (2.12) is equivalent to the representation given by the same formula (2.13) in the space $\mathcal{H}_{-1}$ of homogeneous functions of degree of homogeneity -1, equipped with scalar product

$$(F,G)_{-1} = \frac{-1}{2\pi^4} \int \overline{F(u)}\, \frac{1}{(uv)^2}\, G(v)\delta(u_0 - 1)\delta(v_0 - 1)\delta(u^2)\delta(v^2)d^5u d^5v\ . \tag{2.14}$$

We mention that the integral in (2.14) is in general divergent because of the singularity for $u = v$. It has to be defined by analytic continuation with respect to N of the hermitian form $(F,G)_N$ (in which $-\frac{1}{2\pi^4}(uv)^{-2}$ in the integrand is replaced by $\frac{2^{N+1}\Gamma(-N)}{\pi^{7/2}\Gamma(-N - \frac{3}{2})}(uv)^{-3-N}$) (cf. [14]). The scalar product defined through this analytic continuation is positive-definite if and only if $N(N + 3) < 0$. The normalization is chosen in such a way that $(u_0^N, u_0^N)_N = 1$ ($F = u_0^N$ is the only $SO(4)$ invariant vector in $\mathcal{H}_N$ (up to a factor)). The intertwining operator T which maps $\mathcal{H}_{-1}$ onto $\mathcal{H}_{-2}$ and its inverse are given by

$$\begin{aligned}(TF)(u) &= \frac{-1}{2\pi^2} \int F(v)\delta(v_0 - 1)\delta(v^2)\, \frac{d^5v}{(uv)^2}\ , \\ (T^{-1}F)(v) &= \frac{1}{2\pi^2} \int F(u)\delta(u_0 - 1)\delta(u^2)\, \frac{d^5u}{uv}\ .\end{aligned} \tag{2.15}$$

The action of the five additional generators Γ_a $(a = 0,1,2,3,5)$ of the Lie algebra of $SO_0(4,2)$ in the space $\mathcal{H}_{-2}$ is defined by

$$(\Gamma_a F)(u) = [T(u_a F)](u) = \frac{-1}{2\pi^2}\int \frac{v_a}{(uv)^2} F(v)\delta(v_0 - 1)\delta(v^2)d^5v\ . \tag{2.16}$$

It can be verified by a straightforward computation that these operators satisfy (together with the generators Γ_{ab} of $SO_0(4,1)$) the commutation relations (2.4). In particular,

$$i[\Gamma_a,\Gamma_b] = \Gamma_{ab} = \begin{cases} -i(u_a \frac{\partial}{\partial u_b} - u_b \frac{\partial}{\partial u_a}) & \text{for } a,b = 1,2,3,5 \\ -iu_0 \frac{\partial}{\partial u_b} & \text{for } a = 0,\ b = 1,2,3,5\ . \end{cases} \tag{2.17}$$

It is easily seen also that the operators (2.16) are hermitian with respect to the scalar product (2.12). Some further property of the representation $\mathbf{R}_0$ are given in the Appendix. (In particular, we show that $\mathbf{R}_0$ defined so far as a representation of the Lie algebra of $SO_0(4,2)$ can be in fact integrated to a representation of the group; the global form of the representation coincides with the familiar realization of the conformal group in space-time which leaves invariant the D'Alembert equation $\Box f(x) = (\frac{\partial^2}{\partial x_0^2} - \sum_{j=1}^{3} \frac{\partial^2}{\partial x_j^2})f(x) = 0.$)

2.3. Algebraic Form of Equation (2.2)

In the space $\mathcal{H}$ of functions $\phi(p)$ the operators Γ_a (2.16) assume the form

$$(\Gamma_\mu\phi)(p) = -\frac{2}{\pi^2}\int \frac{q_\mu}{[(p-q)^2]^2}\phi(q)\varepsilon(q_0)\delta(q_0 - 1)d^4q \tag{2.18}$$

$$(\Gamma_5\phi)(p) = -\frac{2}{\pi^2}\int \frac{1}{[(p-q)^2]^2}\phi(q)\varepsilon(q_0)\delta(q_0 - 1)d^4q\ . \tag{2.19}$$

Comparing (2.18) with (2.19) we see that

$$(p_\mu\phi)(p) = (\frac{1}{\Gamma_5}\Gamma_\mu\phi)(p)\ . \tag{2.20}$$

Taking into account that for any analytic function F of Γ_5 we have

$$F(\Gamma_5)(\Gamma_\mu \pm \Gamma_{\mu 5}) = (\Gamma_\mu \pm \Gamma_{\mu 5})F(\Gamma_5 \pm i) \tag{2.21}$$

and using Equations (C.9), (C.10) (see Appendix C) we can verify that for $\lambda = 0$ the operators

$$p_\mu = \frac{1}{\Gamma_5}\Gamma_\mu \tag{2.22}$$

satisfy the identities $[p_\mu,p_\nu] = 0,\ p_\mu p^\mu = 1.$

On the other hand, one can check directly (or by using (2.15)) that

$$(\frac{1}{\Gamma_5}\phi)(p) = -\frac{1}{2\pi^2}\int \frac{1}{(p-q)^2}\phi(q)\varepsilon(q_0)\delta(q^2 - 1)d^4q\ . \tag{2.23}$$

Inserting (2.22) and (2.23) in the quasi-potential Equation (2.2) we find the following algebraic equation for the relativistic Coulomb problem

$$\frac{1}{\Gamma_5}\,[\Gamma_0(E - \frac{1}{\Gamma_5}\,\Gamma_0) + \frac{\alpha}{2}]\phi_E(p) = 0 \,. \tag{2.24}$$

Before going to the solution of Equation (2.24) we will make the following general comments.

(1) The prescription (2.22) for the algebraization of the (free) 4-momentum does not depend on the interaction under consideration.

(2) The simple algebraization of the potential based on Equation (2.23) is peculiar to the case of zero mass exchange. The relativistic Yukawa potential

$$V(p,q) = \frac{g^2}{(p - q)^2 - \mu^2} \tag{2.25}$$

leads already to considerable complications (see Section III.2 of Reference [2]). The reason is that the kernel in the scalar product (2.8) in $\mathcal{H}$ is closely related to the relativistic Coulomb potential. If on the other hand we adapt the scalar product in our representation space to the potential (2.25) for $\mu > 0$, the simplicity of the free Hamiltonian will be lost.

(3) We can use Equations (2.18-20) and (2.23) to solve the inverse problem: given *ad hoc* an infinite-component wave equation in the representation space $\mathcal{H}$ of R_0 (see References [13,15,16]) to reconstruct an equivalent integral equation in momentum space.

3. SOLUTION OF THE COULOMB EIGENVALUE PROBLEM

3.1. Group Theoretical Treatment of the Algebraic Equation

In order to get rid of the inverse powers of Γ_5 in Equation (2.24) we multiply it from the left by $\Gamma_5\Gamma_0^{-1}\Gamma_5$ and put

$$\phi_E = \Gamma_0 f_E \,. \tag{3.1}$$

This leads to the following equation for f_E:

$$[(\Gamma_0 - E\Gamma_5)\Gamma_0 - \frac{\alpha}{2}\,\Gamma_5]f_E = 0 \,. \tag{3.2}$$

First of all we observe that the operators Γ_0, Γ_5 and Γ_{05} generate the Lie algebra of SO(2,1):

$$[\Gamma_0,\Gamma_{05}] = i\Gamma_5,\ [\Gamma_5,\Gamma_0] = i\Gamma_{05},\ [\Gamma_{05},\Gamma_5] = -i\Gamma_0 \,. \tag{3.3}$$

Equation (C.12) of Appendix C shows that for the representation R_0 the Casimir operator of $\underline{SO}(2,1)$ is equal to the Casimir of $\underline{SO}(3)$. Hence, for fixed angular momentum ℓ

$$\Gamma_0^2 - \Gamma_5^2 - \Gamma_{05}^2 = \underline{L}^2 = \ell(\ell + 1) \,. \tag{3.4}$$

Since Equation (3.2) is obviously SO(3) invariant, we will require that f_E is an eigenvector of $\underline{L}^2$, say $f_{E\ell}$.

Equation (3.4) and the positivity of Γ_0 imply that we have to deal with one of the discrete series of unitary representations of SO(2,1) described by Bargmann[17] (see also [14] Chapter 7). Each irreducible representation $R_0^{(\ell)}$ of this series can be realized as a group of coordinate transformations (with a suitable multiplier) in the space $\mathcal{H}_\ell$ of analytic functions on the unit disk

$$D_1 = \{z \in \mathbb{C},\ |z| < 1\} . \tag{3.5}$$

$\mathcal{H}_\ell$ is considered as a Hilbert space with scalar product

$$(g,f)_\ell = \frac{2\ell + 1}{\pi} \int_{D_1} (1 - z\bar{z})^{2\ell}\, \overline{g(z)} f(z) d^2 z . \tag{3.6}$$

The generators of the representation $R_0^{(\ell)}$ are first order differential operators with respect to z:

$$\Gamma_0 = z \frac{d}{dz} + \ell + 1,\ \Gamma_5 = (\ell + 1)z + \frac{1}{2}(z^2 + 1)\frac{d}{dz}$$
$$\Gamma_{05} = i[(\ell + 1)z + \frac{1}{2}(z^2 - 1)\frac{d}{dz}] . \tag{3.7}$$

It is easily seen that the operators (3.7) satisfy the commutation relations (3.3) and the identity (3.4).

Inserting (3.7) in (3.2) we get the following second order (linear) differential equation for $f_{E\ell}(z)$:

$$\{zQ \frac{d^2}{dz^2} + [(\ell + 2 + \frac{\alpha}{2E})Q + (\ell + 1)Q'z + \frac{\alpha}{2E} z] \frac{d}{dz}$$
$$+ (\ell + 1)[(\ell + 1)Q' + \frac{\alpha}{2} z]\}f = 0 \tag{3.8}$$

where

$$Q = \frac{E}{2}(z^2 + 1) - z,\ Q' = Ez - 1 .$$

3.2. Calculation of the Energy Eigenvalues

The eigenvalues of E have to be determined from the condition that $f_{E\ell}$ be regular in the unit disk. The possible singular points of any solution of (3.8) are $z = 0$, $z = \infty$ and

$$z = z_\pm = \frac{1}{E} \pm \frac{1}{E}\sqrt{1 - E^2} . \tag{3.9}$$

Among these four points only two $z = 0$ and $z = z_-$ belong to D_1. They are both "weak singularities" of the differential Equation (3.8) and there are regular solutions f_0 and f_- in the neighborhood of any of them. In order to ensure that these two solutions are analytic continuation of one another, it is necessary to assume that the branch points at $z = z_+$ and $z = z_\infty$ are of the same type (so that one could consider a single-valued solution of (3.8) regular in the cut z-plane with a cut between z_+ and ∞ which does not cross the unit disk).

For $z \to z_+$ the asymptotic form of (3.8) is

$$[A(z - z_+) \frac{d^2}{dz^2} + B \frac{d}{dz} + C]f_+ = 0 \tag{3.10}$$

with $A = \sqrt{1 - E^2}\, z_+$, $B = z_+[\sqrt{1 - E^2}\,(\ell + 1) + \frac{\alpha}{2E}]$. For $z \to z_+$ the singular solution f_+ of (3.10) behaves like $(z - z_+)^{\nu_+}$ where

$$\nu_+ = 1 - \frac{B}{A} = -\ell - \frac{\alpha}{2E\sqrt{1 - E^2}} . \tag{3.11}$$

For $z \to \infty$ Equation (3.8) is equivalent to

$$[z^2 \frac{d^2}{dz^2} + (3\ell + 4 + \frac{\alpha}{2E})z \frac{d}{dz} + 2(\ell + 1)(\ell + 1 + \frac{\alpha}{2E})]f_\infty = 0 . \tag{3.12}$$

The relevant solution of (3.12) is $f_\infty = z^{\nu_\infty}$ with

$$\nu_\infty = -\ell - 1 - \frac{\alpha}{2E} . \tag{3.13}$$

The branch points at $z = z_+$ and $z = \infty$ are of the same type if and only if $\nu_\infty - \nu_+$ is an integer. So, we put

$$\nu_\infty - \nu_+ = \frac{\alpha}{2E} (\frac{1}{\sqrt{1 - E^2}} - 1) - 1 = n - 1 . \tag{3.14}$$

Thus, the eigenvalues E_n of E are determined from the equation

$$\frac{\alpha}{2n} = (E_n + \frac{\alpha}{2n})\sqrt{1 - E_n^2} \tag{3.15}$$

or

$$E_n^3 + \frac{\alpha}{n} E_n^2 - (1 - \frac{\alpha^2}{4n^2})E_n - \frac{\alpha}{n} = 0 . \tag{3.16}$$

Only one of the three real roots of (3.16) satisfies (3.15). It can be written as an expansion in $\alpha_n \equiv \frac{\alpha}{2n}$:

$$\sqrt{1 - E_n^2} = \alpha_n - \alpha_n^2 + \frac{3}{2}\alpha_n^3 - 3\alpha_n^4 + \dots$$

$$E_n = 1 - \frac{1}{2}\alpha_n^2 + \alpha_n^3 - \frac{17}{8}\alpha_n^4 + \dots \quad . \tag{3.17}$$

In order to find the range* of the quantum number n we look at the power series expansion of the solution of Equation (3.8):

$$f(z) = \sum_{\nu=0}^{\infty} f_\nu z^\nu . \tag{3.18}$$

In view of (3.8) the coefficients f_ν satisfy the following recurrence relation

$$(\nu + 1)(\nu + \ell + 2 + \beta)f_{\nu+1} - 2(\nu + \ell + 1)^2 \mathrm{ch}\lambda f_\nu + [(\nu - 1)(\nu + 3\ell + 2 + \beta) + 2(\ell + 1)(\ell + 1 + \beta)]f_{\nu-1} = 0, \; \nu = 0, 1, 2, \dots, \; \beta = \frac{\alpha}{2E} . \tag{3.19}$$

The radius of convergence of the power series (3.18) is determined by the behavior of the coefficients f_ν for large ν. Dividing the left-hand side of (3.19) by

---

* This problem was not touched in Reference [2].

$\nu + 1$ and neglecting the terms of order $\frac{1}{\nu}$ we obtain the following asymptotic form for the recurrence relation

$$(\nu + \ell + 2 + \beta)f_{\nu+1} - \frac{2}{E}(\nu + 2\ell + 1)f_\nu + (\nu + 3\ell + \beta)f_{\nu-1} = 0. \tag{3.20}$$

It corresponds to a first order differential equation which can be obtained by multiplying by z^ν and summing over ν. The result is

$$zQf' + \{(2\ell + 1)Q + \frac{E}{2}[\ell(z^2 - 1) + \beta(z^2 + 1)]\}f = \frac{E}{2}(\ell + 1 + \beta) . \tag{3.21}$$

(We have used the initial conditions $f_{-1} = 0$, $f_0 = f(0) = 1$.) The solution of (3.21) regular (and normalized to 1) for $z = 0$ is

$$f(z) = \frac{\ell + 1 + \beta}{z^{\ell + 1 + \beta}}\left(\frac{z - z_-}{z - z_+}\right)^{\beta/\sqrt{1-E^2}} \frac{1}{[(z - z_+)(z - z_-)]^\ell} \int_0^z \xi^{\ell+\beta}\left(\frac{\xi - z_+}{\xi - z_-}\right)^{\beta/\sqrt{1-E^2}} [(\xi - z_+)(\xi - z_-)]^{\ell-1}d\xi . \tag{3.22}$$

We can define $f(z)$ as analytic single valued function in the cut z-plane with a cut along the real semi axis $z \geq z_+$ provided that

$$\frac{\beta}{\sqrt{1 - E^2}} = \beta + n, \; n = 1, 2, \ldots \tag{3.23}$$

in accordance with (3.15) (β is defined in (3.19)). It is regular for $z = z_-$ only if $n \geq \ell$. For $\ell = 0$ we actually have to require $n \geq 1$; it is easily verified that for $E = 0$, Equation (3.8) has no solution regular for $z = 0$. (This shows that contrary to the Wick-Cutkosky model [6] there is no limit of "maximal binding" in our quasi-potential equation.) The present argument cannot exclude however the values $n = \ell$ for $\ell \geq 1$. We observe that (3.22) gives the exact solution of Equation (3.8) for the s waves ($\ell = 0$) but not for $\ell \geq 1$ (however, it has for all ℓ the correct behavior $(z - z_+)^{\nu_+}$ as $z \to z_+$). We expect that the exact range of the quantum number n is always $n \geq \ell + 1$, which would give the familiar SO(4) degeneracy of the energy levels of the non-relativistic hydrogen atom (as well as of the Wick-Cutkosky model). We mention that the second order term in Equation (3.17) reproduces precisely the Balmer formula for the non-relativistic Coulomb energy levels as it should be in any consistent relativistic generalization of the Coulomb problem.

APPENDIX A

DIFFERENT REALIZATIONS AND PROPERTIES OF THE EXCEPTIONAL REPRESENTATION R_0 OF $SO_0(4,2)$

A. The Set of Conformal Transformations in Space-time as a Global Realization of R_0

Consider the space X of negative frequency solutions

$$f(x) = \frac{1}{(2\pi)^{3/2}} \int \mathfrak{f}(\xi) e^{-ix\xi} \delta_0^+(\xi) d^4\xi, \quad \delta_0^+(\xi) = \theta(\xi_0)\delta(\xi^2) \tag{A.1}$$

of the D'Alembert equation

$$\Box f(x) \equiv \left(\frac{\partial^2}{\partial x_0^2} - \Delta\right) f(x) = 0 \tag{A.2}$$

with scalar product

$$\begin{aligned}(f,g) &= i \int_{x_0=t} \left(\bar{f}(x) \frac{\partial g(x)}{\partial x_0} - \frac{\partial \bar{f}(x)}{\partial x_0} g(x)\right) d^3\underline{x} \\ &= \int \bar{\mathfrak{f}}(\xi) \mathfrak{g}(\xi) \delta_0^+(\xi) d^4\xi .\end{aligned} \tag{A.3}$$

The representation of the conformal group acting in X which leaves Equation (A.2) and the scalar product (A.3) invariant is generated by the following transformations:

(i) Poincaré transformations

$$[U(a,\Lambda)f](x) = f(\Lambda^{-1}(x - a)) \tag{A.4}$$

(ii) Dilations

$$(U(\lambda)f)(x) = \lambda^{-1} f(\lambda^{-1}x) \tag{A.5}$$

(iii) Inversion

$$[U(R)f](x) = \frac{1}{x^2} f\left(\frac{-x}{x^2}\right) . \tag{A.6}$$

The inversion $(Rx)_\mu = -\frac{x_\mu}{x^2}$ does not actually belong to the connected component of the identity of the conformal group, but the set of non-linear transformations

$$[R\{b,1\}Rx]_\mu = \frac{x_\mu - x^2 b_\mu}{d(b,x)}, \quad d(b,x) = 1 - 2bx + b^2x^2 \tag{A.7}$$

belongs to $SO_0(4,2)$ and generates the so-called special conformal transformations

$$[U(R\{-b,1\}R)f](x) = \frac{1}{d(b,x)} f\left(\frac{x_\mu - x^2 b_\mu}{d(b,x)}\right) . \tag{A.8}$$

The (hermitian) infinitesimal operators of the subgroups (A.4), (A.5), and (A.8) are given by

$$P_\mu = i\partial_\mu,\ M_{\mu\nu} = i(x_\mu\partial_\nu - x_\nu\partial_\mu),\ (\partial_\mu \equiv \frac{\partial}{\partial x^\mu})\ ,$$
$$D = -i(1 + x_\mu\partial^\mu),\ K_\mu = i(2x_\mu + 2x_\mu x_\nu\partial^\nu - x^2\partial_\mu)\ . \qquad (A.9)$$

These operators are related to the generators Γ_{ab} and Γ_a used in Section 2.2 by

$$M_{\mu\nu} \Leftrightarrow \Gamma_{\mu\nu},\ P_\mu \Leftrightarrow \Gamma_\mu + \Gamma_{\mu 5},\ D \Leftrightarrow \Gamma_5,\ K_\mu \Leftrightarrow \Gamma_\mu - \Gamma_{\mu 5}\ . \qquad (A.10)$$

This well-known representation of the conformal group (related to the 0-spin 0-mass particles) is equivalent to the representation R_0 defined in Section 2.2. The intertwining operator V which maps $\mathcal{H}$ onto X can be written down explicitly:

$$\mathcal{H} \ni \phi(p) \overset{V}{\rightarrow} f(x) = \frac{1}{\pi}\int D_0^{(-)}(p + x)\varepsilon(p_0)\delta(p^2 - 1)\phi(p)d^4p \qquad (A.11)$$

where

$$D_0^{(-)}(x) = \frac{i}{(2\pi)^3}\int e^{-ix\xi}\delta_0^+(\xi)d^4\xi = \frac{-i}{(2\pi)^2}\frac{1}{(x_0 - i0)^2 - \underline{x}^2} \qquad (A.12)$$

is the Lorentz invariant negative frequency solution of Equation (A.2). (The distribution $D_0^{(-)}(x)$ appears in quantum field theory as the two-point function of a zero mass field.) The realization of the representation R_0 in X displays its irreducibility with respect to the Poincaré subgroup of the conformal group.

B. R_0 As One of the Metaplectic Representations of SU(2,2)

The metaplectic series of unitary representations of SU(2,2) can be constructed in infinitesimal form starting with the 4-dimensional representation (2.5) of the Lie algebra. To do this, we introduce the 4-component operator valued spinor φ satisfying the canonical commutation relations

$$[\varphi^\alpha,\varphi^\beta] = 0,\ [\varphi^\alpha,\tilde{\varphi}_\beta] = \delta^\alpha_\beta,\ \alpha,\beta = 1,2,3,4,\ \tilde{\varphi} = \varphi^*\beta\ ; \qquad (B.1)$$

here β is the hermitian matrix satisfying (2.7) and normalized by the requirements $\det\beta = 1$, $\beta\gamma_0$ is positive definite. It is easy to verify that the set of operators

$$\Gamma_{AB} = \tilde{\varphi}\gamma_{AB}\varphi \qquad (B.2)$$

obeys the commutation relations (2.4) since (B.1) implies that

$$[\Gamma_{AB},\Gamma_{CD}] = \tilde{\varphi}[\gamma_{AB},\gamma_{CD}]\varphi\ . \qquad (B.3)$$

The metaplectic series of the so-called ladder representations of SU(2,2) corresponds to the (star) representation of the canonical commutation relations (B.1) in the Fock space F defined in the following way. There exists a unit vector $|0)$ in F (defined up to a phase factor) for which

$$(\gamma_0 + 1)\varphi|0) = \tilde{\varphi}(\gamma_0 - 1)\ |0) = 0,\ (\Gamma_0\ |0) = |0))\ . \qquad (B.4)$$

The vector $|0)$ so defined is $SU(2) \times SU(2)$ invariant.

In order to label the irreducible representations of the metaplectic series, it is convenient to extend the representation defined by (B.2) to a representation of U(2,2) by introducing a 16th generator,

$$C = \frac{1}{2}\tilde{\varphi}\varphi \ . \tag{B.5}$$

C belongs to the center of the enveloping algebra of the Lie algebra $\underline{U}(2,2)$ and hence, should be a multiple of the identity in each irreducible subspace of F. It is easy to verify that the spectrum of C in F is given by

$$C = \lambda - 1,\ \lambda = 0,\ \pm\frac{1}{2},\ \pm 1,\ \ldots \ . \tag{B.6}$$

It can be proved that for fixed C (or λ) the ladder representation R_λ acting in the corresponding invariant subspace F_λ of F is already irreducible. All elements of the center of the enveloping algebra of the metaplectic series are functions of λ. In particular, the second order Casimir operator C_2 of SU(2,2) is given by

$$C_2 = \frac{1}{2}\Gamma_{AB}\Gamma^{AB} = 3(\lambda^2 - 1) \ . \tag{B.7}$$

It has been shown explicitly in Reference [18] that the metaplectic representations R_λ so defined are equivalent to the representation of the conformal group in space-time, corresponding to zero-mass particles of helicity λ. In particular, for $\lambda = 0$, we recover the representation R_0 described in Section 2.2 and Appendix A.

The ladder representations R_λ are closely related to the two metaplectic representations of the real symplectic group Sp(4,R) in 8-dimension described in References [19,20]. Namely, if $R^{(0)}$ is the single-valued and $R^{(1)}$ the double-valued representation of Sp(4,R) acting in the same Fock space F, then

$$\begin{aligned} R^{(0)} &= \sum_{\lambda=0,\pm1,\pm2,\ldots} \oplus\, R_\lambda \\ R^{(1)} &= \sum_{\lambda=\pm\frac{1}{2},\pm\frac{3}{2},\ldots} \oplus\, R_\lambda \ . \end{aligned} \tag{B.8}$$

More about the different realizations of the ladder representations and their equivalence is said in Appendix to Reference [2]. The term metaplectic and the first mathematical description of the metaplectic representations of Sp(n,R) is due to Weil [21]. (See also Mackey [22].) The description of the metaplectic representations of $\underline{U}(2,2)$ in terms of creation and annihilation operators was first given by Kurşunoglu [23].

C. Quadratic Identities in the Enveloping Algebra of the Metaplectic Representations

We shall collect in this section a set of quadratic identities which hold in the enveloping algebra of the metaplectic representation of $\underline{U}(2,2)$. They can be derived by using (B.1), (B.2), and the identity

$$\sum_{a=0,1,2,3,5} (\gamma_a)^\alpha_\beta (\gamma^a)^\delta_\delta = \delta^\alpha_\beta \delta^\delta_\delta + 2\varepsilon^{\alpha\delta\sigma\tau} B_{\sigma\beta} B_{\tau\delta} \ , \tag{C.1}$$

where $\varepsilon^{\alpha\delta\sigma\tau}$ is the completely antisymmetric unit tensor in 4-dimension

($\varepsilon^{1234} = 1$) and B is defined (up to an irrelevant sign) by

$$B\gamma_{ab}B^{-1} = -{}^t\gamma_{ab}, \ (a,b = 0,1,2,3,5), \ {}^tB = -B, \ (B^{-1})^{\beta\alpha} = \frac{1}{2}\varepsilon^{\alpha\beta\sigma\tau}B_{\sigma\tau} \qquad (C.2)$$

(the superscript t to the left of a matrix stands for transposition).

Each of the metaplectic representations R_λ remains irreducible when restricted to any of the 5-dimensional rotation subalgebras of $\underline{SO}(4,2)$. Hence, their second order Casimir operators are functions of λ only. A direct calculation gives

$$\frac{1}{2}\Gamma_{ab}\Gamma^{ab} = 2(\lambda^2 - 1) = \frac{1}{2}\Gamma_{\mu\nu}\Gamma^{\mu\nu} + \Gamma_\mu\Gamma^\mu \qquad (C.3)$$

(repeated upper and lower indices have to be summed over the range $a,b = 0,1,2,3,5$; $\mu,\nu = 0,1,2,3$). Comparing (C.3) with (B.7), we find

$$\Gamma_\mu\Gamma^\mu = \Gamma_{5\mu}\Gamma^{5\mu} = \lambda^2 + \Gamma_5^2 - 1 . \qquad (C.4)$$

We also have

$$\frac{1}{2}\Gamma_{\mu\nu}\Gamma^{\mu\nu} = \underline{L}^2 - \underline{N}^2 = \lambda^2 - 1 - D^2, \ \underline{LN} = -\lambda\Gamma_5 \qquad (C.5)$$

(with $\Gamma_{ij} = \varepsilon_{ijk}L_k$, $\Gamma_{oj} = N_j$, $i,j,k = 1,2,3$). More generally, the following tensor identities hold:

$$\{\Gamma_{\mu5},\Gamma_\nu\} - \{\Gamma_\mu,\Gamma_{\nu5}\} = 2\Gamma_5\Gamma_{\mu\nu} - \lambda\varepsilon_{\mu\nu\sigma\tau}\Gamma^{\sigma\tau} \qquad (C.6)$$

$$\{\Gamma_{CA},\Gamma^{CB}\} = (\lambda^2 - 1)\delta_A^B (A,B = 0,1,2,3,5,6) . \qquad (C.7)$$

As mentioned before, each of the representations R_λ remains also irreducible when restricted to the Poincaré subgroup generated by P_μ and $M_{\mu\nu}$ (or K_μ and $M_{\mu\nu}$; see (A.10)). This gives

$$P_\mu P^\mu = K_\mu K^\mu = 0, \ \underline{P}\,\underline{L} = P_0\lambda . \qquad (C.8)$$

Equation (C.6) implies

$$P_\mu K_\nu - P_\nu K_\mu = 2(\Gamma_5 - i)\Gamma_{\mu\nu} - \lambda\varepsilon_{\mu\nu\sigma\tau}\Gamma^{\sigma\tau} . \qquad (C.9)$$

The scalar product of P and K is a function of λ and Γ_5:

$$KP = (PK)^* = 2[\lambda^2 + (\Gamma_5 + i)^2] . \qquad (C.10)$$

The Casimir operators of the $\underline{SO}(4)$ subalgebra are expressed in terms of Γ_0 and λ:

$$\sum_{j=1}^{3}(L_j^2 + \Gamma_{j5}^2) = \Gamma_0^2 + \lambda^2 - 1, \ L_j\Gamma_{j5} = \lambda\Gamma_0 . \qquad (C.11)$$

From (C.4) and (C.11), it follows that

$$\underline{L}^2 = \Gamma_0^2 - \Gamma_{05}^2 - \Gamma_5^2 . \qquad (C.12)$$

REFERENCES

[1] Itzykson, C., Kadyshevsky, V. G., and Todorov, I. T., *Three Dimensional Formulation of the Relativistic Two-Body Problem and Infinite Component Wave Equations*, Institute for Advanced Study, Princeton, preprint (1969) and *Phys. Rev.* (to be published).

[2] Itzykson, C., and Todorov, I. T., "An Algebraic Approach to the Relativistic Two-Body Problem" in *Proceedings of the Coral Gables Conference on Fundamental Interactions on High Energy*, T. Gudehus *et al.* editors, Gordon and Breach, New York (1969).

[3] Todorov, I. T., "On the Three Dimensional Formulation of the Relativistic Two-Body Problem", *Lectures Presented at the Theoretical Physics Institute*, University of Colorado, Boulder (1969).

[4] Kadyshevsky, V. G., "Relativistic Equations for the S-Matrix in the p-Representation", I "Unitarity and Causality Conditions"; II, *Soviet Phys. JETP*, 19, 443, 597 (1964).

[5] Nakanishi, N., "A General Survey of the Theory of the Bethe-Salpeter Equation", *Prog. Theor. Phys. Suppl.*, No. 43, 1 (1969).

[6] Wick, G. C., "Properties of the Bethe-Salpeter Wave Functions", *Phys. Rev.*, 96, 1124 (1954).

Cutkosky, R. E., "Solutions of a Bethe-Salpeter Equation", *Phys. Rev.*, 96, 1135 (1954).

[7] Kadyshevsky, V. G., "Quasi-potential Equation for the Relativistic Scattering Amplitude", *Nucl. Phys.*, 136, 125 (1968).

Kadyshevsky, V. G., and Mateev, M. D., "On a Relativistic Quasi-potential Equation in the Case of Particles with Spin", *Nuovo Cimento*, 55A, 233 (1968).

[8] Faustov, R. N., and Helashvili, A. A., "Normalization Condition for Simultaneous Wave Function of the Bound State of Two Particles", *JINR*, Dubna, preprint P2-4345 (1969).

[9] Logunov, A. A., and Tavkhelidze, A. N., "Quasi-optical Approach in Quantum Field Theory", *Nuovo Cimento*, 29, 380 (1963).

Logunov, A. A., Tavkhelidze, A. N., Todorov, I. T., and Khrustalev, O. A., "Quasi-potential Character of the Scattering Amplitude", *Nuovo Cimento*, 30, 134 (1963).

[10] Kyriakopoulos, E., "Dynamical Groups and the Bethe-Salpeter Equation", *Phys. Rev.*, 174, 1846 (1968).

[11] Todorov, I. T., "Discrete Series of Hermitian Representations of the Lie Algebra of U(p,q)", Int. Centre Theoret. Phys., Trieste, preprint IC/66/71 (1966).

[12] Yao, Tsu, "Unitary Irreducible Representations of SU(2,2), I and II", *J. Math. Phys.*, 8, 1931 (1967) and 9, 1615 (1968).

[13] Fronsdal, C., "Infinite Multiplets and the Hydrogen Atom", *Phys. Rev.*, 156, 1665 (1967).

[14] Gel'fand, I. M., Graev, M. I., and Vilenkin, N. Ya., "Integral Geometry and Representation Theory" in *Generalized Functions*, Vol. 5, Academic Press,

New York (1966). See also "Properties and Operations", Appendix B to Vol. 1, Academic Press, New York (1964).

[15] Nambu, Y., "Infinite-component Wave Equations with Hydrogen-like Mass Spectra", *Phys. Rev.*, 160, 1171 (1967).

[16] Barut, A. O., and Kleinert, H., "Current Operators and Majorana Equation for the Hydrogen Atom from Dynamical Groups", *Phys. Rev.*, 157, 1180 (1967).

[17] Bargmann, V., "Irreducible Unitary Representations of the Lorentz Group", *Annals of Math.*, 48, 568 (1947).

[18] Mack, G., and Todorov, I. T., "Irreducibility of the Ladder Representations of U(2,2) When Restricted to Its Poincaré Subgroup", *J. Math. Phys.*, 10, 2078 (1969).

[19] Itzykson, C., "Remarks on Boson Commutation Rules", *Commun. Math. Phys.*, 4, 92 (1967).

[20] Bargmann, V., "Group Representations on Hilbert Spaces of Analytic Functions" in *Lectures at the International Symposium on Analytic Methods in Mathematical Physics, Indiana University 1968*, Gordon and Breach, New York (1970).

[21] Weil, A., "Sur Certains Groupes d'Operateurs Unitaires", *Acta Math.*, 111, 143 (1964).

[22] Mackey, G., "Some Remarks on Symplectic Automorphisms", *Proceedings Amer. Math. Soc.*, 16, 393 (1965).

[23] Kurşunoglu, B., *Modern Quantum Theory*, W. H. Freeman and Co., San Francisco (1962), p. 257.

TENSOR OPERATORS FOR THE GROUP SL(2,C)

by

W. Rühl*

INTRODUCTION

This talk consists of three parts: some selected topics of a purely mathematical theory of irreducible tensor operators, the adaptation of this theory to the decomposition of the current density operators of elementary particle physics restricted to single-particle spaces, and an application of this formalism to a phenomenological analysis of certain scattering experiments.

1. IRREDUCIBLE TENSOR OPERATORS

1.1. Notations and Some Known Facts About the Representations of SL(2,C)

We shall mainly adhere to the notations of Gel'fand and Naimark. [1] In particular we make explicit use of matrices like the following ones

$$a = \begin{pmatrix} a_{11} & a_{12} \\ a_{21} & a_{22} \end{pmatrix} \in SL(2,C); \; u = \begin{pmatrix} u_{11} & u_{12} \\ u_{21} & u_{22} \end{pmatrix} \in SU(2) \equiv K, \; k = \begin{pmatrix} \lambda^{-1} & \mu \\ 0 & \lambda \end{pmatrix}, \; \zeta = \begin{pmatrix} 1 & 0 \\ z & 1 \end{pmatrix}$$

where λ, μ, z, are complex numbers and K stands for "maximal compact subgroup" of SL(2,C). One possibility to realize the principal series of representations of SL(2,C) is on a space of measurable functions $f(z)$ which have finite norm with respect to the scalar product

$$(f_1,f_2) = \int f_1(z)f_2(z)dz \; . \tag{1.1}$$

We denote this space $L^2(Z)$. The group operations are introduced by

$$T_a^{\chi}f(z) = \alpha^{\chi}(z,a)f(z_a) \tag{1.2}$$

with

$$\zeta a = k\zeta_a; \; \zeta_a = \begin{pmatrix} 1 & 0 \\ z_a & 1 \end{pmatrix}, \; k = \begin{pmatrix} \lambda^{-1}(z,a) & \mu \\ 0 & \lambda(z,a) \end{pmatrix},$$

$$\alpha^{\chi}(z,a) = |\lambda(z,a)|^{i\rho-2}\left(\frac{\lambda(z,a)}{|\lambda(z,a)|}\right)^{-m} \tag{1.3}$$

---

* European Organization for Nuclear Research, Geneva, Switzerland.

where ρ is real and m is an integer. We characterize the representation χ by the pairs of numbers

$$\chi = (m,\rho) = \{n_1,n_2\};\ n_{1,2} = \frac{-m}{+2} + \frac{i}{2}\rho \tag{1.4}$$

and use

$$-\chi = (-m,-\rho) \quad \text{if} \quad \chi = (m,\rho) .$$

We call this realization of the principal series the "noncompact picture".

Another realization of the principal series is obtained in a space $L^2_m(K)$ of measurable functions $\varphi(u)$ on K satisfying the constraint

$$\varphi(u(\psi)u) = e^{im\psi}\varphi(u)$$
$$u(\psi) = \begin{pmatrix} e^{i\psi} & 0 \\ 0 & e^{-i\psi} \end{pmatrix} \tag{1.5}$$

which have finite norm with respect to the scalar product

$$(\varphi_1,\varphi_2) = \int \overline{\varphi_1(u)}\varphi_2(u)d\mu(u) . \tag{1.6}$$

We introduce the operators T^χ_a by the definition

$$T^\chi_a\varphi(u) = \alpha^\chi(u,a)\varphi(u_a);\ ua = ku_a,\ k = \begin{pmatrix} \lambda^{-1}(u,a) & \mu \\ 0 & \lambda(u,a) \end{pmatrix} \tag{1.7}$$

and a relation between λ and α^χ as in (1.3). This realization is denoted the "compact picture". We arrive at the principal series in this compact picture if we use the technique of induced representations and induce from one-dimensional unitary representations $\xi(k)$ of the subgroup of triangular matrices k

$$\xi(k) = |\lambda|^{i\rho}\left(\frac{\lambda}{|\lambda|}\right)^{-m}$$

and identify the cosets of this subgroup in $SL(2,C)$ with the cosets of the subgroup $U(1)$ in $SU(2)$ by

$$a = ku .$$

The equivalence of the compact and the noncompact picture is easily established.

Following Gel'fand [1] we consider a set of closed topological vector spaces D_χ which are dense in the Hilbert spaces $L^2_m(K)$. They consist of infinitely differentiable functions $\varphi(u)$ satisfying (1.5) and possess a topology typical for a space of type K in Gel'fand's notation [2]. In the noncompact picture the corresponding spaces D_χ consist of infinitely differentiable functions $f(z)$ (considered as functions of two real variables) which possess an asymptotic expansion

$$f(z) \cong |z|^{i\rho-2}\left(\frac{z}{|z|}\right)^{-m} \sum_{k,\ell=0}^{\infty} C_{k\ell}z^{-k}\bar{z}^{-\ell} \tag{1.8}$$

around $z = \infty$. The topology is simply carried over from the compact picture. The spaces D_χ are invariant subspaces under operation of T^χ_a, and the operators T^χ_a are continuous. We emphasize that in the compact picture the definition of the spaces D_χ is independent of the parameter ρ. One space D_χ may therefore be used simultaneously for the definition (1.7) of operators T^χ_a with fixed m but variable complex ρ. Completing the spaces D_χ for fixed m with respect to the scalar product norm (1.6), we obtain the original Hilbert spaces $L^2_m(K)$. In this fashion we can make $L^2_m(K)$ to carry representations χ for arbitrary complex ρ, which are nonunitary if $\mathrm{Im}\rho \neq 0$. Translating this construction into the noncompact picture we obtain representations in Hilbert spaces $L^2_{\rho_1}(Z)$ with the scalar product defined by

$$(f_1,f_2) = \int \overline{f_1(z)} f_2(z)(1 + |z|^2)^{\rho_1} dz,\ \rho_1 = \mathrm{Im}\rho .$$

For any fixed χ with arbitrary complex ρ there is an isometric mapping from $L^2_{\rho_1}(Z)$ onto $L^2_m(K)$ which intertwines the bounded operators T^χ_a.

The spaces D_χ possess invariant closed subspaces E_χ if both n_1 and n_2 (1.4) are positive integers. In the noncompact picture the spaces E_χ are spanned by polynomials in z and $\bar{z}$ of maximal order $n_1 - 1$, respectively $n_2 - 1$. Therefore

$$\dim E_\chi = n_1 n_2 . \tag{1.9}$$

In addition the space D_χ possesses an invariant subspace if both n_1, n_2 are negative integers. This subspace is denoted F_χ and consists of all functions of D_χ whose momenta

$$\int z^k \bar{z}^\ell f(z) dz \tag{1.10}$$

vanish for all orders

$$0 \leq k \leq -n_1 - 1$$

$$0 \leq \ell \leq -n_2 - 1 .$$

Again we find

$$\dim D_\chi / F_\chi = n_1 n_2 . \tag{1.11}$$

Next we recall Gel'fand's results [1] on bilinear invariant functionals on spaces D_χ. We define such a functional $B(f_1,f_2)$ for two functions $f_{1,2} \in D_{\chi_{1,2}}$ requiring

a) linearity

$$B(\sum_i \alpha_i f_i, \sum_j \beta_j h_j) = \sum_{ij} \alpha_i \beta_j B(f_i,h_j) ; \tag{1.12}$$

b) continuity in each argument;

c) invariance for all $a \in SL(2,C)$

$$B(T^{\chi_1}_a f_1, T^{\chi_2}_a f_2) = B(f_1,f_2) . \tag{1.13}$$

In the case that both D_{χ_1} and D_{χ_2}, $\chi_1 = \{n_1,n_1'\}$, $\chi_2 = \{n_2,n_2'\}$, are such that neither pair consists of nonnegative or nonpositive integers (we call χ regular in such a case) the functional B can be proved to be generated by a homogeneous distribution M(z) in the form

$$B(f_1,f_2) = \int (\int M(z_1) f_1(z_2 + z_1) dz_1) f(z_2) dz_2 \tag{1.14}$$

where M(z) is nontrivial only in the following two cases

$$\chi_1 = -\chi_2 : M(z) = C\delta(z) \tag{1.15}$$

$$\chi_1 = \chi_2 = (m,\rho) : M(z) = C|z|^{-i\rho-2}\left(\frac{z}{|z|}\right)^{+m} . \tag{1.16}$$

Because of (1.15) we call two representations χ_1, χ_2 dual to each other if $\chi_1 = -\chi_2$. The kernel (1.16) serves also as an intertwining operator for a dual pair of representations. A convolution with $M(z)$ can be shown to establish a one-to-one and bicontinuous map from D_χ onto $D_{-\chi}$ if χ is regular, which intertwines T_a^χ and $T_a^{-\chi}$.

In order to treat also the nonregular cases which were so far excluded, we recall the properties of the distribution

$$p_{\sigma,m}(z) = |z|^\sigma\left(\frac{z}{|z|}\right)^m$$

(σ complex, m integral). Considered in its analytic dependence on σ it is a meromorphic function with simple poles at

$$\sigma_n = -2n - 2 - |m| ,\ n = 0, 1, 2, \cdots$$

and the residues

$$\operatorname*{Res}_{\sigma_n} p_{\sigma,m}(z) = \frac{2\pi(-1)^m}{n!(n + |m|)!} \frac{\partial^{\sigma_1}}{\partial z^{\sigma_1}} \frac{\partial^{\sigma_2}}{\partial \bar{z}^{\sigma_2}} \delta(z) \tag{1.17}$$

where

$$\begin{aligned} \sigma_1 &= n + \frac{1}{2}(|m| - m) \\ \sigma_2 &= n + \frac{1}{2}(|m| + m) . \end{aligned} \tag{1.18}$$

Formally the distribution $M(z)$ (1.16) is identical with $p_{-i\rho-2,m}(z)$, only the domains on which they operate differ. However, a look at the compact picture or at the asymptotic expansion (1.8) convinces us that the distribution $p_{-i\rho-2,m}(z)$ can be extended from test functions $f(z)$ with compact support (say) onto D_χ by continuity. Analytic continuations in ρ can still be given a rigorous meaning in the compact picture, where the domain D_χ is independent of the parameter ρ. The results on the analytic structure of $p_{-i\rho-2,m}(z)$ in ρ apply therefore also to $M(z)$, the only source of singularities is the behavior at $z = 0$.

The residue or the constant term in the Laurent expansion of $p_{-i\rho-2,m}(z)$ around a pole give new distribution kernels $M(z)$ which lead to bilinear invariant functionals and corresponding intertwining operators, that establish one-to-one and bicontinuous mappings between the spaces of the following pairs (we set $\chi = \{n,n'\}$)

a) D_χ and $D_{-\chi}$, if one of the numbers n, n' is zero, whereas the other is an arbitrary integer;

b) E_χ and $D_{-\chi}/F_{-\chi}$ if n, n' are both positive integers;

c) F_χ and $D_{-\chi}/E_{-\chi}$ if n, n' are both negative integers;

d) F_χ and $D_{\pm\chi'}$, if n, n' are both negative integers and χ' is defined by $\chi' = \{n,-n'\}$ for $\chi = \{n,n'\}$.

Together with the regular case, this is a complete list of intertwining operators for the spaces D_χ, E_χ, F_χ, and their quotient spaces.

1.2. Trilinear Invariant Functionals

The same relation which exists between intertwining operators and bilinear invariant functionals holds true between irreducible tensor operators and trilinear invariant functionals. Trilinear invariant functionals for three arbitrary representations can be obtained by analytic continuation in the three ρ's from a trilinear invariant functional for three representations of the principal series using essentially the same method as for the bilinear invariant functional. The kernel which generates the trilinear invariant functional for three representations of the principal series is the same as the kernel which was used by Naimark to decompose the tensor product of two representations of the principal series [3]. We start our discussion with his results.

We refer to the noncompact picture. We define a Hilbert space $L^2(Z \times Z)$ of measurable functions $f(z_1,z_2)$ which have finite norm

$$(f_1,f_2) = \int \overline{f_1(z_1,z_2)} f_2(z_1,z_2) dz_1 dz_2 \ . \tag{1.19}$$

This space carries the unitary representation $\chi_1 \times \chi_2$ defined by

$$T_a^{\chi_1 \times \chi_2} f(z_1,z_2) = \alpha^{\chi_1}(z_1,a)\alpha^{\chi_2}(z_2,a) f((z_1)_a,(z_2)_a) \tag{1.20}$$

with the notations (1.3). We call this representation the tensor product of the representations χ_1 and χ_2 which are both assumed to belong to the principal series. The issue solved by Naimark is the decomposition of this tensor product into a direct integral of irreducible representations.

We consider the set of Hilbert spaces $L^2(Z,\chi)$ each of which consists of measurable functions $f(z,\chi)$ in z, which have finite norm with respect to the scalar product

$$(f_1,f_2)_\chi = \int \overline{f_1(z,\chi)} f_2(z,\chi) dz \ . \tag{1.21}$$

χ runs over the principal series, and each $L^2(Z,\chi)$ is assumed to carry the representation χ. We form the direct integral

$$H = \int^{\oplus} L^2(Z,\chi) d\chi \tag{1.22}$$

where $d\chi$ is the Plancherel measure of $SL(2,C)$ normalized as

$$d\chi = (m^2 + \rho^2) d\rho \ . \tag{1.23}$$

We sum over all m and integrate over the real ρ axis, thus we count two representations χ and $-\chi$ out of almost each equivalence class. Actually we want to consider the spaces $L^2(Z,\chi)$ and $L^2(Z,-\chi)$ as isometric images of each other via the intertwining operator, such that the double counting is only a symmetric way of writing

$$H = 2 \sum_m^{\oplus} \int_0^{\infty\,\oplus} d\rho (m^2 + \rho^2) L^2(Z,\chi) \ .$$

The Hilbert space H can be decomposed into two orthogonal subspaces H_+ and H_- which are obtained by restricting the integration (1.22) to even respectively odd m.

We define Naimark's kernel by

$$N(z_1,z_2,z_3|\chi_1,\chi_2,\chi_3) = \frac{1}{8\pi^2} |z_1 - z_2|^{\sigma_3} |z_2 - z_3|^{\sigma_1} |z_3 - z_1|^{\sigma_2} \left(\frac{z_1 - z_2}{|z_1 - z_2|}\right)^{\mu_3} \left(\frac{z_2 - z_3}{|z_2 - z_3|}\right)^{\mu_1} \left(\frac{z_3 - z_1}{|z_3 - z_1|}\right)^{\mu_2} \tag{1.24}$$

if $\sum_i m_i$ is even and by zero if this sum is odd. The parameters σ_i and μ_i are linear combinations of the m_i and ρ_i

$$\sigma_i = -\frac{i}{2} (\rho_1 + \rho_2 + \rho_3 - 2\rho_i) - 1$$
$$\mu_i = +\frac{1}{2} (m_1 + m_2 + m_3 - 2m_i) . \tag{1.25}$$

Naimark has proved the following assertion [3]: The integral transformations

$$f(z,\chi) = \int N(z,z_1,z_2|-\chi,\chi_1,\chi_2) f(z_1,z_2) dz_1 dz_2$$
$$f(z_1,z_2) = \int N(z,z_1,z_2|\chi,-\chi_1,-\chi_2) f(z,\chi) dz d\chi \tag{1.26}$$

which can be made to converge in the sense of the respective image spaces by an appropriate regularization procedure, establish an isometric mapping of $L^2(Z \times Z)$ on H_s, $s = (-1)^{m_1+m_2}$, such that for fixed $f(z_1,z_2)$ and its image $f(z,\chi)$ the vectors

$$T_a^{\chi_1 \times \chi_2} f(z_1,z_2),\ T_a^{\chi} f(z,\chi)$$

are mapped onto each other for all $a \in SL(2,C)$.

Naimark [3] proved this theorem by reducing it to the Plancherel theorem for $SL(2,C)$.

We can now define irreducible tensor operators. Since we want later to continue analytically, we restrict ourselves to the spaces D_χ from the outset. Let three spaces D_{χ_i}, $i = 1, 2, 3$, be given such that $\sum_i m_i$ is even. We consider an operator A on the tensor products $f_2 \times f_1$, $f_{1,2} \in D_{\chi_{1,2}}$,

$$A(f_2 \times f_1) \in D_{\chi_3} .$$

First we require that A has to be continuous in its arguments f_1 and f_2 separately. In the case that all three χ_i are regular we call A an irreducible tensor operator if it satisfies in addition the "covariance" relation

$$A(T_a^{\chi_2} f_2 \times T_a^{\chi_1} f_1) = T_a^{\chi_3} A(f_2 \times f_1) . \tag{1.27}$$

If all three χ_i belong to the principal series, A is, up to a constant, uniquely determined and is given by Naimark's kernel (1.24) in the form

$$A(f_2 \times f_1)(z_3) = \int N(z_3,z_2,z_1|-\chi_3,\chi_2,\chi_1) f_2(z_2) f_1(z_1) dz_2 dz_1 . \tag{1.28}$$

This operator is obviously related with the trilinear invariant functional

$$B(f_3,f_2,f_1) = \int N(z_3,z_2,z_1|\chi_3,\chi_2,\chi_1) f_3(z_3) f_2(z_2) f_1(z_1) dz_3 dz_2 dz_1 . \tag{1.29}$$

If all three $f_i(z_i)$ in (1.29) are in D_{χ_i} this functional can be continued off the principal series.

In particular we are interested in the cases where D_{χ_3} possesses an invariant subspace E_{χ_3} or F_{χ_3}, such that applying an appropriate limiting process to (1.28) yields an operator A whose range is in the invariant subspace. If we keep $\chi_{1,2}$ at regular positions, a nontrivial operator obtained in this fashion with image in E_{χ_3} or in the quotient space D_{χ_3}/F_{χ_3} is denoted a finite irreducible tensor operator. These are the operators of major physical interest.

In the special case that

$$n_3 = n_3' = 2, \dim E_{\chi_3} = 4$$

so that E_{χ_3} carries the four-vector representation, we call A a vector operator or a generalized Dirac matrix. We use the same notation when

$$n_3 = n_3' = -2, \dim D_{\chi_3}/F_{\chi_3} = 4$$

such that the four-vector representation appears on the quotient space. This approach to the generalized Dirac matrices, which are known since the work of Gel'fand and Yaglom [4] and Naimark [1], is due to Wess [5]. Our presentation is an extension of Wess's work but still by no means complete.

1.3. Finite Tensor Operators

If we continue the trilinear invariant functional off the principal series, singularities arise due to the behavior of Naimark's kernel on the manifolds $z_1 = z_2$ etc. We consider first the case that we reach a point

$$\chi_3 = \{n_3, n_3'\},\ n_3, n_3' > 0 \ \text{integral} \ . \tag{1.30}$$

We want to investigate the condition under which the integral (1.28) lies entirely in the invariant subspace E_{χ_3}. With the notation

$$\begin{aligned} A_i &= \frac{1}{2}(\sigma_i + \mu_i) \\ B_i &= \frac{1}{2}(\sigma_i - \mu_i) \end{aligned} \tag{1.31}$$

a necessary condition for this to happen is

$$\left.\begin{aligned} A_3 &= -\nu - 1 \\ B_3 &= -\mu - 1 \end{aligned}\right\} \nu,\mu = 0, 1, 2, \cdots \ . \tag{1.32}$$

If in (1.29) $f_3(z_3)$ lies not entirely in $F_{-\chi_3}$ (replace χ_3 by $-\chi_3$ in (1.29)), the trilinear invariant functional has a pole in σ_3 with the residue

$$\begin{aligned} \operatorname{Res} B(f_3,f_2,f_1) &= \frac{1}{4\pi\nu!\mu!} \int dz_3 f_3(z_3) \\ &\times \int f_2(z)|z - z_3|^{\sigma_1}\left(\frac{z - z_3}{|z - z_3|}\right)^{\mu_1} \frac{\partial^\nu}{\partial z^\nu}\frac{\partial^\mu}{\partial \bar{z}^\mu}|z_3 - z|^{\sigma_2}\left(\frac{z_3 - z}{|z_3 - z|}\right)^{\mu_2} \\ &\times f_1(z)dz \ . \end{aligned} \tag{1.33}$$

In order that the inner integral in (1.33) defines an element of E_{χ_3}, i.e., a polynomial in $z, \bar{z}$, we must implement (1.32) by the requirement

$$\begin{aligned} \nu &\leqq n_3 - 1 \\ \mu &\leqq n_3' - 1 \end{aligned} \tag{1.34}$$

which can be obtained by inspection from

$$\sigma_1 + \sigma_2 = i\rho_3 - 2 = n_3 + n_3' - 2$$

$$\mu_1 + \mu_2 = -m_3 = n_3 - n_3' .$$

As an example we consider the four-vector with

$$n_3 = n_3' = 2$$

in which case ν, μ range over the values 0 and 1 only. We write the inner integral in (1.33) as

$$\int f_2(z) A(\nu,\mu|z,z_3) f_1(z) dz \tag{1.35}$$

and get the following operators $A(\nu,\mu|z,z_3)$ (in a new normalization)

$$\begin{aligned} A(0,0|z,z_3) &= (z - z_3)(\bar{z} - \bar{z}_3) \\ A(1,0|z,z_3) &= (z - z_3)(\bar{z} - \bar{z}_3) \frac{\partial}{\partial z} - (n_1 - 1)(\bar{z} - \bar{z}_3) \\ A(0,1|z,z_3) &= (z - z_3)(\bar{z} - \bar{z}_3) \frac{\partial}{\partial \bar{z}} - (n_1' - 1)(z - z_3) \\ A(1,1|z,z_3) &= (z - z_3)(\bar{z} - \bar{z}_3) \frac{\partial^2}{\partial z \partial \bar{z}} - (n_1 - 1)(\bar{z} - \bar{z}_3) \frac{\partial}{\partial \bar{z}} \\ &\quad - (n_1' - 1)(z - z_3) \frac{\partial}{\partial z} + (n_1 - 1)(n_1' - 1) . \end{aligned} \tag{1.36}$$

The representations χ_1 and χ_2 are restricted by (1.32) to

$$\begin{aligned} m_1 &= -m_2 - 2(\nu - \mu) \\ \rho_1 &= -\rho_2 - 2i(\nu + \mu - 1) \end{aligned} \tag{1.37}$$

which relation we abbreviate as $\chi_1 = (-\chi_2)_{\nu\mu}$.

Finally we consider the case that both n_3, n_3' are negative integers, so that D_{χ_3} possesses an invariant subspace F_{χ_3}. In this case the integrals

$$\int dz_3 z_3^{\ell} \bar{z}_3^{k} \int N(z_3,z_2,z_1|-\chi_3,\chi_2,\chi_1) f_2(z_2) f_1(z_1) dz_2 dz_1 \tag{1.38}$$

can easily be evaluated by elementary methods and shown to be zero for any pair of functions $f_{1,2}$ belonging to regular representations $\chi_{1,2}$ and for all

$$0 \leqq \ell \leqq -n_3 - 1$$

$$0 \leqq k \leqq -n_3' - 1$$

provided A_3 and B_3 (1.31) are not simultaneously integers on the half axis

$$\begin{aligned} A_3 &\leqq -n_3 - 1 \\ B_3 &\leqq -n_3' - 1 . \end{aligned} \tag{1.39}$$

This means that with the sole exception of the cases (1.39) the integral (1.28) defines an element of F_{χ_3}. Only in the exceptional cases (1.39) the components in the finite dimensional space D_{χ_3}/F_{χ_3} are nonzero. If we implement (1.39) by the further condition

$$A_3 \geqq 0, \ B_3 \geqq 0 \tag{1.40}$$

we obtain finite irreducible tensor operators. In fact, it is possible in this case to decompose Naimark's kernel into two intertwining operators that map $D_{\chi_{1,2}}$ each onto $D_{-\chi_{1,2}}$, the finite irreducible tensor operator obtained earlier which maps the tensor product of $D_{-\chi_1}$ and $D_{-\chi_2}$ into $E_{-\chi_3}$, and the intertwining operator from $E_{-\chi_3}$ into D_{χ_3}/F_{χ_3}.

2. CURRENT DENSITY OPERATORS

2.1. Vertex Functions

First we introduce a special realization of a unitary irreducible representation of the group $SL(2,C) \times T_4$ for a particle of mass M and spin S and positive energy. We define a Hilbert space of measurable, vector valued functions on $SL(2,C)$ $\Phi_q(a)$, $-S \leqq q \leqq S$, $S - q$ integral, which have finite norm with respect to the scalar product

$$(\Phi^1,\Phi^2) = \int d\mu(a) \sum_q \overline{\Phi^1_q(a)}\Phi^2_q(a) \tag{2.1}$$

$d\mu(a)$ is the Haar measure on $SL(2,C)$ normalized (in the notations of Section 1.1) as

$$a = \zeta k, \ d\mu(a) = (2\pi)^{-4} dz d\lambda d\mu \ .$$

In addition, we require that the functions $\Phi_q(a)$ be covariant on right cosets of $SU(2)$, i.e.,

$$\Phi_q(ua) = \sum_{q'} D^S_{qq'}(u)\Phi_{q'}(a) \ . \tag{2.2}$$

The matrix D^S describes a unitary irreducible representation of $SU(2)$ of spin S. We call this Hilbert space $L^2(M,S)$. In this space we define the representation $\{M,S\}$ by

$$\begin{aligned} U_a\Phi_q(a_1) &= \Phi_q(a_1 a) \\ U_x\Phi_q(a) &= \exp\{\tfrac{1}{2}\, iM\mathrm{Tr}(\underset{\sim}{x} a^+ a)\}\Phi_q(a) \end{aligned} \tag{2.3}$$

where $\underset{\sim}{x}$ is a two-by-two matrix constructed from the translation four-vector x as

$$\underset{\sim}{x} = \sum_{\mu=0,1,2,3} x_\mu \sigma_\mu \ .$$

Here σ_o is the unit matrix and σ_k, $k = 1, 2, 3$, are the familiar Pauli matrices. We mention that the four-momentum vector used in physics is related with the argument $a \in SL(2,C)$ by

$$\begin{aligned} P_o &= \frac{1}{2}\, M\mathrm{Tr}(a^+ a) \\ P_k &= -\frac{1}{2}\, M\mathrm{Tr}(\sigma_k a^+ a) \ . \end{aligned} \tag{2.4}$$

The unitary irreducible representation $\{M,S\}$ of $SL(2,C) \times T_4$ is unitary but reducible on the subgroup $SL(2,C)$. In order to reduce this representation we embed the space $L^2(M,S)$ in the space $L^2(SL(2,C))$ which carries the right regular representation. A canonical way of doing this is

$$\Phi_q(a) = \int d\mu(u) D^S_{q'}(u) \tilde{\Phi}(u^{-1}a)$$

where $\tilde{\Phi}(a)$, is any element of $L^2(SL(2,C))$. If we apply the Plancherel theorem of $SL(2,C)$ to the right regular representation, we obtain the direct integral decomposition

$$L^2(M,S) = \sum_{m=-2S}^{2S}{}^{\oplus} \int_0^{\infty}{}^{\oplus} d\rho\,(m^2+\rho^2) L^2(\chi) \tag{2.5}$$

where $L^2(\chi)$ carries the principal series representation χ of $SL(2,C)$. As a realization we may use for example the space $L^2_m(K)$ discussed in Section 1. We emphasize that this decomposition is free of degeneracies. If we restrict a current density operator to single particle spaces, say its domain is in $L^2(M_1,S_1)$ and its range in $L^2(M_2,S_2)$, it decomposes together with the two spaces, and it is this decomposition which we are interested in.

In order to define suitable vertex functions for a given current density operator $j_\mu(x)$ acting between the spaces $L^2(M_{1,2},S_{1,2})$, we consider the matrix elements

$$\begin{aligned}\langle\Phi^2|j_\mu(0)|\Phi^1\rangle &= (\Phi^2, j_\mu(0)\Phi^1) \\ &= N_1N_2 \sum_{q_1q_2} \int d\mu(a_1)d\mu(a_2)\overline{\Phi^2_{q_2}(a_2)}\,\Gamma_\mu(a_2,a_1)_{q_2q_1}\Phi^1_{q_1}(a_1)\end{aligned} \tag{2.6}$$

where $\Phi^i \in L^2(M_i,S_i)$. The N_i are normalization constants. The normalization customarily used in physical literature is such that for the matrix element of the electromagnetic current between proton states we have

$$\Gamma_\mu(e,e)_{q_2q_1} = \delta_{\mu o}\delta_{q_1q_2} \text{ times charge of the proton}$$

(e is the unit element of $SL(2,C)$) which is achieved by

$$N = (2S+1)\left(\frac{2M}{(2\pi)^3}\right)^{\frac{1}{2}} 8\pi^2M^2 \ .$$

Of course the domain of $j_\mu(0)$ is not the whole Hilbert space $L^2(M_1,S_1)$ in general, but at least it is not smaller than the space $C^\infty_c(M_1,S_1)$ of infinitely differentiable functions with compact support on $SL(2,C)$ that satisfy the constraint (2.2). Under the Fourier decomposition (2.5) this space goes over into a space of functions (for the realizations $L^2_m(K)$ these functions can be written $\varphi(u,m,\rho)$ satisfying the constraint (1.5)) which are entire in ρ.

The definition (2.6) is not yet unique, we complete it by requiring covariance on right cosets of $SU(2)$ in (2.12). The vertex function $\Gamma_\mu(a_2,a_1)_{q_2,q_1}$ is then a vector valued function. Let us define

$$|a|^2 = \mathrm{Tr}(a^+a) \ .$$

The "four-momentum transfer" q

$$q = p_2 - p_1,\ p_{1,2} = p(a_{1,2})$$

(see (2.4)) lies in the domain

$$q^2 = q_0^2 - q_1^2 - q_2^2 - q_3^2 = M_1^2 + M_2^2 - M_1M_2|a_1a_2^{-1}|^2 \leqq (M_1 - M_2)^2 .$$

From field theory we know that below a "threshold mass" M_{th}

$$-\infty < q^2 < M_{th}^2$$

(M_{th} may equal two pion masses, for example) the vertex function is analytic as a function of the real variables on $SL(2,C) \times SL(2,C)$. In the worst case, namely when

$$M_{th}^2 < (M_1 - M_2)^2$$

there is a finite q^2 interval on which we have no analyticity. But a physicist's intuition lets us expect that in this interval we have at most a finite number of singular points due to additional thresholds with continuity at these points and continuous differentiability in between. The harmonic analysis of the vertex functions is consequently beset with at most a complication due to their behavior if q^2 tends to infinity. We may try to handle this complication by means of a regularization procedure.

In order to formulate the four-vector covariance of the vertex function and the covariance on the right cosets of SU(2) it is advantageous to introduce a basis in $L_m^2(K)$, the "canonical basis". We use the functions

$$\varphi_q^j(u) = (2j+1)^{\frac{1}{2}} D^j_{\frac{1}{2}m,q}(u);\ -j \leqq q \leqq j,\ j = \frac{1}{2}|m| + n,\ n = 0, 1, 2, \cdots \qquad (2.7)$$

where D^S is the same unitary matrix as in (2.2). The orthonormality and completeness of this basis in $L_m^2(K)$ follows from the theorem of Peter and Weyl. This basis lies in the spaces D_χ and a subbasis can be used to span the invariant subspaces E_χ and F_χ. It can be carried over to the noncompact picture where we denote its elements by $f_q^j(z)$. If the operator T_a^χ in D_χ acts on a basis element $f_q^j(z)$ we obtain the "coordinate functions"

$$T_a^\chi f_q^j(z) = \sum_{j'q'} D^\chi_{j'q'jq}(a) f_{q'}^{j'}(z) . \qquad (2.8)$$

In particular we have

$$\begin{aligned} D^\chi_{j_1q_1j_2q_2}(u) &= \delta_{j_1j_2} D^{j_1}_{q_1q_2}(u) \\ D^\chi_{j_1q_1j_2q_2}(d) &= \delta_{q_1q_2} d^\chi_{j_1j_2q}(\eta) \end{aligned} \qquad (2.9)$$

where the matrix d is defined by

$$d = \begin{pmatrix} e^{\frac{1}{2}\eta} & 0 \\ 0 & e^{-\frac{1}{2}\eta} \end{pmatrix},\ \eta \geqq 0 . \qquad (2.10)$$

Finally, we switch from the vector labels $\mu = 0, 1, 2, 3$ to components with respect to the canonical basis in the space E_χ, $\chi = (0,-4i)$, which carries the vector representation

$$\Gamma^J_Q(a_2,a_1)_{q_2q_1} : J = Q = 0 \quad \text{and} \quad J = 1,\ Q = +1,\ 0,\ -1$$

$$\Gamma^0_0 = \pi^{\frac{1}{2}}\,\Gamma_0,\ \Gamma^1_0 = -(\tfrac{\pi}{3})^{\frac{1}{2}}\,\Gamma_3,\ \Gamma^1_{\pm 1} = \pm(\tfrac{\pi}{6})^{\frac{1}{2}}(\Gamma_1 \mp i\Gamma_2)\ . \tag{2.11}$$

Then the covariance properties of the vertex functions are expressed by the formulae

$$\Gamma^J_Q(u_2a_2,u_1a_1)_{q_2q_1} = \sum_{q_1'q_2'} D^{S_2}_{q_2q_2'}(u_2)D^{S_1}_{q_1'q_1}(u_1^{-1})\Gamma^J_Q(a_2,a_1)_{q_2'q_1'} \tag{2.12}$$

$$\Gamma^J_Q(a_2a^{-1},a_1a^{-1})_{q_2q_1} = \sum_{\substack{J'Q' \\ J'\leq 1}} D^{(0,-4i)}_{JQJ'Q'}(a)\Gamma^{J'}_{Q'}(a_2,a_1)_{q_2q_1}\ . \tag{2.13}$$

2.2. The Decomposition of a Vertex Function with Covariance in the Principal Series

In (2.12), (2.13) the covariance was formulated in so general terms that we may immediately modify these equations and study vertex functions which transform as a representation χ of the principal series. To avoid confusion we add the label χ to the arguments of the vertex functions, the coordinate functions in (2.13) $D^{(0,-4i)}$ are replaced by D^χ. The main tool of the Fourier decomposition of the vertex function obtained in this fashion is Naimark's theorem. At the end we continue in χ analytically until we arrive at the point $(0,-4i)$ again.

For χ in the principal series complex conjugation maps D_χ onto $D_{-\chi}$ (independently of the two pictures), in particular

$$\overline{f^J_q(z)^\chi} \in D_{-\chi}\ . \tag{2.14}$$

Denoting analytic continuations of the complex conjugate off the principal series by $(\cdots)^*$, we have from (2.14) and (1.2)

$$(f^j_q(z)^\chi)^* \in D_{-\chi}$$

$$(T^\chi_a f^j_q(z)^\chi)^* = T^{-\chi}_a (f^j_q(z)^\chi)^*\ . \tag{2.15}$$

The unitarity of the principal series representations implies

$$(D^\chi_{j_1q_1j_2q_2}(a))^* = D^\chi_{j_2q_2j_1q_1}(a^{-1})\ . \tag{2.16}$$

The bilinear invariant functional (1.14), (1.15) enables us to introduce a matrix calculus by

$$B((f^{j_2}_{q_2})^*, Af^{j_1}_{q_1}) = \langle \chi_2;j_2q_2|A|\chi_1;j_1q_1\rangle \tag{2.17}$$

provided

$$f^{j_1}_{q_1} \in D_{\chi_1},\ f^{j_2}_{q_2} \in D_{\chi_2},\ AD_{\chi_1} \subset D_{\chi_2}\ .$$

For $A = \mathbb{1}$ this gives

$$\langle \chi;j_2q_2|\mathbf{1}|\chi;j_1q_1\rangle = \delta_{j_1j_2}\delta_{q_1q_2} \tag{2.18}$$

whereas $A = T^{\chi}_{a}$ leads us back to the coordinate functions (2.8)

$$\langle \chi;j_2q_2|T^{\chi}_{a}|\chi;j_1q_1\rangle = D^{\chi}_{j_2q_2j_1q_1}(a) \ . \tag{2.19}$$

A similar notation can be used for the trilinear invariant functional (1.29) for the representations $\chi_1, -\chi_2, -\chi_3$

$$B((f^J_Q)^*,(f^{j_2}_{q_2})^*,f^{j_1}_{q_1}) = \langle \chi_2;j_2q_2|A^J_Q(\chi_3)|\chi_1;j_1q_1\rangle \tag{2.20}$$

for any

$$f^{j_1}_{q_1} \in D_{\chi_1}, \ f^{j_2}_{q_2} \in D_{\chi_2}, \ f^J_Q \in D_{\chi_3}$$

and χ_3 in the principal series, say. The linearity and continuity of the functional implies

$$B((f^J_Q)^*,(T^{\chi_2}_{a_2^{-1}}f^{j_2}_{q_2})^*,T^{\chi_1}_{a_1^{-1}}f^{j_1}_{q_1}) = \sum_{j_1'q_1'j_2'q_2'} D^{\chi_1}_{j_1'q_1'j_1q_1}(a_1^{-1})D^{\chi_2}_{j_2q_2j_2'q_2'}(a_2)$$
$$\times \langle \chi_2;j_2'q_2'|A^J_Q(\chi_3)|\chi_1;j_1'q_1'\rangle \ .$$

With (2.16), (2.19) and matrix calculus we can continue this equation

$$= \langle \chi_2;j_2q_2|T^{\chi_2}_{a_2}A^J_Q(\chi_3)T^{\chi_1}_{a_1^{-1}}|\chi_1;j_1q_1\rangle \ . \tag{2.21}$$

From (2.15), (2.16) and the invariance of the trilinear functional we have

$$\langle \chi_2;j_2q_2|T^{\chi_2}_{a^{-1}}A^J_Q(\chi_3)T^{\chi_1}_{a}|\chi_1;j_1q_1\rangle = \sum_{J'Q'} D^{\chi_3}_{JQJ'Q'}(a)\langle \chi_2;j_2q_2|A^{J'}_{Q'}(\chi_3)|\chi_1;j_1q_1\rangle \ . \tag{2.22}$$

Comparing (2.21), (2.22) with (2.12), (2.13) we recognize that the vertex function $\Gamma^J_Q(a_2,a_1|\chi)_{q_2q_1}$ has the same covariance properties as the matrix element

$$\langle \chi_2;s_2q_2|T^{\chi_2}_{a_2}A^J_Q(\chi)T^{\chi_1}_{a_1^{-1}}|\chi_1;s_1q_1\rangle$$

where χ_1 and χ_2 are arbitrary. This fact suggests that we decompose the vertex functions into such matrix elements.

In fact we define a Fourier transform by

$$M(\chi_2,\chi_1;\chi) = \int d\mu(a_2a_1^{-1}) \sum_{JQ} \Gamma^J_Q(a_2,a_1|\chi)_{q_2q_1}$$
$$\times \langle \chi_2;s_2q_2|T^{\chi_2}_{a_2}A^J_Q(\chi)T^{\chi_1}_{a_1^{-1}}|\chi_1;s_1q_1\rangle^* \tag{2.23}$$

when all three $\chi_{1,2}$ and χ are in the principal series. The left-hand side of (2.23) can be verified to be independent of q_1 and q_2. The main tool in the inversion of this Fourier transformation is Naimark's theorem in the form (note that the product $\chi \times \chi_2$ is decomposed into $\int^{\oplus} d\chi_1$)

$$\sum_{j_1 q_1} \int d\chi_1 \langle \chi_2; j_2' q_2' | A^{J'}_{Q'}(\chi) | \chi_1; j_1 q_1 \rangle^* \langle \chi_2; j_2 q_2 | A^J_Q(\chi) | \chi_1; j_1 q_1 \rangle = \delta_{j_2 j_2'} \delta_{q_2 q_2'} \delta_{JJ'} \delta_{QQ'} \tag{2.24}$$

and the Plancherel theorem for SL(2,C) in a similar matrix version

$$\frac{1}{2} \int d\chi \sum_{J_1 Q_1 J_2 Q_2} D^\chi_{J_1 Q_1 J_2 Q_2}(a_1) (D^\chi_{J_1 Q_1 J_2 Q_2}(a_2))^* = \delta(a_1, a_2) \ . \tag{2.25}$$

Here $\delta(a_1, a_2)$ is the delta-function on SL(2,C) normalized with respect to the Haar measure (see (2.1)). From (2.12) and (2.13) we have

$$\begin{aligned} &\frac{1}{2} \int d\chi_1 d\chi_2 M(\chi_2, \chi_1; \chi) \langle \chi_2; S_2 q_2 | T^{\chi_2}_{a_2} A^J_Q(\chi) | \chi_1; S_1 q_1 \rangle \\ &= \frac{1}{2} \int d\chi_1 d\chi_2 \sum_{j_1 j_2 q_1' q_2'} \langle \chi_2; j_2 q_2' | T^{\chi_2}_{a_2} A^J_Q(\chi) | \chi_1; j_1 q_1' \rangle \\ &\times \int d\mu(a_2') \sum_{J'Q'} \langle \chi_2; j_2 q_2' | T^{\chi_2}_{a_2'} A^{J'}_{Q'}(\chi) | \chi_1; j_1 q_1' \rangle^* \Gamma^{J'}_{Q'}(a_2', e | \chi)_{q_2 q_1} \end{aligned}$$

where we exploited the covariance to set $a_1' = e$. Inserting the two formulae (2.24), (2.25) into this expression we obtain the desired result $\Gamma^J_Q(a_2, e|\chi)_{q_2 q_1}$. By means of covariance we can extend this result to the general inversion formula

$$\Gamma^J_Q(a_2, a_1|\chi)_{q_2 q_1} = \frac{1}{2} \int d\chi_1 d\chi_2 M(\chi_2, \chi_1; \chi) \times \langle \chi_2; S_2 q_2 | T^{\chi_2}_{a_2} A^J_Q(\chi) T^{\chi_1}_{a_1^{-1}} | \chi_1; S_1 q_1 \rangle \ . \tag{2.26}$$

It does not make much sense to discuss the convergence of (2.23) and (2.26) using the information on the vertex function supplied by field theory. We mention only that a sufficient condition for the proper convergence of both (2.23) and (2.26) is infinite differentiability and rapid decrease (that is, faster decrease than any power of $|a|$) of $\Gamma^J_Q(a, e|\chi)_{q_2 q_1}$ on SL(2,C). We call such vertex functions "smooth".

2.3. The Decomposition of the Four-vector vertex Function

For a smooth vertex function we write (2.23), (2.26) in the form

$$\begin{aligned} \Gamma^J_Q(a_2, a_1|\chi)_{q_2 q_1} &= \frac{1}{2} \int d\chi_1 d\chi_2 \int d\mu(a_2' a_1'^{-1}) \sum_{J'Q'} \\ &\times \Gamma^{J'}_{Q'}(a_2', a_1'|\chi)_{q_2 q_1} \langle \chi_2; S_2 q_2 | T^{\chi_2}_{a_2} A^J_Q(\chi) T^{\chi_1}_{a_1^{-1}} | \chi_1; S_1 q_1 \rangle \\ &\times \langle \chi_2; S_2 q_2 | T^{\chi_2}_{a_2'} A^{J'}_{Q'}(\chi) T^{\chi_1}_{a_1'^{-1}} | \chi_1; S_1 q_1 \rangle^* \end{aligned} \tag{2.27}$$

which we want to continue in χ. Because of the smoothness of the vertex function we may handle the expression (2.27) rather freely without facing problems of convergence. When we reach the point

$$\chi = \{n, n'\}, \ n = n' = 2$$

we define $\Gamma_Q^J = 0$ for $J > 1$. The first bracket $\langle\cdots\rangle$ in (2.27) with J restricted to $J \leq 1$ is finite everywhere except that a pole occurs whenever

$$\chi_1 = (\chi_2)_{\nu\mu}$$

(see (1.37), remember that χ_2 has been replaced by $-\chi_2$ in (2.20)). The bracket $\langle\cdots\rangle^*$ of (2.27) is in turn zero everywhere except at the same position, where it assumes a finite limit. Both the residue of the pole and the finite limit are matrix elements of vector operators (1.36). A careful analysis shows that the integration over χ_1 and χ_2 picks up just the residue of the pole, and of the two integrations only one is left. One of the two representations $\chi_{1,2}$ is pushed off the principal series, for convenience we choose χ_1. The smoothness of the vertex function accounts for the nonunitarity (polynomial increase) of χ_1. With the new variable $\lambda = \frac{1}{2} i\rho$ and $S = \min(S_1,S_2)$ we get

$$\begin{aligned}\Gamma_Q^J(a_2,a_1)_{q_2q_1} &= \frac{8i}{\pi} \sum_{\nu,\mu=0,1} \sum_{m=-2S}^{2S} \int_{-i\infty}^{+i\infty} d\lambda \\ &\times M_{\nu\mu}(m,\lambda)\langle\chi;S_2q_2|T^\chi_{a_2} A_Q^J(\nu,\mu)T^{\chi_{\nu\mu}}_{a_1^{-1}}|\chi_{\nu\mu};S_1q_1\rangle\end{aligned} \tag{2.28}$$

and the Fourier transform

$$\begin{aligned}M_{\nu\mu}(m,\lambda) &= \sum_{JQ} \int d\mu(a_2a_1^{-1})\Gamma_Q^J(a_2,a_1)_{q_2q_1}(-1)^J(2J+1) \\ &\times \langle\chi;S_2q_2|T^\chi_{a_2} A_Q^J(1-\mu,1-\nu)T^{\chi_{1-\mu,1-\nu}}_{a_1^{-1}}|\chi_{1-\mu,1-\nu};S_1q_1\rangle^* .\end{aligned} \tag{2.29}$$

As long as χ is in the principal series $(\cdots)^*$ means the complex conjugate.

Formula (2.29) for the Fourier transform can easily be simplified. We set first

$$\begin{aligned}a_2 &= u_2d(\eta_2)a \\ a_1 &= u_1d(-\eta_1)a\end{aligned} \qquad \eta_{1,2} \geq 0,\ \eta = \eta_1 + \eta_2$$

($d(\eta_{1,2})$ as in (2.10)) and can decompose the measure correspondingly

$$d\mu(a_2a_1^{-1}) = (4\pi)^{-1}d\mu(u_1)d\mu(u_2)\mathrm{sh}^2\eta d\eta .$$

Inserting this into (2.29) we obtain

$$\begin{aligned}M_{\nu\mu}(\chi) &= [4\pi(2S_1+1)(2S_2+1)]^{-1} \int_0^\infty d\eta\,\mathrm{sh}^2\eta \\ &\times \sum_{JQJ_1J_2q_1q_2} (-1)^J(2J+1)\Gamma_Q^J(d(\eta_2),d(\eta_1)^{-1})_{q_2q_1} \\ &\times d^{\chi^*}_{S_2J_2q_2}(\eta_2)d^{(\chi^*)_{1-\mu,1-\nu}}_{J_1S_1q_1}(\eta_1)\langle\chi^*;J_2q_2|A_Q^J(1-\mu,1-\nu)|(\chi^*)_{1-\mu,1-\nu};J_1q_1\rangle\end{aligned} \tag{2.30}$$

where we used the notation

$$\chi^* = (m,-\rho) \quad \text{for} \quad \chi = (m,\rho) .$$

In (2.30) we may set η_1 or η_2 equal to zero in which case one of the two d-functions drops out. We recall that by (2.4) the matrix $d(\eta)$ corresponds to the four-momentum

$$p = (M\mathrm{ch}\eta, 0, 0, -M\mathrm{sh}\eta) .$$

The vertex function entering (2.30) has therefore been brought into a "collinear" frame of inertia. In these frames it is easy to express vertex functions by some conventional kind of form factors. The momentum transfer q^2 is

$$q^2 = M_1^2 + M_2^2 - 2M_1M_2\mathrm{ch}\eta .$$

In physical applications the vertex functions have to be regularized to fulfill the assumption of smoothness. One of the basic premises in standard applications to physics is that the removal of the regularization can be accounted for by a mere change of the integration contours in (2.28). Typical statements arrived at in such applications involve asymptotic expansions of vertex functions. The derivation of such asymptotic expansions is always based on the following "Weyl symmetry relations" which reflect the existence of an intertwining operator for two representations χ and $-\chi$:

$$d^{\chi}_{j_1 j_2 q}(\eta) = \beta^{j_1}(-\lambda)\beta^{j_2}(\lambda) d^{-\chi}_{j_1 j_2 q}(\eta) \tag{2.31}$$

$$\begin{aligned} \langle \chi; S_2 q_2 | A^J_Q(\nu,\mu) | \chi_{\nu\mu}; S_1 q_1 \rangle &= -\beta^{S_2}(-\lambda)\beta^{S_1}(\lambda_{\nu\mu}) \\ &\quad \times \langle -\chi; S_2 q_2 | A^J_Q(1-\nu, 1-\mu) | (-\chi)_{1-\nu,1-\mu}; S_1 q_1 \rangle \end{aligned} \tag{2.32}$$

$$M_{\nu\mu}(\chi) = -\beta^{S_2}(\lambda)\beta^{S_1}(-\lambda_{\nu\mu}) M_{1-\nu,1-\mu}(-\chi) \tag{2.33}$$

with

$$\beta^S(\lambda) = \frac{\Gamma(S+1+\lambda)}{\Gamma(S+1-\lambda)} . \tag{2.34}$$

We close this section with the remark that the principal series is not sufficient for an expansion of vertex functions with tensorial covariance of higher rank, e.g., with a covariance like that of antisymmetric or symmetric traceless tensors of rank two. In addition, we have then contributions from a "discrete" series. Details on the material presented in this section can be found in [6].

3. PHENOMENOLOGICAL ANALYSIS OF THE ELECTROMAGNETIC VERTEX FUNCTION OF HADRONS

3.1. Asymptotic Expansions of Form Factors

We assume that the Fourier transformation (2.29) yields an analytic function of λ. Moreover, we assume (only for simplicity) that the singularities closest to the imaginary axis are simple poles. At present there is in fact no justification of this hypothesis other than an *a posteriori* verification of its implications by experiment. The situation is even worse than in the formally related case of Regge poles, since nonrelativistic quantum mechanics cannot serve as a heuristic guide.

We set $a_1 = e$, $a_2 = d(\eta)$ in (2.28) and have

$$\Gamma^J_Q(d(\eta),e)_{q_2q_1} = \frac{8i}{\pi} \sum_{\nu,\mu} \sum_m \int_{-i\infty}^{+i\infty} d\lambda M_{\nu\mu}(\chi) \times \sum_{J_2} d^\chi_{S_2J_2q_2}(\eta)\langle\chi;J_2q_2|A^J_Q(\nu,\mu)|\chi_{\nu\mu};S_1q_1\rangle \equiv \Gamma^J_Q(\eta)_{q_2q_1} . \tag{3.1}$$

Obviously J_2 is restricted to $|J_2 - S_1| \leqq 1$. The Weyl symmetry (2.31) of the coordinate function can be made explicit if we introduce "coordinate functions of the second kind" (or e-functions) by

$$d^\chi_{j_1j_2q}(\eta) = e^\chi_{j_1j_2q}(\eta) + \beta^{j_1}(-\lambda)\beta^{j_2}(\lambda)e^{-\chi}_{j_1j_2q}(\eta) . \tag{3.2}$$

These e-functions are defined uniquely by their asymptotic property

$$\overline{\lim_{\mathrm{Re}\,\lambda\leqq 0,|\lambda|\to\infty}} |\lambda e^{-\lambda\eta} e^\chi_{j_1j_2q}(\eta)| < \infty \quad \text{for} \quad \eta > 0 .$$

The Weyl symmetry (2.32), (2.33) of both the Fourier transforms and the matrix elements of the vector operators $A(\nu,\mu)$ allows us to rewrite (3.1) as

$$\Gamma^J_Q(\eta)_{q_2q_1} = \frac{16i}{\pi} \sum_{\nu,\mu} \sum_m \int_{-i\infty}^{+i\infty} d\lambda M_{\nu\mu}(\chi) \times \sum_{J_2} e^\chi_{S_2J_2q_2}(\eta)\langle\chi;J_2q_2|A^J_Q(\nu,\mu)|\chi_{\nu\mu};S_1q_1\rangle . \tag{3.3}$$

If $M_{\nu\mu}(\chi)$ has the properties assumed at the beginning, the asymptotic behavior of $\Gamma^J_Q(\eta)_{q_2q_1}$ in η for $\eta \to \infty$ is

$$\Gamma^J_Q(\eta) \cong -32 \operatorname*{Res}_{\lambda=\lambda'} M_{\nu'\mu'}(m',\lambda) \times \sum_{J_2} e^{(m',\lambda')}_{S_2J_2q_2}(\eta)\langle m',\lambda';J_2q_2|A^J_Q(\nu',\mu')|\chi'_{\nu\mu};S_1q_1\rangle$$

where the dominant pole is assumed to appear in the Fourier transform labelled ν', μ', m' and at the position λ'. The asymptotic behavior of the e-function in η is

$$e^\chi_{j_1j_2q}(\eta) = Ce^{(\lambda-1-|q+\frac{1}{2}m|)\eta}(1 + O(e^{-2\eta})) \tag{3.5}$$

where C is independent of η.

3.2. Electromagnetic Form Factors of the Nucleon

We want now to investigate in detail the case that the asymptotic behavior of the form factors is caused by a simple pole in the Fourier transforms. We neglect the other singularities, mention, however, that the known analytic structure of the form factors is not reproduced by a finite number of poles.

As long as we consider only one process, elastic electron proton scattering, for example, the physical meaning of such pole is difficult to describe. Quite the same situation arises in the case of Regge poles. The interpretation of the Regge pole as an exchanged object is connected with the possibility to identify this Regge pole in a whole class of processes and to characterize it by a set of quantum numbers. A meaningful interpretation of the poles we are considering here necessitates the simultaneous discussion of a whole class of processes as well. Since we shall do this in the next section we postpone the further discussion.

In an actual physical application we have to extend the group SL(2,C) to parity, time reversal, and isospin. In addition, we know that the electromagnetic current is conserved and that its restriction to single-particle spaces is selfadjoint. The group extensions are in fact trivial generalizations, but current conservation imposes a subsidiary condition on the Fourier transforms which deserves a detailed study. We can show that it implies linear difference equations whose coefficients depend on the masses, the spins, and the intrinsic parities. In the case of the nucleon form factors [7] the isospin invariance is taken into account by treating the isoscalar and isovector parts of the current independently but in an analogous fashion. Since $S_1 = S_2 = \frac{1}{2}$ in this case, the representations of SL(2,C) occurring have $m = \pm 1$. The following symmetry relations

$$\begin{aligned} M_{00}(1,\lambda) &= M_{00}(-1,\lambda) \\ M_{01}(-1,\lambda) &= -M_{10}(1,\lambda) \\ M_{11}(1,\lambda) &= M_{11}(-1,\lambda) \end{aligned} \tag{3.6}$$

are due to parity invariance; Weyl symmetry relations imply

$$\begin{aligned} M_{01}(-1,\lambda) &= -M_{10}(1,-\lambda) \\ M_{00}(1,\lambda) &= (\lambda + \tfrac{1}{2})(\lambda - \tfrac{3}{2})M_{11}(-1,-\lambda) \ . \end{aligned} \tag{3.7}$$

Current conservation and time reversal invariance yield finally

$$M_{11}(1,\lambda) = M_{11}(1,-\lambda - 1) \ . \tag{3.8}$$

Due to the selfadjointness all $M_{\nu\mu}(\chi)$ are real for real λ.

The unnormalized form factors of Sachs type [7] are given by

$$\begin{aligned} \Gamma_0^0(\eta)_{\pm\frac{1}{2},\pm\frac{1}{2}} &= e\sqrt{\pi}\ \mathrm{ch}\ \frac{1}{2}\ \eta G_E(q^2) \\ \Gamma_0^1(\eta)_{\pm\frac{1}{2},\pm\frac{1}{2}} &= e\sqrt{\frac{\pi}{3}}\ \mathrm{sh}\ \frac{1}{2}\ \eta G_E(q^2) \\ \Gamma_{\pm 1}^1(\eta)_{\mp\frac{1}{2},\pm\frac{1}{2}} &= e\sqrt{\frac{2\pi}{3}}\ \mathrm{sh}\ \frac{1}{2}\ \eta G_M(q^2) \ . \end{aligned} \tag{3.9}$$

The most appealing ansatz seems to be a dominant pole in M_{01} or M_{10} of the isovector current (we denote such pole "isovector class one") at $\lambda = \lambda_1$ on the real axis close to $-\frac{1}{2}$. It implies

$$\left.\begin{aligned} G_E &\cong C_E\left(-\frac{q^2}{M^2}\right)^{\lambda_1-\frac{3}{2}} \\ G_M &\cong C_M\left(-\frac{q^2}{M^2}\right)^{\lambda_1-\frac{3}{2}} \end{aligned}\right\} q^2 \to -\infty$$

$$C_E : C_M = (\lambda_1 - \tfrac{1}{2})^{-1} \ . \tag{3.10}$$

We can describe by this ansatz a positive definite proton magnetic form factor, a negative neutron magnetic form factor, and a proton electric form factor which has to change sign before the asymptotic domain is reached (somewhere between 5 and 10 GeV^2). These findings are in agreement with the experimental data.

3.3. Form Factors for the Electroproduction of Nucleon Resonances

Inelastic scattering of electrons off a proton serves to analyze the electromagnetic transition matrix elements for a proton going into nucleon resonances. We consider an irreducible representation of SL(2,C) extended by parity and isospin. We call a set of resonances a tower, if their spin-parity and isospin quantum numbers allow us to fit them into one such representation. Towers are therefore labelled by the invariants of the extended SL(2,C). Neither need all places in such representations be occupied by resonances, nor is the number of resonances occupying one state in a representation or the number of towers to which one resonance belongs bounded by one (we allow for an arbitrary "representation mixing"). We substantiate this definition by the assumption that the Fourier transforms of the electromagnetic transition elements from the proton to the resonances of one tower exhibit a pole at the same position. The quantum numbers and the SL(2,C) invariants of the tower of resonances are coupled with the corresponding quantum numbers of the proton tower via the vector operator and isospin Clebsch-Gordon coefficients. In particular is the location of the pole in the Fourier transforms identical with the SL(2,C) invariant λ of the tower of resonances. Since as we mentioned an infinite set of poles is necessary to reproduce the known analytic properties of the form factors, one resonance has contributions in an infinite set of towers, we have "infinite representation mixing".

The concept thus arrived at is best compared with Barut's notion of dynamical groups [8]. It deviates mainly from it by weakening most of its premises. In particular we need not specify

a) The noncompact group. Such group must always include SL(2,C) as a subgroup because of relativistic invariance. We use the minimal group SL(2,C) itself.

b) The representations by their unitarity, irreducibility, degeneracy, etc. We note that a dynamical group model based on a simple Lie group which is strictly bigger than SL(2,C) implies Fourier transforms on SL(2,C) which exhibit sequences of equally spaced poles. This behavior is analogous to the reduction of a Toller pole into infinite families of Regge poles.

c) The form of the current.

We lose by this generalization a global representation of the form factors and are left only with asymptotic expansions. However, the generality of our ansatz lets us hope that our scheme might prove useful for a phenomenological analysis of the electron scattering data. The tower hypothesis requires an experimental verification.

We want to illustrate finally how this tower hypothesis correlates data for different production processes. We consider a tower with isospin $\frac{1}{2}$ and spin parity content $S^P = \frac{1^\pm}{2}, \frac{3^\pm}{2}, \cdots$ which is connected with the proton via the isovector part of the electromagnetic current, and whose SL(2,C) invariant λ corresponds to the position of the isovector class one pole (see Section 3.2). This is justified by the fact that the proton itself fits into this tower. Other candidates for this tower are

$$N(1518),\ S^P = \frac{3}{2}-,\quad N(1680),\ S^P = \frac{5}{2}-,\quad N(1688),\ S^P = \frac{5}{2}+ .$$

If such pole dominates the form factors, we obtain for the ratio of the cross sections in the laboratory frame and fixed electron scattering angle

$$\lim_{\substack{q^2\to-\infty \\ \theta \text{ fixed}}} \left(\frac{d\sigma}{d\Omega}\right)_{\text{res}} : \left(\frac{d\sigma}{d\Omega}\right)_{\text{elast}} = \text{const.}$$

where the const. does not depend on θ. Its value remains unknown, since we did not specify the residues corresponding to the different members of one tower in our model. We can use this additional freedom to adjust these residues for current conservation without restricting the mass spectrum.

3.4. Conclusion

The use which can be made of our mathematical formalism is certainly not restricted to the analysis of the phenomenology of electromagnetic processes as sketched in Section 3. The priority given to this application is historical and due to the simple fact that this application is formally the simplest one. Further details on this kind of application are contained in the original articles [9].

REFERENCES

[1] Gel'fand, I. M., Graev, M. I., and Vilenkin, N. Ya., *Generalized Functions*, Vol. 5: "Integral Geometry and Representation Theory", New York (1966). Naimark, M. A., *Linear Representations of the Lorentz Group*, London (1964).

[2] Gel'fand, I. M., and Shilov, G. E., *Generalized Functions*, Vol. 2: "Spaces of Fundamental and Generalized Functions", New York (1968).

[3] Naimark, M. A., *Amer. Math. Soc. Transl. Ser.2*, 36, 101 (1964).

[4] Gel'fand, I. M., and Yaglom, A. M., *Zhur. Eksper. i Teor. Fiz.*, 18, 703 (1948).

[5] Wess, J., *Lectures in Theoretical Physics*, Vol. 10B, edited by A. O. Barut and W. E. Brittin, New York (1968), p. 325.

[6] Rühl, W., *Nuovo Cimento*, 63A, 1131, 1163 (1969), and CERN preprint, TH 1125.

[7] Källén, G., *Elementary Particle Physics*, Reading, Mass. (1964).

[8] Barut, A. O., and Kleinert, H., *Phys. Rev.*, 161, 1464 (1967).

[9] Rühl, W., *Nucl. Phys.*, B11, 505 (1969) and, with J. Kupsch, *Nuovo Cimento*, 64A, 991 (1969).

LIE ALGEBRAS OF LOCAL CURRENTS AND THEIR REPRESENTATIONS*

by

G. A. Goldin and D. H. Sharp**

1. INTRODUCTION

In these lectures, our aim is to describe some problems concerning representations of infinite dimensional Lie algebras, whose solution would be of considerable interest to physicists. These problems arise quite generally in trying to implement the "current algebra" approach to elementary particle physics. However, the specific topics we shall discuss here have to do with recent suggestions that one might be able to write relativistic theories of hadrons exclusively in terms of local observables such as currents [1-4].

The talks are organized as follows. First, we shall try to explain briefly how our approach fits in with what physicists usually call "current algebra". Secondly, we shall rewrite ordinary non-relativistic quantum mechanics in terms of local currents, and present the mathematical framework for discussing representations of the current algebra thus obtained. This discussion will provide a non-trivial example where the idea of working exclusively with local currents can be carried out in an explicit and mathematically rigorous way.

Next, we shall display a representation of the current algebra for a non-relativistic system having infinitely many degrees of freedom. This representation is obtained by taking the limit of a theory with N identical non-interacting bosons in a volume V, as the number of particles and the volume become infinite, while the average density (N/V) remains fixed. Finally, we shall briefly discuss a relativistic model for charged scalar mesons based on local currents, and mention a few of the many questions which remain unanswered in the non-relativistic and relativistic theories.

2. BACKGROUND [5]

The "currents" which usually appear in relativistic current algebras are the weak and electromagnetic currents of the strongly interacting particles or, as they are called, the hadrons.

---

* Work supported in part by the National Science Foundation and the U.S. Atomic Energy Commission.

** Department of Physics, University of Pennsylvania, Philadelphia, Pennsylvania.

The existence and properties of these currents are inferred from experimental studies of the hadronic weak and electromagnetic interactions. While we have been familiar with the basic properties of the electromagnetic four-vector current $J^{\mu}_{EM}(x)$ for quite a while, the nature of the vector and axial vector currents which play a fundamental role in the weak interactions has become reasonably clear only within the past fifteen years or so.

One of the interesting consequences of the approximate SU(3) invariance of the strong interactions is that it allows a certain unification in the description of the weak and electromagnetic currents. This is achieved by combining the various parts of the electromagnetic and vector weak currents into a single object having eight components, which we write as

$$F^{\mu}_{j}(x); \; \mu = 0,1,2,3, \; j = 1,\dots,8 \; . \tag{2.1}$$

This "vector octet" of currents behaves like a 4-vector under Lorentz transformations and transforms like an octet under SU(3) rotations. The pieces of the axial vector weak currents can likewise be combined into a second eight-component object

$$F^{5\mu}_{j}(x) \; , \tag{2.2}$$

which is an axial vector and which also transforms like an SU(3) octet.

In the vector octet, for example, the strangeness-conserving part of the vector weak current is proportional to $F^{\mu}_{1} + iF^{\mu}_{2}$, the electromagnetic current $J^{\mu}_{EM}(x) = e(F^{\mu}_{3}(x) + \frac{1}{\sqrt{3}} F^{\mu}_{8}(x))$, while $F^{\mu}_{4}(x),\dots,F^{\mu}_{7}(x)$ are related to the strangeness-changing weak currents.

The space integrals of the time components of the local current densities $F^{\mu}_{j}(x)$ and $F^{5\mu}_{j}(x)$ define a set of charges, $F_{j}(x_{o})$ and $F^{5}_{j}(x_{o})$. For $j = 1,2,3$, $F_{j}(x_{o}) = I_{j}$, which is the isotopic spin; the hypercharge $Y = \frac{2}{\sqrt{3}} F_{8}$, and the electric charge $Q = \int \overset{o}{J}_{EM}(x)d^{3}x = e(I_{3} + \frac{1}{2} Y)$. We remark that the charges F_{1}, F_{2}, F_{3}, and F_{8} arise from conserved currents and are thus constants of the motion, whereas the other charges may vary with time.

The local currents $F^{\mu}_{j}(x)$ and $F^{5\mu}_{j}(x)$ and their associated charges are the basic objects of study in "current algebra".

The fundamental hypothesis of current algebra, due to Gell-Mann [6,7], states that the time components of the physical vector and axial vector octet currents satisfy the <u>equal-time</u> commutation relations:

$$[F^{o}_{k}(\vec{x}), F^{o}_{\ell}(\vec{y})]|_{x^{o}=y^{o}} = i\delta(\vec{x} - \vec{y}) f_{k\ell m} F^{o}_{m}(\vec{x}) \tag{2.3a}$$

$$[F_k^{\circ}(\vec{x}),F_{\ell}^{5\circ}(\vec{y})]|_{x^\circ=y^\circ} = i\delta(\vec{x}-\vec{y})f_{k\ell m}F_m^{5\circ}(\vec{x}) \tag{2.3b}$$

$$[F_k^{5\circ}(\vec{x}),F_{\ell}^{5\circ}(\vec{y})]|_{x^\circ=y^\circ} = i\delta(\vec{x}-\vec{y})f_{k\ell m}F_m^{\circ}(\vec{x}) \ , \tag{2.3c}$$

where the numbers $f_{k\ell m}$ are the structure constants of SU(3). We remark that these commutation relations define an infinite-dimensional Lie algebra of local currents when integrated with a suitable class of testing functions.

Integration of Equations (2.3) over $\vec{x}$ and $\vec{y}$ leads to the equal-time charge algebra

$$[F_k(x^\circ),F_\ell(x^\circ)] = if_{k\ell m}F_m(x^\circ) \tag{2.4a}$$

$$[F_k(x^\circ),F_\ell^5(x^\circ)] = if_{k\ell m}F_m^5(x^\circ) \tag{2.4b}$$

$$[F_k^5(x^\circ),F_\ell^5(x^\circ)] = if_{k\ell m}F_m(x^\circ) \ . \tag{2.4c}$$

This weaker version of Gell-Mann's hypothesis is what has actually been used in many of the most successful applications of current algebra, as in the derivation of the famous Adler-Weisberger relation [8,9].

To the physicist, Gell-Mann's hypothesis is very beautiful. The reason for this is that it captures so much of what we really think is correct in our understanding of the weak and electromagnetic interactions of hadrons in the form of simple, possibly exact, relationships between experimentally observable quantities. For example, this idea allows one to formulate the notion of universality of strength of the weak interactions in a way that does not require a detailed description of how the hadronic weak current is built up out of particle fields. Furthermore, the commutation relations (2.3) and (2.4) specify a mathematical sense in which the group SU(3) × SU(3) acts in the strong interactions, even though it is not an invariance group.

These ideas of Gell-Mann are the foundation on which we would like to build. An obvious extension of Equations (2.3) is to try to find the commutation relations satisfied by the other components of the octet currents, and to extract the physics contained in them. But we wish to discuss the possibility that one can go further, and write complete relativistic theories in which all of the fundamental dynamical variables in the theory are local observables, such as the vector and axial vector currents mentioned above.

To clarify the question, let us recall the canonical field theory of neutral scalar mesons. As discussed in Todorov's lectures [10], one has fields $\varphi(\vec{x},t)$ and $\pi(\vec{x},t)$ which satisfy the equal-time commutation relations

$$[\varphi(\vec{x},t),\pi(\vec{y},t)] = i\delta(\vec{x}-\vec{y}) \ . \tag{2.5}$$

It is assumed that $\varphi(x)$ and $\pi(x)$ form a complete set of operators in the sense that the manifold of all states available to the system spans a single irreducible representation of the local algebra (2.5). The dynamics of the theory is contained in the Hamiltonian

$$H = \int d^3\vec{x}[\pi^2(x) + \nabla\varphi(x) \cdot \nabla\varphi(x) + \mu^2\varphi^2(x)] + H_I \ , \tag{2.6}$$

where H_I is the interaction Hamiltonian, usually taken to be a polynomial in $\varphi(x)$ Thus H is explicitly a function of $\varphi(x)$ and $\pi(x)$. We are asking whether one can repeat this pattern using as "coordinates" local observables such as the currents themselves, with a local current algebra replacing Equation (2.5) and with the Hamiltonian an explicit function of the currents.

Remarks. (i) The analogue of the canonical commutation relations in a theory based on currents is an equal-time current algebra, such as Equations (2.3). In studying the mathematical structure of local current algebras, one is already studying relationships between observable quantities which are subject, in principle to direct experimental tests.

(ii) Another familiar point is that, among the hundred-odd known hadrons, there are presently no candidates to play the role of "elementary particle", quarks not yet having been observed. Since relativistic theories have traditionally been written in terms of canonical fields whose quanta may be considered as the building blocks of matter, one may be at a loss, when presented with the hadron spectrum, to know where to start.

Local currents treat all particles on an equal footing in the sense that, if one starts with the physical current, and postulates various commutation relations between its components, one does not have to say anything at the beginning about what kinds of particles are present in the theory. All of the different charged particles will make their contribution to the electromagnetic current, for example, but instead of trying to specify at the outset how the current is constituted in terms of particle fields, one can learn this in the process of solving the theory. Thus one might hope that the currents could define a theory in which no hadron plays a special role.

(iii) We expect that in theories written in terms of currents, the local currents themselves will be fields which satisfy Wightman's axioms [11].

(iv) It is hardly necessary to emphasize how far we are today from being able to implement these ideas in situations of immediate relevance to high energy particle physics. We face not only the problem of writing down the correct current commutation relations and the proper Hamiltonian; we have not even identified with any degree of certainty a complete set of local currents in terms of which to describe the hadron system.

To explore the basic ideas, we therefore take various canonical field theory models and rewrite them in terms of currents, obtaining a current algebra and a formula for the Hamiltonian as a function of the currents. Once one has abstracted these relationships from the underlying field theory, one is entitled to take them as a new starting point for the description of the physical system. In the following, we outline some results on representations of the current algebras which arise in non-relativistic quantum mechanics and in a relativistic model for charged scalar mesons.

3. NON-RELATIVISTIC CURRENT ALGEBRA

3.1. n-Particle Representations of the Current Algebra [12-16]

Our starting point is the second-quantized formulation of the quantum mechanics of a system of spinless particles. In this formalism we introduce a Hilbert space $\mathcal{H} = \bigoplus_{n=0}^{\infty} \mathcal{H}_n$, where $\mathcal{H}_n$ is the Hilbert space of symmetric (or antisymmetric) L^2 functions of n vector variables. An element $\Psi \in \mathcal{H}$ may be written as $\Psi = (\Psi_0, \Psi_1, \ldots)$ with $(\Psi,\Psi) = \sum_{n=0}^{\infty} (\Psi_n,\Psi_n) < \infty$.

For the commutation relations $[\psi(\vec{x}),\psi^*(\vec{y})]_- = \delta(\vec{x} - \vec{y})$, the equations

$$(\psi(\vec{x})\Psi)_n(\vec{x}_1,\ldots,\vec{x}_n) = \sqrt{n+1}\, \Psi_{n+1}(\vec{x}_1,\ldots,\vec{x}_n,\vec{x})$$

and

$$(\psi^*(\vec{x})\Psi)_n(\vec{x}_1,\ldots,\vec{x}_n) = \frac{1}{\sqrt{n}} \sum_{j=1}^{n} \delta(\vec{x} - \vec{x}_j)\Psi_{n-1}(\vec{x}_1,\ldots,\hat{\vec{x}}_j,\ldots,\vec{x}_n) \tag{3.1}$$

define operator-valued distributions $\psi(\vec{x})$ and $\psi^*(\vec{x})$ in the Hilbert space of symmetric functions. Likewise fields satisfying anticommutation relations, $[\psi(\vec{x}),\psi^*(\vec{y})]_+ = \delta(\vec{x} - \vec{y})$, are defined by the equations

$$(\psi(\vec{x})\Psi)_n(\vec{x}_1,\ldots,\vec{x}_n) = \sqrt{n+1}\, \Psi_{n+1}(\vec{x}_1,\ldots,\vec{x}_n,\vec{x})$$

and

$$(\psi^*(\vec{x})\Psi)_n(\vec{x}_1,\ldots,\vec{x}_n) = \frac{(-1)^{n+1}}{\sqrt{n}} \sum_{j=1}^{n} (-1)^{j+1}\delta(\vec{x} - \vec{x}_j)\Psi_{n-1}(\vec{x}_1,\ldots,\hat{\vec{x}}_j,\ldots,\vec{x}_n) \tag{3.2}$$

in the Hilbert space of antisymmetric functions.

Defining the number density of particles as

$$\rho(\vec{x}) = \psi^*(\vec{x})\psi(\vec{x}) ,$$

and the particle flux density by

$$\vec{J}(\vec{x}) = \frac{1}{2i} [\psi^*(\vec{x})\nabla\psi(\vec{x}) - (\nabla\psi^*(\vec{x}))\psi(\vec{x})] , \tag{3.3}$$

one can obtain by direct calculation from either (3.1) or (3.2) that

$$(\rho(f)\Psi)_n = \sum_{j=1}^{n} f(\vec{x}_j)\Psi_n$$

$$(J(\vec{g})\Psi)_n = \frac{1}{2i} \sum_{j=1}^{n} [\vec{g}(\vec{x}_j) \cdot \nabla_j + \nabla_j \cdot \vec{g}(\vec{x}_j)]\Psi_n \tag{3.4}$$

for the smeared currents $\rho(f) = \int \rho(\vec{x})f(\vec{x})d\vec{x}$ and $J(\vec{g}) = \int \vec{J}(\vec{x}) \cdot \vec{g}(\vec{x})d\vec{x}$.

Restricted to $\mathcal{H}_n$, Equation (3.4) defines an irreducible representation, called the n-particle representation, of the non-relativistic current algebra [1]:

$$\begin{aligned}[\rho(f),\rho(g)] &= 0\\ [\rho(f),J(\vec{g})] &= i\rho(\vec{g}\cdot\nabla f)\\ [J(\vec{g}),J(\vec{h})] &= iJ(\vec{h}\cdot\nabla\vec{g}-\vec{g}\cdot\nabla\vec{h})\ .\end{aligned}\tag{3.5}$$

In the representation (3.4), the number operator N^{op} is a super-selecting operator.

3.2. Exponentiating the Lie Algebra of Currents [12,13]

The currents in (3.5) are in general unbounded operators; thus they are defined only on a dense domain D which may depend on the testing function. Under these circumstances, current commutators might not always make sense. Therefore we look for a group, which we can represent by unitary operators.

Let $\vec{\varphi}^{\vec{g}}_t\colon \mathbb{R}^s \to \mathbb{R}^s$ denote the flow for time t by the vector field $\vec{g}$; i.e.,

$$\frac{\partial\vec{\varphi}^{\vec{g}}_t}{\partial t}(\vec{x}) = \vec{g}(\vec{\varphi}^{\vec{g}}_t(\vec{x}))$$

with $\vec{\varphi}^{\vec{g}}_{t=0}(\vec{x}) = \vec{x}$. If $\vec{g}$ has components in Schwartz' space S, then $\vec{\varphi}^{\vec{g}}_t$ exists and is C_∞ for all t.

It turns out that the correct objects to define are $U(f) = e^{i\rho(f)}$ and $V(\vec{\varphi}^{\vec{g}}_t) = e^{itJ(\vec{g})}$ where

$$U(f_1)V(\vec{\psi}_1)U(f_2)V(\vec{\psi}_2) = U(f_1 + f_2\circ\vec{\psi}_1)V(\vec{\psi}_2\circ\vec{\psi}_1)\ .\tag{3.6}$$

One can prove this by studying the n-particle representation, which becomes

$$\begin{aligned}U(f)\Psi_n &= \exp[i\textstyle\sum_{j=1}^{n} f(\vec{x}_j)]\Psi_n\\ V(\vec{\psi})\Psi_n(\vec{x}_1,\dots,\vec{x}_n) &= \Psi_n(\psi(\vec{x}_1),\dots,\psi(\vec{x}_n))\prod_{j=1}^{n}\sqrt{\det\left(\frac{\partial\psi^k}{\partial x^\ell}(\vec{x}_j)\right)}\ .\end{aligned}\tag{3.7}$$

From (3.7) one can verify (3.4) and hence (3.5) using Stone's theorem; therefore (3.6) is the correct group law to study.

Thus we must consider representations of the semidirect product $S \wedge K$, where S is Schwartz' space, and K is the group of C_∞ diffeomorphisms from $\mathbb{R}^s \to \mathbb{R}^s$ generated by the flows $\vec{\varphi}^{\vec{g}}_t$ under composition. K may be appropriately topologized. It may be pointed out that S is needed in order to be able to take successive derivatives in (3.5).

3.3. The Gel'fand-Vilenkin Formalism [12,13]

The Gel'fand-Vilenkin formalism [17] is suitable for the study of representations of groups such as $S \wedge K$, in which an abelian subgroup is a nuclear space. Such groups also occur in relativistic models [1-4,12]. We assume familiarity with the topology of S and remark that S' denotes the continuous dual of S, with (F,f) the value of $F \in S'$ at $f \in S$.

A <u>cylinder set</u> in S' is a set of the form $\{F \in S' | ((F,f_1),\ldots,(F,f_n)) \in A\}$ for $A \subseteq \mathbb{R}^n$. A is called the <u>base</u> of the cylinder set.

A <u>cylindrical measure</u> μ on S' is a countably additive normalized measure μ on the σ-algebra generated by all cylinder sets with Borel base.

An important result is expressed in the following <u>Theorem</u> (Bochner's theorem for nuclear spaces):

If $L(f)$ is a continuous functional on S, with $L(0) = 1$, which satisfies the "positivity condition"

$$\sum_{j,k=1}^{n} \bar{C}_k C_j L(f_j - f_k) \geq 0 \tag{3.8}$$

for $f_j \in S$ and $C_j \in \mathbb{C}$, then there exists a unique cylindrical measure μ such that

$$L(f) = \int_{S'} e^{i(F,f)} d\mu(F) . \tag{3.9}$$

If U is a strongly continuous cyclic representation of S in H with cyclic vector Ω, we can let $L(f) = (\Omega, U(f)\Omega)$ define a cylindrical measure μ according to Equation (3.9). Then H can be realized as $L^2_\mu(S')$ with $\Omega(F) \equiv 1$, and

$$U(f)\Psi(F) = e^{i(F,f)}\Psi(F) . \tag{3.10}$$

If $U(f)V(\vec{\psi})$ is a representation of $S \wedge K$, with $\Omega \in H$ cyclic for U, then μ is <u>quasi-invariant for K</u> in the following sense: if we define $(\vec{\psi}^*F,f) = (F,f \circ \vec{\psi})$ and $\mu^{\vec{\psi}}(X) = \mu(\vec{\psi}^*X)$, then $\mu^{\vec{\psi}}$ and μ have the same sets of measure zero. Along with (3.10), we have

$$V(\vec{\psi})\Psi(F) = \chi_{\vec{\psi}}(F)\Psi(\vec{\psi}^*F)\sqrt{\frac{d\mu^{\vec{\psi}}}{d\mu}}\,(F) \tag{3.11}$$

where $\frac{d\mu^{\vec{\psi}}}{d\mu}(F)$ is the Radon-Nikodym derivative.

The "multiplier" $\chi_{\vec{\psi}}(F)$ is a complex-valued function of modulus one. While $\chi_{\vec{\psi}}(F) \equiv 1$ is always a possibility, one can obtain nontrivial inequivalent representations with the same μ from different families of χ's. The χ's satisfy

$$\chi_{\vec{\psi}_2}(F)\chi_{\vec{\psi}_1}(\vec{\psi}_2^*F) = \chi_{\vec{\psi}_1\circ\vec{\psi}_2}(F) . \tag{3.12}$$

Many deep parallels with Mackey's theory [18] of representations of semidirect products of locally compact groups inhere in the Gel'fand-Vilenkin formalism.

For the n-particle representation (3.7), μ is concentrated on the set $F = \{F_{\vec{x}_1} + \ldots + F_{\vec{x}_n} ; \vec{x}_j \neq \vec{x}_k\}$, where $(F_{\vec{x}},f) = f(\vec{x})$, with $d\mu(F_{\vec{x}_1} + \ldots + F_{\vec{x}_n}) \propto e^{-\vec{x}_1^2}\ldots e^{-\vec{x}_n^2} d\vec{x}_1 \ldots d\vec{x}_n$. In the symmetric case, $\chi_{\vec{\psi}}(F) \equiv 1$, while in the antisymmetric case, this is no longer true. The two cases are unitarily inequivalent in more than one spatial dimension. This method of describing particle statistics is discussed in detail in [13-15].

3.4. Representations of the Non-Relativistic Current Algebra in the "N/V" Limit [19]

The n-particle representations of $S \wedge K$ are of course a mere restatement of the ordinary quantum mechanics of n identical particles; i.e., a system of finitely many degrees of freedom. Here we see how this reformulation leads to some particularly simple expressions in the limit of infinitely many noninteracting identical particles at constant average density.

To consider N bosons in a volume V, we impose periodic boundary conditions on the wave functions $\Psi(\vec{x}_1,\ldots\vec{x}_n)$, which are symmetric with respect to interchange of particle coordinates. This corresponds to a representation of $C_\infty(T^s) \wedge K(T^s)$ where T^s is the s-torus, $C_\infty(T^s)$ is topologized like a nuclear space, and $K(T^s)$ is the group of C_∞ diffeomorphisms from $T^s \to T^s$.

We know that the state of lowest energy is $\Omega_{N,V}(\vec{x}_1,\ldots,\vec{x}_N) = \left(\frac{1}{\sqrt{V}}\right)^N$. Thus $L_{N,V}(f) = (\Omega_{N,V}, e^{i\rho(f)}\Omega_{N,V})$ becomes

$$[\frac{1}{V}\int d\vec{x} e^{if(\vec{x})}]^N = [1 + \frac{1}{V}\int d\vec{x}[e^{if(\vec{x})} - 1]]^N .$$

Setting $\bar{\rho} = N/V$ and taking the limit as $N,V \to \infty$, one obtains

$$L(f) = \exp[\bar{\rho}\int (e^{if(\vec{x})} - 1)d\vec{x}] . \tag{3.13}$$

One can check that if $L(f)$ is given by Equation (3.13) it is continuous, positive, and satisfies $L(0) = 1$. Thus, $L(f)$ is the Fourier transform of a cylindrical measure μ in S', and defines a representation of S.

By the same procedure, one can obtain

$$L(f,\vec{\psi}) \equiv (\Omega, U(f)V(\vec{\psi})\Omega) = \exp[\bar{\rho}\int (e^{if(\vec{x})}\sqrt{J_{\vec{\psi}}(\vec{x})} - 1)d\vec{x}] , \tag{3.14}$$

where $J_{\vec{\psi}}(\vec{x}) = \det(\frac{\partial\psi^k}{\partial x^j}(\vec{x}))$ is the Jacobian of $\vec{\psi}$. From (3.14), one can compute all of the n-point ground-state expectation functions of the currents. For example,

$$\begin{aligned} &<\rho(f)> = \bar{\rho}\int f(\vec{x})d\vec{x} , \\ &<J(\vec{g})> = 0 , \\ &<\rho(f)\rho(g)> = <\rho(fg)> + <\rho(f)> <\rho(g)> , \\ &<\rho(f)J(\vec{g})> = -\frac{1}{2i} <\rho(\vec{g} \cdot \nabla f)> , \\ &<J(\vec{g})J(\vec{h})> = \frac{1}{4} <\rho(\nabla \cdot \vec{g}\nabla \cdot \vec{h})> , \end{aligned} \tag{3.15}$$

and so on. Equation (3.14) is equivalent to (3.13) together with the commutation relations (3.5) and the equation

$$(\nabla\rho + 2i\vec{J})(\vec{x})\Omega = 0 , \tag{3.16}$$

which is true for every N,V. Since the kinetic energy piece of the Hamiltonian density can be written in terms of currents as [1,14,15]

$$H(\vec{x}) = \frac{1}{8M} (\nabla\rho - 2i\vec{J})(\vec{x}) \frac{1}{\rho(\vec{x})} (\nabla\rho + 2i\vec{J})(\vec{x}) \ , \tag{3.17}$$

Equation (3.16) implies that $H\Omega = 0$. We shall discuss apparently singular Hamiltonians such as (3.17) in Section (3.5).

It is also possible to carry out an "N/V" limit for non-interacting particles satisfying Fermi statistics. The results are described in [19].

3.5. "Singular" Hamiltonians

The Hamiltonian density (3.17) seems to contain the singular expression $\rho^{-1}(\vec{x})$. It has been shown [14,15] that in an irreducible n-particle representation of (3.5) the factor $(\nabla\rho + 2i\vec{J})(\vec{x})$ appearing in $H(\vec{x})$ is proportional to $\rho(\vec{x})$. Thus the factor $\rho^{-1}(\vec{x})$ is explicitly cancelled in (3.17) with the result that $H(f) = \int f(\vec{x})H(\vec{x})d\vec{x}$ is actually a well-defined operator in Hilbert space. Here we shall indicate how the quantity $\rho^{-1}(\vec{x})$ can be given a direct mathematical definition.

Suppose we are given a Hilbert space $\mathcal{H}$, an operator valued distribution $\rho(\vec{x})$ in $\mathcal{H}$ and a dense domain D for ρ with $\rho(f)D \subseteq D$ for all $f \in \mathcal{S}$. Define $\mathcal{V}$ to be the linear span of $\{f(\vec{x})\rho(\vec{x})\Phi | \Phi \in D, f \in \mathcal{O}_M\}$, where $\mathcal{O}_M$ denotes the real-valued C_∞ functions which, together with all derivatives, are of polynomial growth at ∞. Thus $\mathcal{V}$ is a family of vector-valued distributions. It may well be the case that for distinct choices of Φ and f, e.g., Φ_1, Φ_2, and f_1, f_2, one can have $f_1(\vec{x})\rho(\vec{x})\Phi_1 = f_2(\vec{x})\rho(\vec{x})\Phi_2$. Then $\rho^{-1}(\vec{x}): \mathcal{V} \times \mathcal{V} \to \mathcal{S}'$ is given by

$$(f(\vec{x})\rho(\vec{x})\Phi, \rho^{-1}(\vec{x})g(\vec{x})\rho(\vec{x})\Psi) = (\Phi, f(\vec{x})\rho(\vec{x})g(\vec{x})\Psi) \quad , \tag{3.18}$$

extended sesqui-linearly to $\mathcal{V} \times \mathcal{V}$. It is now an easy lemma to show that $\rho^{-1}(\vec{x})$ is well-defined. One should note that $\rho^{-1}(\vec{x})$ is <u>not</u> well-defined by the requirement that "$\rho^{-1}(\vec{x}): \mathcal{V} \to \mathcal{V}$" be given by $f(\vec{x})\rho(\vec{x})\Phi = f(\vec{x})\Phi$. Thus $\rho^{-1}(\vec{x})$ is a map from $\mathcal{V} \times \mathcal{V} \to \mathcal{S}'$, although of course it is not an operator-valued distribution in $\mathcal{H}$.

Let $K(\vec{x})$ be a (not necessarily Hermitian) operator-valued distribution in $\mathcal{H}$ on D, with $K(\vec{x})D$ and $K^*(\vec{x})D$ contained in $\mathcal{V}$. Then K is <u>related</u> to ρ in a certain sense, and one can define the matrix elements of $H(\vec{x}) = K^*(\vec{x})\rho^{-1}(\vec{x})K(\vec{x})$ by

$$(\Psi, H(\vec{x})\Phi) = (K(\vec{x})\Psi, \rho^{-1}(\vec{x})K(\vec{x})\Phi) \quad . \tag{3.19}$$

In the "N/V" example above, $\vec{K}(\vec{x}) = (\nabla\rho + 2i\vec{J})(x)$ is related to ρ by the commutation relations together with Equation (3.16).

4. A RELATIVISTIC MODEL FOR CHARGED SCALAR MESONS [2,20]

The charged scalar model was originally defined [2] in terms of the operators

$$\begin{aligned} j_\mu(x) &= i[\varphi^*(x)\partial_\mu\varphi(x) - (\partial_\mu\varphi^*(x))\varphi(x)] \\ S(x) &= \varphi^*(x)\varphi(x) \\ \dot{S}(x) &= \varphi^*(x)\pi^*(x) + \pi(x)\varphi(x) \ , \end{aligned} \tag{4.1}$$

where $\partial_0\varphi = \pi^*$. The fields are assumed to satisfy the canonical equal-time commutation relations

$$[\varphi(\vec{x}),\pi(\vec{y})] = [\varphi^*(\vec{x}),\pi^*(\vec{y})] = i\delta(\vec{x}-\vec{y}) \ , \tag{4.2}$$

with all of the other commutators vanishing. This leads to the current commutation relations

$$[j_0(f),j(\vec{g})] = -2iS(\vec{g}\cdot\nabla f) \tag{4.3a}$$

$$[S(f),\dot{S}(g)] = 2iS(fg) \tag{4.3b}$$

$$[j(\vec{g}),\dot{S}(f)] = 2ij(f\vec{g}) \ . \tag{4.3c}$$

All of the other commutators vanish.

Setting $K_\mu(x) = \partial_\mu S(x) - ij_\mu(x)$, the energy-momentum tensor in this model is (without interactions)

$$\theta_{\mu\nu}(x) = \frac{1}{4}K^*_\mu \frac{1}{S} K_\nu + \frac{1}{4}K^*_\nu \frac{1}{S} K_\mu - g_{\mu\nu}[\frac{1}{4}K^*_\alpha \frac{1}{S} K^\alpha - m^2 S] \ . \tag{4.4}$$

Let us emphasize that we do not actually know any representations in which (4.1) and (4.3) together make literal sense. We are indeed considering a situation where, having guessed the current algebra, it is taken as the fundamental starting point of a theory based solely on currents.

One may choose to look at the subalgebra of (4.1) consisting of j_μ and S. It is then consistent to represent S by a multiple of the identity [20]: $S(\vec{x}) = \frac{c}{2}I$. Of course $\dot{S}$ then equals zero, and the commutation relations (4.3b) and (4.3c) must be abandoned. If $\dot{S} = 0$, Equation (4.4) implies that $[H,S(f)] = 0$, so this is at least a consistent model. It is in fact Sugawara's model [4] for the case of a trivial internal symmetry group. The Hamiltonian density becomes [20]

$$H(\vec{x}) = \frac{1}{2c}[j_0(\vec{x})j_0(\vec{x}) + \vec{j}(\vec{x})\cdot\vec{j}(\vec{x})] + m^2 \tag{4.5}$$

which is the same as in the Sugawara model.

The choice $S(\vec{x}) = cI$ might at first be regarded as natural for it makes unambiguous sense out of $\frac{1}{S(\vec{x})}$ in (4.4) and the Hamiltonian becomes bilinear in the currents. But we know that products of distributions at a point rarely make mathematical sense, while we have seen in the non-relativistic model how the "inverse of an operator-valued distribution" <u>can</u> make sense when appropriately sandwiched between vector-valued distributions. In fact (4.4) may be <u>less</u> singular than a bilinear expression; the factor $\frac{1}{s}$ might <u>cancel</u> something in the numerator.

5. QUESTIONS

Now it is time to reveal the extent of our ignorance by mentioning a few of the questions to which we don't have answers.

A <u>complete</u> classification of the irreducible representations of $S \wedge K$ would presumably amount to solving the many-body problem, at least in the "N/V" limit, and is therefore very likely a forlorn hope. However, any examples of representations beyond those mentioned would be extremely interesting. To construct such examples, it would be helpful to know something about the measurability of the orbits in S' under the action of K.

We would like to have a way to determine the functional L(f) in the N/V limit directly, without first having to start from the form of the functional in a box. Preliminary results in this direction have been obtained, using functional differential equations [19,21]. Furthermore, one would like to have techniques for the <u>approximate</u> determination of L(f), in view of the fact that it is unlikely that this functional can be calculated exactly in most situations of practical interest.

Finally, we reiterate that we have no concrete representations of the charged scalar algebra, or any other interesting local, relativistic current algebras, at this time. To construct such representations may be a crucial step in extending the results described here to the domain of particle physics.

6. ACKNOWLEDGMENTS

The authors wish to thank the staff of the Battelle Memorial Institute for hospitality extended to them during the summer of 1969 when these lectures were prepared. It is also our pleasure to thank Professors V. Bargmann, G. Mackey, E. Stein and other participants in the 1969 Rencontres for numerous helpful and interesting discussions.

7. REFERENCES

[1] Dashen, R. F., and Sharp, D. H., *Phys. Rev.*, 165, 1857 (1968).

[2] Sharp, D. H., *Phys. Rev.*, 165, 1867 (1968).

[3] Callan, C. G., Dashen, R. F., and Sharp, D. H., *Phys. Rev.*, 165, 1883 (1968).

[4] Sugawara, H., *Phys. Rev.*, 170, 1659 (1968).

[5] The presentation of the material in the first half of this section follows S. L. Adler and R. F. Dashen, Chapter I in *Current Algebras*, Benjamin, N. Y., (1968). For further background material one can consult other chapters in the Adler-Dashen book as well as B. Renner, *Current Algebras and Their Applications*, Pergamon, N. Y., (1968), and the lectures of L. Michel and L. O'Raifeartaigh in these Proceedings.

[6] Gell-Mann, M., *Phys. Rev.*, 125, 1067 (1962).

[7] Gell-Mann, M., *Physics*, 1, 63 (1964).

[8] Adler, S. L., *Phys. Rev. Letters*, 14, 1051 (1965).

[9] Weisberger, W. I., *Phys. Rev. Letters*, 14, 1047 (1965).

[10] Todorov, I. T., lectures on quantum field theory. (Not reproduced here)

[11] For reviews of axiomatic field theory see: R. F. Streater and A. S. Wightman, *PCT, Spin and Statistics and All That*, Benjamin, N. Y., (1964), and R. Jost, *The General Theory of Quantized Fields*, American Mathematical Society (1963).

[12] Goldin, G., Ph.D. Thesis, Pinceton University (1968), unpublished.

[13] Goldin, G., "Non-Relativistic Current Algebras as Unitary Representations of Groups", *J. Math. Phys.*, (to be published).

[14] Grodnik, J., and Sharp, D. H., "Representations of Local Non-Relativistic Currents", *Phys. Rev.*, (to be published).

[15] Grodnik, J., and Sharp, D. H., "Description of Spin and Statistics in Non-Relativistic Quantum Theories Based on Local Currents", *Phys. Rev.*, (to be published).

[16] Grodnik, J., Ph.D. Thesis submitted to the University of Pennsylvania (1969), (unpublished).

[17] Gel'Fand, I., and Vilenkin, N., "Applications of Harmonic Analysis", Vol. 4 in *Genera. Fun.*, Academic Press, N. Y. (1964).

[18] Mackey, G., *Ann. Math.*, 55, 101 (1952).

[19] Goldin, G., Grodnik, J., Powers, R. T., and Sharp, D. H., "Non-Relativistic Current Algebra in the N/V Limit", (to be published).

[20] Dicke, A., and Goldin, G., (to be published).

[21] Grodnik, J., and Sharp, D. H., (to be published).

INFINITE DIMENSIONAL LIE ALGEBRAS AND CURRENT ALGEBRA*

by

Robert Hermann**

ABSTRACT

The "current algebras" of elementary particle physics and quantum field theory are interpreted as infinite dimensional Lie algebras of a certain definite kind. The possibilities of algebraic structure and certain types of representations of these algebras by differential operators on manifolds are investigated, in a tentative way. The Sugawara model is used as a typical example. A general differential geometric method (involving jet spaces) for defining currents associated with classical field theories is presented. In connection with the abstract definition of current algebras as modules, a purely module-theoretic definition of a "differential operator" is presented and its properties are studied.

---

* This research was supported by the Office of Naval Research. Reproduction in whole or part is permitted for any purpose by the United States Government.

** Institute for Advanced Study, Princeton, New Jersey 08540

1. INTRODUCTION

In the sense used in this paper, "current algebra" means a program of studying elementary particle physics and quantum field theory from the viewpoint of Lie algebra theory. Specifically, we are concerned with the existence and mathematical properties of certain infinite dimensional Lie algebras whose representations might serve to define the states of physically interesting field-theoretic dynamical systems. As proposed by M. Gell-Mann [6], this study seems to offer the simplest and most natural method for understanding the observed elementary particle symmetries and using them to derive further, deeper facts about the elementary particles. We refer to the books by Adler and Dashen [1] and Renner [16] for further motivation concerning the "physics" of current algebras. Here, we will mainly be concerned with various mathematical questions which are suggested by the broad program. This paper will report on work in progress.

To give a quick idea of what is involved, proceed as follows: Choose the following range of indices;

$$1 \leq a,b \leq n; \qquad 1 \leq i,j \leq 3$$

Let $x = (x_i)$, $y = (y_i)$ denote 3-vectors, i.e., elements of R^3; Consider "symbols" $v_a(x)$ satisfying relations of the following form:

$$[v_a(x),v_b(y)] = c_{abc}v_c(x)\delta(x - y) + d_{abci}\partial_i(v_c(x)\delta(x - y)) + \dots \qquad (1.1)$$

(The terms ... will mean terms involving higher order derivatives.)

Now, the "Lie algebra" defined symbolically by (1.1) can be defined in a more precise mathematical way as follows. Introduce the set of C^∞, real-valued functions $f\colon R^3 \to R$, denoted by F. Since such functions can be added, multiplied, and multiplied by real scalars, F is a commutative, associative algebra, with the real numbers, R, as field of scalars. For $f \in F$, introduce the following symbol:

$$v_a(f) = \int v_a(x)f(x)dx \quad . \qquad (1.2)$$

Then, the rules (1.1) transcribe following the usual calculational rules for generalized functions into the following expressions:

$$[v_a(f_1),v_b(f_2)] = c_{abc}v_c(f_1f_2) - d_{abc}v_c(\partial_i(f_1)f_2) + \dots \quad . \qquad (1.3)$$

We can now give mathematical structure to these formulas. Let Γ be the real vector space spanned by the symbols $v_a(f)$. Then (1.3) defines a skew-symmetric, real bilinear map $\Gamma \times \Gamma \to \Gamma$ that defines a Lie algebra like structure on Γ. (We do not necessarily require that it satisfy the Jacobi identity; typically, however, a quotient algebra will satisfy the Jacobi identity. See Section 6 for further comments on this point.)

Further, Γ is an F-module, with multiplication by an $f \in F$ defined as follows:

$$f(v_a(f')) = v_a(ff') \quad . \tag{1.4}$$

Now, the bracket [,] defined by (1.3) is not an arbitrary R-bilinear map. Roughly, it involves a differential expression in the F-module structure. To make this precise, we will, in Section 2, give an abstract algebraic definition of a "differential operator" purely within the category of F-modules.

Now, in the "currents" of Lagrangian quantum field theory, one finds among the "$v_a(x)$" expressions labeled as follows:

$$v^{\alpha}_{\mu}(x), \; 1 \leq \alpha,\beta \geq m; \; 0 \leq \mu,\nu \leq 3 \tag{1.5}$$

α is an "internal symmetry" index μ is a "space-time" index. Typically, these objects are determined--at least in a formal way--by well-known formulas from the Lagrangian and the Lie algebra of an internal symmetry transformation group. (See [9] for a discussion of the algebraic properties of these rules.) For example, for the "Sugawara model", [3, 9, 18, 21], the following relations are satisfied:

$$[V^{\alpha}_0(x), \; V^{\beta}_0(y)] = c_{\alpha\beta\gamma}V^{\gamma}_0(x)\delta(x - y) \tag{1.6}$$

$$[V^{\alpha}_j(x), \; V^{\beta}_j(y)] = 0 \tag{1.7}$$

$$[V^{\alpha}_0(x), \; V^{\beta}_i(y)] = c_{\alpha\beta\gamma}V^{\alpha}_i(x)\delta(x - y) + \lambda\delta_{\alpha\beta}\partial^x_i\delta(x - y) \quad . \tag{1.8}$$

In (1.5-1.7), "$c_{\alpha\beta\gamma}$" are the structure constants of a semisimple compact Lie algebra (with respect to a Lie algebra basis that is orthonormal with respect to the Killing form), and λ is a free parameter.

2. DIFFERENTIAL OPERATORS ON MODULES

As indicated in the introduction, in order to have a "definition" of current algebras as mathematical objects, independently of their usual association with quantum field theory, it is desirable to have a definition of "differential operator" valid for arbitrary modules. (There is, in the mathematical literature, a definition for sections of vector-bundles. See [15].) Indeed, this is a question of independent mathematical interest. In this section we will give such a definition.*

Let F be an arbitrary commutative, associative algebra with the real numbers as field of scalars, and with an identity element denoted by "1". Denote

---

* This definition is also known to M. Atiyah.

F-modules by $\Gamma, \Gamma', \ldots$. What is desired, for each integer $r \geq 0$, is a "functor" assigning to each pair (Γ, Γ') another F-module $D^r(\Gamma, \Gamma')$, which may be thought of as the "r-th order differential operators from Γ to Γ'." We will, in fact, define $D^r(\Gamma, \Gamma')$ by induction on r.

First, for $r = 0$, let $D^0(\Gamma, \Gamma')$ be the set of F-linear maps: $\Gamma \to \Gamma'$, i.e. an element $D \in D^0(\Gamma, \Gamma')$ is an R-linear map: $\Gamma \to \Gamma'$ such that:

$$D(f\gamma) = fD(\gamma) \quad \text{for } f \in F, \gamma \in \Gamma \tag{2.1}$$

Suppose now that D is an arbitrary R-linear map: $\Gamma \to \Gamma'$. Define an R-bilinear map: $F \times \Gamma \to \Gamma'$ as follows.

$$D(f, \gamma) = D(f\gamma) - fD(\gamma) \quad \text{for } f \in F, \gamma \in \Gamma \tag{2.2}$$

For fixed $f \in F$, define D_f as a R-linear map: $\Gamma \to \Gamma'$ as follows

$$D_f(\gamma) = D(f, \gamma) \quad . \tag{2.3}$$

Definition

Suppose that $D^{r-1}(\Gamma, \Gamma')$ is defined. Then, $D^r(\Gamma, \Gamma')$ consists of the R-linear maps $D: \Gamma \to \Gamma'$ such that, for each $f \in F$, the map D_f belongs to $D^{r-1}(\Gamma, \Gamma')$.

We must now show that $D^r(\Gamma, \Gamma')$ defined in this way has the usual properties one would expect to justify calling it the "F-module of r-th order differential operators".

Theorem 2.1

If $D \in D^r(\Gamma, \Gamma')$, $D' \in D^s(\Gamma', \Gamma'')$, then

$$D'D \in D^{r+s}(\Gamma, \Gamma'') \quad .$$

Proof. Proceed by induction on $r + s$. For $r + s = 0$, it is evident. Let

$$D'' = D'D \quad .$$

Then,

$$D''_f(\gamma) = D''(f\gamma) - fD''(\gamma) = D'(D(f\gamma)) - fD'D(\gamma)$$

$$= D'(D_f(\gamma) + fD(\gamma)) - fD'D(\gamma) = D'(D_f(\gamma)) + D'_fD(\gamma) \quad .$$

This proves the following basic formula:

$$(D'D)_f = D'D_f + D'_fD \quad . \tag{2.4}$$

By induction hypothesis, the right hand side of (2.4) belongs to $D^{r+s-1}(\Gamma,\Gamma')$, hence $D'D$ belongs to $D^{r+s}(\Gamma,\Gamma')$.

Now, let us determine $D^1(F,\Gamma')$. (Note that F may be considered as an F-module.) Given $D \in D^1(F,\Gamma')$, set

$$f_1 = D(1) \quad . \tag{2.5}$$

Define $D' \in D(F,\Gamma')$ as follows:

$$D'(f) = D(f) - f_1 = D_f(1) \tag{2.6}$$

Theorem 2.2

D' *is a derivation of* F, *into* Γ', *i.e.*

$$D'(ff') = D'(f)f' + fD'(f') \quad \text{for} \quad f, f' \in F \quad . \tag{2.7}$$

Proof. By assumption, D_f is a zero-th order operator, i.e., an F-linear map: $F \rightarrow \Gamma$, hence:

$$D_f(f') = D_f(1)f' \quad . \tag{2.8}$$

Then,

$$D_f(f') = (D(f) - fD(1))f' = D_f(1)f' \quad .$$

But also,

$$D_f(f') = D(ff') - fD(f') = D_{ff'}(1) + ff'D(1) - f(D_{f'}(1) + f'D(1)) \quad .$$

Combining these two formulas gives:

$$D_{ff'}(1) = D_f f' D_f(1) + fD_{f'}(1) \quad . \tag{2.9}$$

In view of (2.6), this proves (2.7).

Theorem 2.3

$D^1(F,\Gamma')$ is a direct sum of the subspace $D^0(F,\Gamma')$ and the space of derivations of F into Γ', i.e., an "inhomogeneous" first order operator can be written in a unique way as a sum of a zero-th order operator and a "homogeneous" first order operator.

Proof. Theorem 2.2 shows that the sum of these two spaces spans $D^1(F,\Gamma')$. We must show that they have no non-zero elements in common. Suppose then that $D \in D^0(F,\Gamma')$ is a derivation. Then,

$$D(ff') = fD(f') + f'D(f) = ff'D(1) = 2ff'D(1) \quad ,$$

forcing $D(1) = 0$, which forces $D = 0$.

Suppose now that Γ, Γ' are F-modules, and that $D: \Gamma \to \Gamma'$ is a differential operator. For $\gamma \in \Gamma$, set:

$$D^{\gamma}(f) = D(f\gamma) \quad . \tag{2.10}$$

Thus, D^{γ} can be considered on an R-linear map: $F \to \Gamma'$.

Theorem 2.4

If $D \in D^{r}(\Gamma,\Gamma')$, then, for fixed γ, D^{γ} belongs to $D(F,\Gamma')$.

Proof. Again, by induction on r. For $f' \in F$,

$$(D^{\gamma})_f(f')^{\gamma} = D^{\gamma}(ff') - fD^{\gamma}(f') = D(ff'\gamma) - fD(ff'\gamma)$$

$$= D_f(f'\gamma) = (D_f)^{\gamma}(f'), \text{ i.e. } (D_f)^{\gamma} = (D^{\gamma})_f \quad . \tag{2.11}$$

By induction hypothesis, since $D_f \in D^{r-1}(\Gamma,\Gamma')$, then $(D_f)^{\gamma} \in D^{r-1}(F,\Gamma')$, hence (2.11) proves that $(D^{\gamma})_f \in D^{r-1}(F,\Gamma')$, which shows that $D^{\gamma} \in D^{r}(F,\Gamma')$.

Definition

$D \in D^{1}(\Gamma,\Gamma')$ is a *homogeneous first order differential operator* if, for each $\psi \in \Gamma$, $D^{\psi} \in D^{1}(F,\Gamma')$ is a derivation of F into Γ'.

Now we turn to the description of $D^{2}(F,\Gamma')$. Given $f \in F$, by Theorem 2.2 there is a derivation: $F \to \Gamma'$ such that

$$D_f(f') = X_f(f') + D_f(1)f' \quad . \tag{2.12}$$

But,

$$D_f(f') = D(ff') - fD(f') \quad .$$

Hence,

$$D(ff') = X_f(f') + D_f(1)f' + fD(f') \quad . \tag{2.13}$$

Set $f' = 1$:

$$D(f) = D_f(1) + fD(1) \quad . \tag{2.14}$$

Thus,

$$D(ff') = X_f(f') + D_f(1)f' + fD(f') = X_{f'}(f) + D_{f'}(1)f + f'D(f) \quad .$$

Subtracting,

$$X_f(f') - X_{f'}(f) = f'(D_f(1) - D(f)) + f(D(f') - D_{f'}(1))$$

$$= \text{using (2.14)} \quad f'fD(1) - ff'D(1) = 0 \quad .$$

i.e.

$$X_f(f') = X_{f'}(f) \quad . \tag{2.15}$$

Theorem 2.5

$$X_{ff'} = fX_{f'} + f'X_f \ , \quad \text{for} \quad f, f' \in F \quad . \tag{2.16}$$

Proof.

$$X_{ff'}(f'') = \text{using (2.15)}, \ X_{f''}(ff') = X_{f''}(f)f' + fX_{f''}(f')$$

$$= X_f(f'')f' + fX_{f'}(f'') = (f'X_f + fX_{f'})(f'') \quad .$$

This proves (2.16).

Remark. Let $V(F,\Gamma')$ denote the F-module of derivations of F into Γ'. Then, (2.16) says that the map $:f \to X_f:$ defined by D determines an element of $V(F,V(F,\Gamma'))$.

We can now leave as an exercise to the reader showing that the decomposition (2.12) characterizes second order differential operators. One can also proceed further to study higher order operators by the same methods.

3. ALGEBRAIC STUDY OF SCHWINGER TERMS

Consider the "Sugawara model" commutation relations, (1.6-1.8). The second term on the right hand side of (1.8) is, of course, called a "Schwinger term". We will now attempt an analysis, in the language of Section 2, of this particular sort of "Schwinger term".

Let Γ be an F-module. Suppose that [,] is an R-bilinear map $:\Gamma \times \Gamma \to \Gamma:$ of the following form:

$$[\gamma_1,\gamma_2] = [\gamma_1,\gamma_2]_0 + \lambda D(\gamma_1,\gamma_2) \ , \quad \text{for} \quad \gamma_1,\gamma_2 \in \Gamma \tag{3.1}$$

where $[\ ,\ ]_0$ is an F-bilinear map $:\Gamma \times \Gamma \to \Gamma:$ which is a Lie algebra structure, and where D is a skew-symmetric, R-bilinear map $:\Gamma \times \Gamma \to \Gamma:$ that is a homogeneous first order differential operator. λ is a real parameter.

We will now investigate the validity of the Jacobi identity for [,] assuming that it is true for $[\ ,\ ]_0$. For $\gamma_1, \gamma_2, \gamma_3 \in \Gamma$ set:

$$T(\gamma_1,\gamma_2,\gamma_3) = [\gamma_1,[\gamma_2,\gamma_3]] - [[\gamma_1,\gamma_2],\gamma_3] - [\gamma_2,[\gamma_1,\gamma_3]] \tag{3.2}$$

$$= [\gamma_1,[\gamma_2,\gamma_3]] + \gamma_1 - [\gamma_2,[\gamma_1,\gamma_3]] + [\gamma_3,[\gamma_1,\gamma_2]] \quad . \tag{3.3}$$

Thus, (3.2) exhibits the relation of T to the "Jacobi identity", while (3.3) indicates how T is formed by permuting 1, 2, and 3 in the expression $[\gamma_1,[\gamma_2,\gamma_3]]$. Then, the following formula holds:

$$T = \frac{1}{2} \epsilon_{ijk}[\gamma_i,[\gamma_j,\gamma_k]] \quad . \tag{3.4}$$

We will now compute this explicitly, using (3.1).

$$[\gamma_1,[\gamma_2,\gamma_3]] = [\gamma_1,[\gamma_2,\gamma_3]_0 + \lambda D(\gamma_2,\gamma_3)]$$

$$= [\gamma_1,[\gamma_2,\gamma_3]_0]_0 + \lambda D(\gamma_1,[\gamma_2,\gamma_3]_0)$$

$$+ \lambda[\gamma_1,D(\gamma_2,\gamma_3)]_0 + \lambda^2 D(\gamma_1,D(\gamma_2,\gamma_3)) \quad . \tag{3.5}$$

Combining (3.4) and (3.5), together with the fact that the Jacobi identity is valid for $[\ ,\ ]_0$, gives the following formula:

$$T = \frac{1}{2} \lambda \, \epsilon_{ijk}[D(\gamma_i,[\gamma_j,\gamma_k]_0) + [\gamma_i,D(\gamma_j,\gamma_k)]_0 + \lambda D(\gamma_i,D(\gamma_j,\gamma_k))] \tag{3.6}$$

Then, if :T = 0: for all λ, we have

$$\epsilon_{ijk}(D(\gamma_i,[\gamma_j,\gamma_k]_0) + [\gamma_i,D(\gamma_j,\gamma_k)]_0) = 0 \tag{3.7}$$

$$\epsilon_{ijk}D(\gamma_i,D(\gamma_j,\gamma_k)) = 0 \quad . \tag{3.8}$$

Condition (3.7) is a cocycle-type condition. (See [8] for an explanation of the relation between the "deformation" of Lie algebra structures and Lie algebra cohomology theory.) It is not too clear what is the "general" meaning of condition (3.8), although certain simple ways of satisfying it can be readily presented.

Let us attempt to solve relations (3.7-3.8) with a special Ansatz which may be thought of as a general case of the Sugawara conditions (1.6-1.8). Namely, let us suppose that there is a fixed element labeled "γ_0" of Γ such that:

$$[\Gamma,\gamma_0]_0 = 0 \quad . \tag{3.9}$$

Suppose also that there is a homogeneous 1-differential operator $d\colon \Gamma \times \Gamma \to F$ such that:

$$D(\gamma_1,\gamma_2) = d(\gamma_1,\gamma_2)\gamma_0 \ , \quad \text{for } \gamma_1,\gamma_2 \in \Gamma \quad , \tag{3.10}$$

$$d(\gamma_1,\gamma_2) = -d(\gamma_2,\gamma_1) \tag{3.11}$$

$$d(\gamma_0,\gamma) = 0 \ , \quad \text{for} \quad \gamma \in \Gamma \quad . \tag{3.12}$$

Then, (3.12) guarantees that (3.8) is satisfied. (3.7) is the only condition that needs to be taken into account. Note that, in view of (3.9) and (3.11), (3.7) takes the following form:

$$d(\gamma_1,[\gamma_2,\gamma_3]_0) - d([\gamma_1,\gamma_2]_0,\gamma_3) - d(\gamma_2,[\gamma_1,\gamma_3]_0) = 0, \tag{3.13}$$

$$\text{for} \quad \gamma_1,\gamma_2,\gamma_3 \in \Gamma \quad .$$

4. THE SYMBOL OF DIFFERENTIAL OPERATORS ON VECTOR BUNDLES

We now aim to put the conditions found in Section 3 for the existence of "current algebras" on a slightly different foundation. Let M be a manifold. (See [14] for the notations and ideas of differential geometry to be used here.) Let F be the algebra of C^∞ real valued functions on M. As is well known, differential-geometric ideas can be described in two "languages", that of F-modules and that of vector bundles over M. It is important to be able to pass back and forth between them. The "symbol" of a differential operator expresses the operator-defined generally in the F-module language of Section 2 in terms of vector bundles.

Let $\pi: E \to M$ be a map between manifolds that defines E as a vector bundle over M. Let $\Gamma(E)$ denote the space of cross-section map: $M \to E$. Such cross-section maps can be added (because the fibers of π are vector spaces) and multiplied by functions in F, i.e. $\Gamma(E)$ is an F-module.

Suppose $D \in D^r(\Gamma(E),\Gamma(E'))$. Given a point $p \in M$, we will define the *symbol of* D *at* p, denoted by $\sigma(p,D)$: as an element of the fiber of a vector bundle defined over M, which depends on r.

For $r = 0$, proceed as follows. D is then an F-linear map: $\Gamma(E) \to \Gamma(E')$.

Lemma 4.1

If $\gamma \in \Gamma(E)$ vanishes at p, so does $D(\psi)$.

Proof. Suppose first that γ can be written as: $f\gamma_1$, where $\gamma_1 \in \Gamma(E)$, $f \in F$, and $f(p) = 0$. Then,

$$D(\gamma) = D(f\gamma_1) = fD(\gamma_1) \quad ,$$

hence

$$D(\gamma)(p) = f(p)D(\gamma_1)(p) = 0 \quad .$$

Using the local product structure for the vector bundle and a partition of unity for M, one sees that an arbitrary $\gamma \in \Gamma(E)$ that vanishes at p can be written as the sum of elements of the form $:f\gamma_1:$, hence the lemma is proved.

Let $E(p) = \pi^{-1}(p)$, $E'(p) = \pi^{1-1}(p)$ denote the fiber of the vector bundles over E. Then, the point-evaluation map defines R-linear map: $\Gamma(E) \to E(p)$, $\Gamma(E') \to E'(p)$. Lemma 4.1 shows that D (an element of $D^0(\Gamma(E), \Gamma(E'))$ passes to the quotient to define a linear map which we define as $\sigma(p,D)$ of $E(p) \to E'(p)$.

Now, suppose $r = 1$, and $D \in D^1(\Gamma(E), \Gamma(E'))$. For $f \in F$, define $D_f \in D^0(\Gamma(E),\Gamma(E'))$ as in Section 2. For $p \in M$, let M_p^* denote the vector space of cotangent vectors at p, i.e. M_p^* is the dual space to the tangent space M_p to M at p. Then, $df(p)$, the value at p of the differential of f, is an element of M_p^*.

Lemma 4.2

If $f(p) = 0$ and $df(p) = 0$, then $\sigma(p,D_f) = 0$.

Proof. For $\gamma \in \Gamma(E)$, recall that

$$D_f(\gamma) = D(f\gamma) - fD(\gamma) \quad .$$

Thus, since $f(p) = 0$,

$$D_f(\gamma)(p) = D(f\gamma)(p) \quad .$$

As we have proved, the map $f \to D(f\gamma)$ is a first order differential operator on f. Hence, if also $df(p) = 0$, then all first order derivatives of f vanish at p, hence: $D_f(\gamma)(p) = 0$. From the definition of $\sigma(p,D_f)$, we see that it is zero.

Thus, let $\theta \in M_p^*$, $v \in E(p)$. Let $f \in F$ be a function which vanishes at p, such that:

$$df(p) = \theta \quad .$$

Thus, we see from Lemma 4.2 that $\sigma(p,D_f)(v) \in E'(p)$ only depends on θ. Let us denote this element as follows:

$$\sigma(p,D)(\theta,v) = \sigma(p,D_f)(v) \tag{4.1}$$

It is readily seen that (4.1) defines $\sigma(p,D)$ as bilinear map: $M_p^* \times E(p) \to E'(p)$. This map is the symbol of D at p.

One can continue inductively to define the symbol of an r-th order operator. It is a multilinear map,

$$\sigma(p,D) = M_p^* \circ \ldots \circ M_p^* \times E(p) \to E'(p) \quad .$$

(See [7, 15].) Here, $\circ$ denotes "symmetric tensor product". However, for our immediate purpose in discussing "Schwinger terms" that only involve first order derivatives of delta functions--it suffices to deal with the cases $r = 0$ or 1, hence we will restrict our attention to these cases.

5. THE SYMBOL ASSOCIATED WITH CURRENT ALGEBRAS

Suppose now that M is a manifold; that $\pi: E \to M$ is a vector bundle over M; that F = the algebra of C^∞, real valued functions on M; and that $\Gamma(E)$ is the F-module of C^∞ cross-sections of E. Suppose that [,] is an R-bilinear, first-order differential operator $:\Gamma(E) \times \Gamma(E) \to \Gamma(E)$: on $\Gamma(E)$ that makes $\Gamma(E)$ into a "current algebra". (Thus, in the situation suggested by quantum field theory, M will be R^3, which can be identified with a space-like hypersurface in R^4, the manifold of space-time.) Let us suppose that:

$$[\gamma_1,\gamma_2] = D_0(\gamma_1,\gamma_2) + \lambda D_1(\gamma_1,\gamma_2) \quad , \quad \text{for} \quad \gamma_1,\gamma_2 \in \Gamma(E) \quad , \tag{5.1}$$

where D_0 and D_1 are zero and first order homogeneous differential operators: $\Gamma(E) \times \Gamma(E) \to \Gamma(E)$, and where λ is a real parameter. (Notice that we are changing our notations slightly from those used in Section 3. To make the identification, change $[\gamma_1,\gamma_2]_0$ to $D_0(\gamma_1,\gamma_2)$, $D(\gamma_1,\gamma_2)$ to $D_1(\gamma_1,\gamma_2)$). Let us also suppose that $D_0(\gamma_1,\gamma_2)$ satisfies the Jacobi identity; i.e.,

$$D_0(\gamma_1,D_0(\gamma_2,\gamma_3)) = D_0(D_0(\gamma_1,\gamma_2),\gamma_3) + D_0(\gamma_2,D_0(\gamma_1,\gamma_3)) \tag{5.2}$$

$$D_0(\gamma_1,\gamma_2) = -D_0(\gamma_2,\gamma_1) \tag{5.3}$$

Now, for $p \in M$, the symbol $\sigma(p,D_0)$ is a bilinear map: $E(p) \times E(p) \to E(p)$. The conditions (5.2-5.3) pass to the quotient to define analogous conditions on the symbol. Namely, they express the fact that $\sigma(p,D_0)$ for each $p \in M$ defines a Lie algebra structure on the fiber $E(p)$, i.e. E is a "bundle of Lie algebras". Let us then denote the Lie algebra bracket defined on $E(p)$ by $\sigma(p,\pi_0)$ by the notation: $[\ ,\ ]_p$.

Now, to express the fact that (5.1) defines a Lie algebra structure on $\Gamma(E)$ for each λ, one must impose condition (3.7) and (3.8). The symbol at p of the differential operator D_1 may be defined as follows:

$$\sigma(p,D_1)(\theta,v_1,v_2) = D_1(f\gamma_1,\gamma_2)(p) \quad , \tag{5.4}$$

where $f \in F$ satisfies: $f = 0$, $df(p) = 0$; $v = \gamma_1(p)$, $v_2 = \gamma_2(p)$.

Suppose now that γ_1, γ_2, γ_3 are elements of $\Gamma(E)$, with:

$$\gamma_i(p) = v_i \quad .$$

Then (3.7) implies the following conditions:

$$\epsilon_{ijk}[D_1(f\gamma_i,D_0(f\gamma_j,\gamma_k)) + D_0(f\gamma_i,D_1(f\gamma_j,\gamma_k))] = 0 \quad . \tag{5.5}$$

In turn, this implies the following condition on the symbol:

$$\epsilon_{ijk}([v_i\sigma(p,D_1)(\theta,v_j,v_k)]_p - \sigma(p,D_1)(\theta,[v_j,v_k]_p,v_i) = 0 \quad . \tag{5.6}$$

In turn, (5.6) readily interpretable in terms of Lie algebra cohomology, namely, the following result holds.

Theorem 5.1

For each $\theta \in M_p^*$ consider the skew-symmetric bilinear map

$$\omega_\theta = (v_1,v_2) \to \sigma(p,D_1)(\theta,v_1,v_2) \tag{5.7}$$

of $E(p) \times E(p) \to E(p)$ as a 2-cocycle associated with the adjacent representation of the Lie algebra structure defined on $E(p)$ by the bracket $[\ ,\ ]_p$. Then, condition (5.6) expresses the fact that ω_θ is a 2-cocycle.

This result illustrates the general technique one may use. Now, let us turn to consideration of more special sets of current algebras, immediate generalization of the Sugawara model relation, (1.7-1.9).

6. MODELS WITH C-NUMBER SCHWINGER TERMS

Suppose now that $F = F(R^3)$, the C^∞, real valued functions $:x \to f(x):$ of a real 3-vector x. Let Γ be an F-module. Suppose that $\mathcal{G}$ is a real subspace of Γ which has a real Lie algebra structure, denoted by $:[\ ,\ ]:$. Also suppose that there is an element, denoted by "1", of Γ, which is linearly independent from $\mathcal{G}$. (Thus, the multiples F1 are the "c-number" in the title of this section.) Let us suppose that there is an algebra structure for Γ, whose bracket is also denoted by $[\ ,\ ]$ such that:

$$[fX,f'Y] = ff'[X,Y] + B_i(X,Y)(\partial_i(f)f' - \partial_i(f')f)1 \ , \quad \text{for } X,\ Y \in \mathcal{G}, f,f' \in F \quad . \tag{6.1}$$

Here, the B_i are symmetric, bilinear maps: $\mathcal{G} \times \mathcal{G} \to R$. Again, notice that the Sugawara model relations, (1.7-1.9), are of this form. Our aim in this section is to investigate the conditions for Jacobi-identity type relations.

Thus, for $X,\ Y,\ Z \in G, f, f', f'' \in F$ set

$$T(X,Y,Z;\ f,f',f'') = [fX,[f'Y,f''Z]] - [f'Y,[fX,f''Z]] - [[fX,f'Y],f''Z] \quad . \tag{6.2}$$

Now,

$$[fX,[f'Y,f''Z]] = [fX,f'f''[Y,Z] + B_i(X,Y)(\partial_i(f')f'' - \partial_i(f''f')]$$

$$= ff'f''[X,[Y,Z]] + B_i(X,[Y,Z])(\partial_i(f)f'f'' - \partial_i(f'f'')f) \quad . \qquad (6.3)$$

Now, our goal in this section is not to derive the sort of condition considered in Section 5, but, a more general case that we can explain as follows.

Notice that T given by (6.2) is always a multiple of the element "1" of Γ. Now, we are ultimately interested in linear representations of the [,]-algebra structure on Γ, i.e. assignment of linear operators to elements of Γ in which the bracket [,] goes over into operator commutator. In order that this be possible, it is not essential that the Jacobi identity be satisfied, i.e., T be zero, in Γ, but that it be zero modulo a certain ideal zero. Now, for the sake of physical applications, it is desirable that all elements of the form $f1$, where f is a compact support function in F such that $:\int f(x)dx = 0:$, go over into the zero operator. Putting these remarks together, we see that it is desirable that T satisfy the following condition:

$$\int T(X,Y,Z;\ f,f',f'')(x)dx = 0 \ ,$$

$$\text{for } X,Y,Z \in G;\ f,f',f'' \text{ compact support functions} \quad . \qquad (6.4)$$

We shall call condition (6.4) the *up-to-a divergence Jacobi identity*. Presumably, the general symbol-type condition derived in Section 5 can be generalized to deal with this condition, but in this case it is just as easy to proceed directly; the general conditions will be investigated in a later publication.

In fact, notice from (6.3) that after integrating by parts

$$\int [fX,[f'Y,f''Z](x)dx = (\int ff'f''(x)dx)[X,[Y,Z]]$$

$$+ 2B_i(X,[Y,Z]) \int (\partial_i(f)f'f'')(x)dx$$

Hence,

$$\int T(X,Y,Z;\ f,f',f'')(x)dx = 2(B_i(X,[Y,Z]) \int \partial_i(f)f'f''dx$$

$$- 2B_i(Y,[X,Z]) \int \partial_i(f')ff''dx + 2B_i(Z,[X,Y]) \int \partial_i(f'')ff'dx$$

$$= \text{, after integrating by parts, } 2B_i(X,[Y,Z]) \int \partial_i(f)f'f''dx$$

$$+ 2B_i(Y,[X,Z]) \int (f'\partial_i(f)f'' + f'f\partial_i(f'')dx$$

$$+ 2B_i(Z,[X,Y]) \int \partial_i(f')f''dx \quad .$$

Thus, in order that (6.4) be satisfied, we must have the following relations.

$$B_i(X,[Y,Z]) + B_i(Y,[X,Z]) = 0 \ , \quad \text{for } X,Y,Z \in G \quad . \qquad (6.5)$$

Now, the skew-symmetry of the [,] bracket on Γ requires that B_i be a symmetric real valued form on $\mathcal{G} \times \mathcal{G}$. Thus, condition (6.5) requires that B_1, B_2, B_3 be symmetric bilinear forms on $\mathcal{G}$ that are invariant under the adjoint representation. For example, the Killing form on $\mathcal{G}$ is a candidate. More generally, it is known that each second order Casimir operator for $\mathcal{G}$ corresponds to such a form [13]. Thus, we see that the calculations of this section provide a general method for constructing one class of "current algebras" which satisfy the Jacobi identity up to a divergence. In fact, by specializing $\mathcal{G}$ and the form B_i suitably one obtains the Sugawara model relations, (1.7-1.9). (There the $\mathcal{G}$ is non-semisimple the direct sum of an abelian ideal and a subalgebra. It would perhaps be interesting to discuss the physical situations whose G itself is semisimple.)

Remark. In summary, we have provided in Sections 2-6 samples (without a definitive discussion) of the sort of work that must be done in order to classify "current algebras", from a purely algebraic point of view.

7. GENERAL REMARKS ABOUT DYNAMICS

What we have done so far is, a-priori, without great physical interest, since we as yet do not know enough data to make a Lorentz invariant theory. So far, we have been dealing with "currents" $\gamma_a(x)$ that are "functions" of a space point x. What is needed is some method for constructing objects $\gamma_a(x,t)$ that depend on space-time points in a Lorentz covariant manner.

Now, it is typical of the "current algebra" approach to physics that one approaches quantum field theory from the "Heisenberg picture" point of view. Thus, instead of regarding $\gamma_a(x,t)$ as "functions" of space time points, one ought to introduce test functions $F = F(R^3)$, as before, and objects of the following form:

$$\gamma_a(f,t) = \int \gamma_a(x,t) f(x) dx \quad .$$

Thus, if Γ denotes the Γ-module spanned by the $\gamma_a(f)$, one might expect to see "dynamics" defined by curves $t \to \gamma_a(f,t)$ in Γ, defined by differential equations, say of the form

$$\frac{\partial}{\partial t} \gamma_a(f,t) = [h, \gamma_a(f,t)] \quad , \tag{7.1}$$

where h is an element of Γ (the "Hamiltonian") and where [,] is an algebra structure on Γ of the "current algebra" type.

Unfortunately, this hope is too simple minded. In model situations, (say the Sugawara model) h is of the following formal form:

$$h = \int h_{ab}\gamma_a(x)\gamma_b(x)dx \quad . \tag{7.2}$$

Now, the bracket of something quadratic of the form (7.2) with $\gamma_a(f)$ goes "outside" of Γ.

In fact, what is required is a construction of the following type: Imbedded Γ as a submodule of an F-module Γ', and find an $h \in \Gamma'$ and a bracket [,] in Γ' so that the "dynamics" is given by (7.1).

In the next few sections we will sketch the construction of such a Γ' in a general situation suggested by the Sugawara model, namely, we will attempt to define "polynomial" objects like (7.2) in a consistent algebraic way.

8. POLYNOMIALS OF CURRENTS

To see what is involved mathematically in carrying out the construction of the F-module Γ' suggested in Section 7, consider the following Sugawara model type of commutation relations:

$$[v_a(x), v_b(y)] = c_{abc}v_c(x)\delta(x - y) - d_{abi}\partial_i^x\delta(x - y) \quad . \tag{8.1}$$

Introduce the symbols $:v_a(f):$ for $f \in F = F(R^3)$, as follows:

$$v_a(f) = \int v_a(x)f(x)dx \quad . \tag{8.2}$$

Let Γ be the F-module spanned by the $v_a(f)$. Then, the bracket in Γ is defined, consistently with (8.1) and (8.2), as follows:

$$[v_a(f), v_b(f')] = c_{abc}v_c(ff') + \frac{1}{2} d_{abi}(\partial_i(f)f' - \partial_i(f')f) \quad . \tag{8.3}$$

Now, introduce new objects of the following sort:

$$v_{ab}(f) = \int f(x)v_a(x)v_b(x)dx$$

$$v_{abc}(f) = \int f(x)v_a(x)v_b(x)v_c(x)dx \tag{8.4}$$

and so forth.

Also, introduce the "partial derivatives" $\partial_i v_a(x), \partial_{ij}v_a(x), \ldots,$ so that the following algebraic rules are satisfied:

$$(\partial_i v_a)(f) = \int \partial_i v_a(x)f(x)dx = -\int v_a(x)\partial_i f(x)dx = -v_a(\partial_i(f)) \tag{8.5}$$

$$(\partial_i\partial_j v_a)(f) = v_a(\partial_j\partial_i(f)) \tag{8.6}$$

and so forth.

Now,

$$[v_{ab}(f),v_c(y)] = \int f(x)[v_a(x)v_b(x),v_c(y)]dx = \int f(x)([v_a(x),v_c(y)]v_b(x)$$
$$+ v_a(x)[v_b(x),v_c(y)])dx \int f(x)([c_{acd}v_d(x)\delta(x-y) - d_{aci}\partial_i^x\delta(x-y)]v_b(x)$$
$$+ v_a(x)[c_{bcd}v_d(x)\delta(x-y) - d_{bci}\partial_i^x\delta(x-y)])dx = f(y)c_{acd}v_d(y)v_b(y)$$
$$+ d_{aci}(\partial_i(f)(y)v_b(y) + f(y)\partial_i v_b(y)) + f(y)c_{bcd}v_a(y)v_d(y)$$
$$+ d_{bci}(\partial_i(f)(y)v_a(y) + f(y)\partial_i v_a(y)) = \partial_i(f)(y)(d_{aci}v_b(y) + d_{bci}v_a(y))$$
$$+ f(y)(c_{acd}v_d(y)v_b(y) + c_{bcd}v_a(y)v_d(y) + d_{aci}\partial_i v_b(y) + d_{bci}\partial_i v_a(y)) \quad . \tag{8.7}$$

In particular, for $f,f' \in F$,

$$[v_{ab}(f),v_c(f')] = d_{aci}v_b(\partial_i(f)f') + d_{bci}v_a(\partial_i(f)f') + c_{acd}v_{db}(ff')$$
$$+ c_{bcd}v_{ad}(ff') + d_{aci}\partial_i(v_b)(ff') + d_{bci}(\partial_i v_a)(ff') \quad . \tag{8.8}$$

Now, introduce Γ' as the vector space spanned by all "polynomials" of the following form:

$$v_{a_1\ldots a_r}(f) = \int v_{a_1}(x)\ldots v_{a_r}(x)f(x)dx \quad . \tag{8.9}$$

One can calculate commutation relations of the following form:

$$[v_{a_1\ldots a_r}(f),v_{b_1\ldots b_s}(f')] \quad , \tag{8.10}$$

using the calculations that led into (8.8) as a pattern. Notice that again Γ' is an F-module (multiply $f \in F$ by $v_{a_1}\ldots v_{a_r}(f')$ to get $v_{a_1\ldots a_r}(ff')$, and the formula for the bracket (8.10) will be of the type that we have called "current algebra" bracket, i.e., will involve differential operator: $\Gamma' \times \Gamma' \to \Gamma'$. (Notice that Γ' is some sort of generalization "universal enveloping algebra" of a Lie algebra.)

Thus we have explained the algebraic background of the work of Sommerfield and Sugawara [18,21]. They showed that a Lorentz invariant dynamical theory could be obtained in which the energy-momentum tensor was a second degree polynomial in the currents. Of course, actually solving these equations is enormously difficult, with no kind of a procedure or approximation method available, and the whole theory is, as of right now, therefore useless from the view point of the practical physicist. However, there is an important point of principle involved. The Sugawara model--and others that one may construct using the generalized procedure sketched here--is a theory in which the dynamics is determined completely by the currents. If one believes that the "currents", and not the "fields", are the basic mathematical and/or physical objects involved in the interaction and classification of elementary particles, then a theory in which the equations of motion can be expressed strictly in terms of the currents is very attractive.

In the Sugawara model, these equations of motion have a very interesting classical analogue. Let G be a Lie group, whose Lie algebra is that described by the structure constants "c_{abc}" appearing in the current algebra commutation relations. Bardacki and Halpern, for special choices of G, and the author in general, have shown [3, 9] that the Lagrangian which gives rise in the simplest way (it is still unknown whether there are other Lagrangians which also do so) to the Sugawara model has as its classical externals the space of *harmonic maps:* $R^4 \to G$, in the sense of Eels and Sampson [5].

We will briefly explain what is involved here. Eels and Sampson define the concept of a harmonic map $\varphi: N \to M$ between two Riemannian manifolds. In this case, the system of differential equations defining φ is a system of elliptic partial differential equations--in general, non-linear--which, as the name indicates, generalize the concept of "harmonic function". (In fact, the harmonic map $\varphi: R^n \to R$, with the Euclidean metric on R^n and R, are the harmonic functions in the usual sense.)

Now, their definition makes perfectly good sense in the case either N or M or both are pseudo-Riemannian manifolds. For example, take $N = R^4$, space-time, with the Lorentz metric, and take $M = G$, a compact, semisimple Lie group, with the bi-invariant metric defined by the Killing form on G. Then, the differential equations defining the harmonic maps are identical with Sugawara's, and form a non-linear, hyperbolic system. Unfortunately, very little seems to be known about such systems. Perhaps their possible usefulness as equations for elementary particles will stimulate some relevant mathematics.

9. CURRENTS AS FUNCTIONS ON JET SPACES

Up to this point, all of our efforts have gone into explaining independently of quantum field theory the mathematical nature of currents. In fact, one of the most useful features of current algebra theory is the fact that it throws a new, more algebraic and geometric light on the more traditional aspects of quantum field theory. In this section, we will explain how currents arise in the context of classical field theory.

First we must explain briefly the differential geometric notion of a "jet" of a mapping. (See [12, 15] for more details.) Let E and M be manifolds, and let $\pi: E \to M$ be a mapping of E onto M. Let N be another manifold. The ordered set (E, M, π, N) is said to define a (local product) fiber space if each point p of M has a neighborhood U in which $\pi: \pi^{-1}(U) \to U$ looks like the

Cartesian projection map: $U \times N \to U$. Then, a "fiber space" is a "globalization" of a product space.

Let $\Gamma(E)$ denote the space of cross-section maps, i.e., $\psi \in \Gamma(E)$ is a map: $M \to E$ such that:

$$\pi\psi(p) = p \quad \text{for} \quad p \in M \quad ,$$

i.e., $\psi(p) \in E(p) = \pi^{-1}(p)$, the "fiber" of E over p.

Now, if E were the product $M \times N$, it should be clear that the elements of $\Gamma(E)$ can be written precisely in the form:

$$p \to (p, \psi'(p)) \quad ,$$

where ψ' is a map: $M \to N$. Then the notion of "cross-section" is a "globalization" of the idea of mapping between two spaces.

Suppose now that ψ, ψ' are two elements of $\Gamma(E)$, and p is a point of M. Let us say that ψ and ψ' *agree to the first order at* p *if:*

a) $\psi(p) = \psi'(p)$

b) In terms of a local product structure in a neighborhood U of p, with ψ, ψ' identified with maps: $U \to N$, the partial derivatives of ψ and ψ' of first order agree at p.

Definition. Consider the following equivalence relation on $M \times \Gamma(E)$: (p,ψ) is equivalent to (p',ψ') if and only if

a) $p = p'$

b) ψ and ψ' agree to the first order at p. (9.1)

Then, $J^1(E)$, the *manifold of first order jets of cross-sections*, is defined as the quotient of $M \times \Gamma(E)$ by the equivalence relation given by (9.1).

As shown in [11] and [12], the manifold $J^1(E)$ is the appropriate one for consideration of the calculus of variation problems underlying quantum field theory. For example, a "Lagrangian" is just a real-valued function: $J^1(E) \to R$.

If $\psi \in \Gamma(E)$, denote by $j^1(\psi)$ (its "one-jet") as a mapping: $M \to J^1(E)$ defined as follows:

$j^1(\psi)(p)$ = equivalence class to which the point (p,ψ) belongs. (9.2)

Then, if $L: J^1(E) \to R$ is a "Lagrangian", if dx is a volume element form for M, if $\psi \in \Gamma(E)$, then:

$$L(\psi) = \int_M L(j^1(\psi)(x))dx \tag{9.3}$$

is the value assigned by L to the cross-section ψ.

In order to establish the equivalence with the more usual formulas of field theory, we must introduce coordinate systems for M and E. Suppose that $M = R^4$. Let $x = (x_\mu)$, $0 \le \mu, \nu \le 3$, be Euclidean coordinates for R^4. Suppose also that (φ_a), $1 \le a, b \le n$, is a coordinate system for the fiber N.

We will define a coordinate system, that we will label $(x_\mu, \varphi_a, \varphi_{b\mu})$ for $J^1(E)$ in the following way. Suppose that (p,ψ) is an element of $M \times \Gamma(E)$. We will define the values of these functions on this point:

a) x_μ are the Euclidean coordinates of the point p.

b) φ_a are the φ-coordinates of the point $x(p)$.

c) $\varphi_{a\mu}$ are the derivatives $\dfrac{\partial\varphi_a(x)}{\partial x_\mu}$ of the function $x \to \varphi_a(x)$ which determine ψ locally as a map: $R^4 \to N$.

Thus, the Lagrangian L becomes a function $L(x,\varphi_a,\varphi_{a\mu})$ of the indicated variables. If $\psi \in \Gamma(E)$, with functions $x \to (\varphi_a(x))$ defining ψ locally, then (9.2) takes the following more classical form:

$$L(\psi) = \int L(x,\varphi_a(x), \partial_\mu\varphi_a(x))dx \quad . \tag{9.4}$$

Suppose we are given such a Lagrangian L. Define functions L_a, $L_{a\mu}$ on $J^1(E)$ as follows:

$$L_a = \frac{\partial L}{\partial\varphi_a}$$

$$L_{a\mu} = \frac{\partial L}{\partial\varphi_{a\mu}} \quad .$$

Then, a cross-section determined by functions $:x \to \varphi_a(x):$ is an *extremal* if it satisfies the following differential equations (called the *Euler-Lagrange equations*):

$$\frac{\partial}{\partial x_\mu}(L_{a\mu}(x,\varphi(x),\partial\varphi(x)) = L_a(x,\varphi(x),\partial\varphi(x)) \quad . \tag{9.5}$$

We now proceed to show how "currents" may be defined. Suppose X is a vector field on the manifold E (see [14] for differential geometric notions, such as vector field) of the following form:

$$X = A_\mu(x)\frac{\partial}{\partial x_\mu} + A_a(x,\varphi)\frac{\partial}{\partial\varphi_a} \quad , \tag{9.6}$$

where A_μ, A_a are functions of the indicated variables. (Geometrically, vector fields of the form (9.6) generate one parameter groups of transformations of E that act on M and permute the fibers of E over this action; they may be called "fiber space automorphisms".)

We can now define a "prolonged" vector field X' on $J^1(E)$, by the following formula:

$$X' = A_\mu\frac{\partial}{\partial x_\mu} + A_a\frac{\partial}{\partial\varphi_a} + \left(\frac{\partial A_a}{\partial x_\mu} - \varphi_{a\nu}\frac{\partial A_\nu}{\partial x_\mu} + \frac{\partial A_a}{\partial\varphi_b}\varphi_{b\mu}\right)\frac{\partial}{\partial\varphi_{a\mu}} \quad . \tag{9.7}$$

This prolongation process is a Lie algebra homomorphism: $V(E) \to V(J^1(E))$, i.e.

$$[X,Y]' = [X'Y'] \quad , \tag{9.8}$$

if X, Y are vector fields on h of form (9.5).

Suppose now that $f \in F(M)$. Then,

$$(fX)' = fX' + \frac{\partial f}{\partial x_\mu} (A_a - \varphi_{a\nu} A_\nu) \frac{\partial}{\partial \varphi_{a\mu}} \quad . \tag{9.9}$$

Thus, if L is a Lagrangian,

$$(fX')(L) = fX'(L) + \frac{\partial f}{\partial x_\mu} (A_a - \varphi_{a\nu} A_\nu) L_{a\mu} \quad . \tag{9.10}$$

In particular, suppose that

$$A_\mu = 0; \; X'(L) = 0 \quad . \tag{9.11}$$

This means, geometrically, that if the one parameter group of automorphisms of E generated by X map the fibers of E into themselves, and the group is a one-parameter group of "internal symmetries" of the Lagrangian L, then

$$(fX)'(L) = \frac{\partial f}{\partial x_\mu} (A_a L_{a\mu}) \quad . \tag{9.12}$$

Now, $A_a L_{a\mu} = V_\mu^X$ is the very familiar formula in quantum field theory for the "vector current" generated by a one-parameter group of symmetries.

In general then, we might associate to each vector field X of form (9.7) the following set of functions on $J^1(E)$:

$$V_\mu^X = A_a L_{a\mu} - L_{a\mu} \varphi_{a\nu} A_\nu \quad . \tag{9.13}$$

This method of defining "currents" in classical field theories may be compared to the now-classical work of Belinfante and Rosenfeld [4, 17]. Now that we have seen that "currents" at the level of classical field theory may be interpreted as functions on the jet spaces, the road is open to use the current commutation relations of quantum field theory to define a "Poisson bracket" operation for functions on the jet spaces. However, we will not pursue this topic here. Instead, we will turn to the study of another related connection between "current algebras" and differential geometry.

10. REPRESENTATIONS OF GAUGE ALGEBRAS BY DIFFERENTIAL OPERATORS

Now we turn to the question of representing current algebras in a natural geometric way--as differential operators on manifolds. This corresponds, roughly, to finding their physical consequences as *classical* dynamical systems. The problem of realizing them irreducibly as operators on Hilbert space is related to their consequences in quantum mechanics, and is a much more difficult (and still unsolved) technical problem. (See the work of Araki, Streater and Wulfsohn [2, 19, 20].)

Now, part of our "grand design" is to see how "current algebras" arise in a natural geometric way. Indeed, I feel that this study will have interesting repercussions in "pure" differential geometry. (Of course, differential geometry used to be not unrelated to events in physics. However, there has been a period of introspection in the last twenty years, and now most of the active workers in this field know nothing of these roots.) Lie algebras first arose in mathematics, in the works of S. Lie, as Lie algebras of differential operators on finite dimensional manifolds. It is still an interesting, unsolved mathematical problem to classify the possible ways a given Lie algebra can so act. In the next few sections we will treat a fragment of this problem for the sorts of Lie algebras (or their generalizations, i.e., algebras satisfying the Jacobi identity up to a divergence) encountered in current algebra theory. In this section, we will treat the simplest case--where the "current algebra" contains no "Schwinger terms", hence is what might be called a "gauge algebra". Precisely, let us adopt the following definition.

Definition. Let F be a commutative, associative algebra over the real numbers, and let Γ be an F-module. A real Lie algebra structure [,] on Γ is said to define a *gauge algebra* if the bilinear map $(\gamma_1,\gamma_2) \to [\gamma_1,\gamma_2]$ of $\Gamma \times \Gamma \to \Gamma$ is also F-linear.

This concept is most useful when combined with the idea of a "free" F-module.

Definition. Let V be a real subspace of Γ, and let

$$\alpha: V \otimes F \to \Gamma$$

be the linear map constructed as follows:

$$\alpha(v)(f) = fv \,, \quad \text{for } v \in V,\ f \in F \quad . \tag{10.1}$$

($\otimes$ denotes the tensor product defined with the real numbers as ground field.) Then, Γ is said to be a *free F-module* with *basis space* V if the map α defined by (10.1) is an isomorphism.

The modules which arise as "current algebras" in physical situations are usually also "free". If this is the case, and if V is a basis space, let us use the following notation; suggested by the physicists' notation:

$$v(f) = fv = \alpha(v \otimes f) \,, \quad \text{for } f \in F,\ v \in V \quad . \tag{10.2}$$

Let $\mathcal{G}$ be a real Lie algebra.

Definition. An F-module Γ is a *free gauge algebra based on the Lie algebra of charges* $\mathcal{G}$ if the following conditions are satisfied:

a) Γ is a free F-module, with basis subspace V.

b) Γ is a gauge Lie algebra, in the sense defined above.

c) With respect to the Lie algebra bracket [,] defined by b), V is a Lie subalgebra of the real Lie algebra Γ.

d) $\mathcal{G}$ is isomorphic, as a Lie algebra, to the Lie subalgebra V.

Now, let us suppose that $F = F(R^3)$. Let $\pi: M \to R^3$ be a map from a manifold M to R^3 that defines M as a fiber space over R^3. If $f \in F = F(R^3)$ is a function on the base space, R^3, then $f \to \pi^*(f)$ defines an imbedding of F or a subalgebra of F(M). In turn, this enables us to consider the tensor fields on M as F-modules. For example, if X is a vector field on M, i.e., an element of V(M), and $f \in F$, we denote by "fX" the product of the function $\pi^*(f)$ and the vector field X.

Now, suppose that Γ is a free gauge Lie algebra with the basis Lie subalgebra V. Thus, the following commutation rules are satisfied:

$$[v_1(f_1),v_2(f_2)] = [v_1,v_2](f_1f_2) \ , \quad \text{for} \quad v_1,v_2 \in V;\ f_1,f_2 \in F \ . \tag{10.3}$$

We will now attempt to find a homomorphism h of the Lie algebra Γ-defined by the commutation rules (10.3) - into the Lie algebra V(M) of vector fields on the manifold M. In fact, we will restrict ourselves, at this point at least, to the search for h of the following form:

$$h(v(f)) = fX_v + \partial_i(f)X_v^i \ . \tag{10.4}$$

For $v \in V$, X_v and X_v^i are vector fields on M. The map $v \to X_v$ and X_v^i then define linear mappings: $V \to V(M)$. We will also suppose that

$$X_v(\pi^*(f)) = 0 = X_v^i(\pi^*(f)) \ , \quad \text{for} \quad v \in V,\ f \in F(= F(R^3)) \ . \tag{10.5}$$

Comparing (10.3-10.5), we can readily write down the conditions that h be a Lie algebra homomorphism:

$$\begin{aligned} h([v_1(f_1),v_2(f_2)]) &= [h(v_1(f_1),h(v_2(f_2))] = [f_1X_{v_1} + \partial_i(f_1)X^i_{v_1},f_2X_{v_2} \\ &+ \partial_i(f_2)X^i_{v_2}] = f_1f_2[X_{v_1},X_{v_2}] + \partial_i(f_1f_2[X^i_{v_1},X_{v_2}] + f_1\partial_i(f_2)[X_{v_1},X^i_{v_2}] \\ &+ \partial_i(f_1)\partial_j(f_2)[X^i_{v_1},X^j_{v_2}] = h([v_1,v_2](f_1f_2)) = f_1f_2X_{[v_1,v_2]} \\ &+ \partial_i(f_1f_2)X^i_{[v_1,v_2]} \end{aligned}$$

Thus, the conditions that h be a homomorphism read as follows:

$$[X_{v_1},X_{v_2}] = X_{[v_1,v_2]} \tag{10.6}$$

$$[X^i_{v_1}, X^i_{v_2}] = 0 \tag{10.7}$$

$$[X^i_{v_1}, X_{v_2}] = X^i_{[v_1,v_2]} \quad . \tag{10.8}$$

(Condition (10.8) can be analyzed further in terms of the cohomology of the Lie algebra V but we will not go into that here.)

In summary, we have presented in this section a geometric method for realizing gauge Lie algebras by means of differential operators. Of course, the method can be generalized considerably beyond what has been presented in this section. What we have done amounts to an illustrative example. Our main immediate goal is to lead into the work of the next section on "Schwinger terms".

11. Schwinger Terms for Gauge Lie Algebras

Continue with the notations of Section 10. Let Γ be a free gauge Lie algebra, with basis subalgebra V, i.e., the commutation relations for the Lie algebra structure on Γ take the form (10.3). Let M be, as in Section 10, a fiber space over R^3.

Now, let $D^1(M)$ denote the Lie algebra of first order inhomogeneous differential operators on M. Let us modify the definition (10.4) of h, to define the linear mapping $h'\colon \Gamma \to D^1(M)$, as follows:

$$h'(v(f)) = fX_v + \partial_i(f)X^i_v + fk_v \,, \quad \text{for } f \in F,\ v \in V \quad . \tag{11.1}$$

In (11.1), X_v, X^i_v are vector fields on M; k_v is a zero-th order differential operator on M, i.e., a function on M. Thus, $v \to k_v$ defines a linear mapping of $V \to F(M) = D^0(M)$.

Let us also suppose that conditions (10.5) are satisfied. (They mean, geometrically, that the vector fields X_v, X^i_v are tangent to the fibers of the map $\pi\colon M \to R^3$.) Then, for $v_1, v_2 \in V$; $f_1, f_2 \in F$,

$$\begin{aligned}[h'(v_1(f_1)), h'(v_2(f_2))] &= f_1f_2[X_{v_1}, X_{v_2}] + \partial_i(f_1)f_2[X^i_{v_1}, X_{v_2}] \\ &+ f_1\partial_i(f_2)[X_{v_1}, X^i_{v_2}] + \partial_i(f_1)\partial_j(f_2)[X^i_{v_1}, X^j_{v_2}] - f_1(f_2X_{v_2}(k_{v_1}) \\ &+ \partial_i(f_2)X^i_{v_2}(k_{v_1})) \quad . \end{aligned} \tag{11.2}$$

Let us suppose, as in Section 10, that V is a Lie algebra. Thus, using (10.3), Γ can be made into a Lie algebra, with a bracket denoted by [,]. Let Γ' be the direct sum of Γ and F itself. Define a "new" bracket for Γ', denoted by [,]', by the following formula:

$$[v_1(f_1),v_2(f_2)]' = [v_1,v_2](f_1f_2)$$

$$+ \beta_i(v_1,v_2)\partial_i(f_1)f_2 - \beta_i(v_2,v_1)f_1\partial_i(f_2) \quad . \tag{11.3}$$

(β_i are bilinear maps: $V \times V \to R$. Then,

$$h'([v_1(f_1),v_2(f_2)]') = f_1f_2X_{[v_1,v_2]} + \partial_i(f_1f_2)X^i_{[v_1,v_2]}$$

$$+ \beta_i(v_1,v_2)\partial_i(f_1)f_2 - \beta_i(v_2,v_1)f_1\partial_i(f_2) \quad . \tag{11.4}$$

Let us now equate (11.2) and (11.4). This imposes the following conditions:

$$[X_{v_1},X_{v_2}] = X_{[v_1,v_2]} \tag{11.5}$$

$$[X^i_{v_1},X^j_{v_2}] = 0 \tag{11.6}$$

$$[X^i_{v_1},X_{v_2}] = X^i_{[v_1,v_2]} \tag{11.7}$$

$$X_{v_1}(k_{v_2}) = X_{v_2}(k_{v_1}) \tag{11.8}$$

$$X^i_{v_1}(k_{v_2}) = \beta_i(v_1,v_2) \tag{11.9}$$

What we have done now is to find the conditions, namely (11.5-11.9), that the set of first order differential operators, of the form (11.1), satisfy the commutation relations whose "abstract" structure relations are given by (11.3). Now we turn to the study of more specific structures of this sort, which arise in the study of the current algebras of quantum field theory.

12. THE CURRENT ALGEBRAS OF QUANTUM FIELD THEORY

Let us now change notations slightly. Choose the following range of indices, together with the corresponding summation conventions

$$1 \le a,\ b \le m;\ 1 \le i,\ j \le 3;\ 0 \le \mu,\ \nu \le 3 \quad .$$

Let $x = (x_i)$, $y = (y_i)$ denote elements of R^3. Let $F = F(R^3)$ be the commutative, associative algebra of real-valued functions on R^3.

Consider objects that are labeled as follows:

$$v^a_\mu(x)$$

Typically, they are "currents" associated with a Lie algebra of symmetries of a physical system. One aspect of the "current algebra" approach to quantum field theory is an attempt to construct Lie algebras from these objects, and investigate how these abstract Lie algebras are realized in terms of physical systems. In this final section of this paper, I will rework some of the ideas in a previous paper of mine [10], in the algebraic language developed here.

First of all, for the currents constructed from "Noether's theorem" (essentially equivalent to the material presented in Section 9), using the most common sort of Lagrangians, the "time" components of the current satisfy the following commutation relations:

$$[v_0^a(x), v_0^b(y)] = c_{abc} v_0^c(x)\delta(x - y) \quad . \tag{12.1}$$

Here, the "c_{abc}" are structure constants of a semisimple Lie algebra $\mathcal{G}$.

Second, postulate the following time-space commutation relations:

$$[v_0^a(x), v_j^b(y)] = c_{abc} v_i^c(x)\delta(x - y) - \partial_j^x(v_{ij}^{ab}(x)\delta(x - y)) \quad . \tag{12.2}$$

In (12.2) the $v_{ij}^{ab}(x,y)$ are objects that are model dependent.

Let us put the commutation relations (12.1-12.2) into "module" form. Introduce

$$v_0^a(f) = \int v_0^a(x) f(x) dx$$

$$v_i^a(f) = \int v_i^a(x) f(x) dx$$

$$v_{ij}^{ab}(f_1, f_2) = \int v_{ij}(x) f_1(x) f_2(x) dxdy. \tag{12.3}$$

Then, (12.1) - (12.2) take the following form:

$$[v_0^a(f_1), v_0^b(f_2)] = c_{abc} v_0^c(f_1 f_2) \tag{12.4}$$

$$[v_0^a(f_1), v_j^b(f_2)] = c_{abc} v_i^c(f_1 f_2) + v_{ij}^{ab}(\partial_j(f_1), f_2) \quad . \tag{12.5}$$

Let us now try to find realizations of the commutation relations (12.4-12.5). Let Γ be an F-module. We will construct a mapping of the objects $v_0^a(f), v_1^a(f)$ into the space $D_0(\Gamma,\Gamma)$ of F-linear mappings: $\Gamma \to \Gamma$.

Set:

$$\rho(v_0^a(f)) = fA^a + \partial_i(f)A_j^a \quad , \tag{12.6}$$

where A_a, A_a^j are operators in $D_0(\Gamma,\Gamma)$. Thus, following the pattern described in Section 10, it is readily verified that the following conditions are equivalent to (12.4):

$$[A^a, A^b] = c_{abc} A^c \quad , \tag{12.7}$$

$$[A_i^a, A_j^b] = 0 \quad , \tag{12.8}$$

$$[A_i^a, A^b] = c_{abc} A_j^c \quad . \tag{12.9}$$

Now, let us attempt to satisfy (12.5) by means of the following assignment:

$$\rho(v_i^a(f)) = fB_i^a \quad , \tag{12.10}$$

with $B_i^a \in D_0(\Gamma,\Gamma)$. Then,

$$[\rho(v_0^a(f_1)),\rho(v_i^b(f_2))] = [f_1 A^a + \partial_j(f_1)A_j^a, f_2 B_i^b]$$

$$= f_1 f_2 [A^a, B_i^b] + \partial_j(f_1) f_2 [A_j^a, B_i^b] \quad .$$

Then, we see that ρ will be a representation of the commutation relations (12.5) provided that:

$$[A^a, B_i^b] = c_{abc} B_i^c \tag{12.11}$$

$$v_{ij}^{ab}(f_1,f_2) = \partial_j(f_1) f_2 [A_j^a, B_i^b] \quad . \tag{12.12}$$

To obtain a model having common features of the "Sugawara model", one can further require that the operators $[A_j^a, B_i^b]$ commute with the operators A^a, B_i^b. A method for an explicit realization of these operators in terms of differential operators has been presented in [10], to which we refer for further details. The next step in this program would be to search for more general (possibly even the most general) realizations of this sort, a task we will attempt in volume III of [12].

REFERENCES

[1] Adler, S. and Dashen, R., *Current Algebras*, W. A. Benjamin, New York (1968).

[2] Araki, H., *Factorizable Representations of Current Algebra*, preprint, (1969).

[3] Bardacki, H., and Halpern, M., *Phys. Rev.*, 172, 1542 (1968).

[4] Belinfonte, F., *Physica*, 7, 449-474 (1940).

[5] Eels, J. and Sampson, J., "Harmonic Mappings of Riemannian Manifolds", *Amer. J. Math*, 86, 109-160 (1964).

[6] Gell-Mann, M. and Ne'eman, Y., *The Eightfold Way*, W. A. Benjamin, New York (1964).

[7] Goldschmidt, H., "Existence Theorems for Analytic Linear Partial Differential Equations", *Ann. of Math.*, 86, 246-270 (1967).

[8] Hermann, R., "Analytic Continuation of Group Representations", *Comm. in Math. Phys.*; "Part I", 2, 251-270 (1966); "Part II", 35, 53-74 (1966); "Part III", 3, 75-97 (1966); "Part IV", 51, 131-156 (1967); "Part V", 5, 157-190 (1967); "Part VI", 6, 205-225 (1967).

[9] Hermann, R., *Phys. Rev.*, 177, 2449 (1969).

[10] Hermann, R., "Current Algebras, the Sugawara Model, and Differential Geometry" to appear *J. Math. Phys.*

[11] Hermann, R., *Lie Algebras and Quantum Mechanics*, to appear W. A. Benjamin, New York.

[12] Hermann, R., *Vector Bundles for Physicists*, to appear, W. A. Benjamin, New York.

[13] Hermann, R., *Lie Groups for Physicists*, W. A. Benjamin, New York (1966).

[14] Hermann, R., *Differential Geometry and the Calculus of Variations*, Academic Press, New York (1968).

[15] Palais, R., *Global Analysis*, W. A. Benjamin, New York (1968).

[16] Renner, B., *Current Algebras and their Applications*, Pergamon Press, London (1968).

[17] Rosenfeld, L., "Sur le tenseur d'impulsion energie", *Acad. Roy. Belgique, CL. Sci. Mem. Coll.*, 18, (FASC 6), 30 pp (1940).

[18] Sommerfeld, C., *Phys. Rev.*, 176, 2019 (1968).

[19] Streater, R., *Nuovo Cimento*, 53, 487-495 (1968).

[20] Streater, R. and Wulfsohn, A., *Nuovo Cimento*, 57, 330-339 (1968).

[21] Sugawara, H., *Phys. Rev.*, 170, 1659 (1968).

ATTENDEES

1969 RENCONTRES

Battelle Seattle Research Center
Seattle, Washington

Dr. Valentine Bargmann
Department of Physics
Palmer Physical Laboratory
Princeton University
P. O. Box 708
Princeton, New Jersey 08540

Dr. Arno Böhm
Department of Physics
University of Texas, Austin
Austin, Texas 78712

Dr. Michael Boon
Institut Battelle
7, route de Drize
1227 Carouge/Geneve
Switzerland

Prof. Leon Ehrenpreis
Courant Institute of Mathematical Sciences
New York University
251 Mercer Street
New York, New York 10012

Dr. Dimitri I. Fotiadi
Centre de Physique Theorique
Ecole Polytechnique
17, rue Descartes
75 Paris V
France

Dr. M. L. Glasser
Battelle Memorial Institute
505 King Avenue
Columbus, Ohio 43201

Dr. Gerald A. Goldin
Department of Physics
David Rittenhouse Laboratory
University of Pennsylvania
Philadelphia, Pennsylvania 19104

Dr. Irvin Grodsky
Department of Physics
Cleveland State University
Cleveland, Ohio

Professor Robert Hermann
Department of Mathematics
Northwestern University
Evanston, Illinois 60201

Mr. Roger E. Howe
Department of Mathematics
University of California, Berkeley
Berkeley, California 94720

Professor Meyer Jerison
Division of Mathematical Sciences
Purdue University
Lafayette, Indiana 47907

Dr. J. E. Keizer
Battelle Memorial Institute
505 King Avenue
Columbus, Ohio 43201

Professor B. Kostant
Department of Mathematics
Massachusetts Institute of Technology
Cambridge, Massachusetts 02139

Dr. D. L. Lessor
Pacific Northwest Laboratories
Battelle Memorial Institute
P. O. Box 999
Richland, Washington 99352

Professor H. Leutwyler
CERN, Theory Division
1211 Geneve 23
Switzerland

Professor George W. Mackey
Department of Mathematics
Harvard University
Cambridge, Massachusetts 02138

Professor Louis Michel
Institute des Hautes Etudes
Scientifiques
35, route de Chartres
91-Bures-Sur-Yvette
France

Mr. William Montgomery
School of Theoretical Physics
Institute for Advanced Studies
64-65 Merrion Square
Dublin, Ireland

Professor Calvin C. Moore
Department of Mathematics
University of California, Berkeley
Berkeley, California 94720

Dr. Robert D. Ogden
Department of Mathematics
DePaul University
2323 N. Seminary
Chicago, Illinois 60614

Professor L. O'Raifeartaigh
School of Theoretical Physics
Dublin Institute for Advanced Studies
64-65 Merrion Square
Dublin 2, Ireland

Dr. Ryszard Raczka
Instytut Badan Jadrowych
Hoza 69
Warsaw, Poland

Dr. Stephen J. Rallis
School of Mathematics
Institute for Advanced Study
Princeton, New Jersey 08540

Dr. W. Rühl
Theoretical Physics Division
CERN, 1211 Geneve 23
Switzerland

Professor David H. Sharp
Department of Physics
University of Pennsylvania
Philadelphia, Pennsylvania 19104

Professor E. M. Stein
Department of Mathematics
Princeton University
Princeton, New Jersey 08540

Dr. Ernest A. Thieleker
Applied Mathematics Division
Argonne National Laboratory
Argonne, Illinois 60439

Professor Ivan T. Todorov
Institute for Advanced Study
Princeton, New Jersey 08540

Dr. Per Tomter
Department of Mathematics
University of California, Berkeley
Berkeley, California 94720

Dr. Michael J. Westwater
School of Natural Science
Institute for Advanced Study
Princeton, New Jersey 08540

Beschaffenheit der Manuskripte
Die Manuskripte werden photomechanisch vervielfältigt; sie müssen daher in sauberer Schreibmaschinenschrift mit ausreichend großer Type geschrieben sein. Handschriftliche Formeln bitte nur mit schwarzer Tusche eintragen. Notwendige Korrekturen sind bei dem bereits geschriebenen Text entweder durch Überkleben des alten Textes vorzunehmen oder aber müssen die zu korrigierenden Stellen mit weißem Korrekturlack abgedeckt werden. Die reproduktionsfähigen Abbildungen (in Originalgröße) sollen in den Text eingeklebt werden. Falls das Manuskript oder Teile desselben neu geschrieben werden müssen, ist der Verlag bereit, dem Autor bei Erscheinen seines Bandes einen angemessenen Betrag zu zahlen. Die Autoren erhalten 50 Freiexemplare.

Zur Erreichung eines möglichst optimalen Reproduktionsergebnisses ist es erwünscht, daß bei der vorgesehenen Verkleinerung der Manuskripte der Text auf einer Seite in der Breite möglichst 18 cm und in der Höhe 26,5 cm nicht überschreitet. **Entsprechende Satzspiegelvordrucke werden vom Verlag gern auf Anforderung zur Verfügung gestellt.**

Manuskripte, in englischer, deutscher oder französischer Sprache abgefaßt, sind einzureichen bei: Springer-Verlag, 6900 Heidelberg, Postfach 1780.

Cette série a pour but de donner des informations rapides, de niveau élevé, sur des développements récents en physique, aussi bien dans la recherche que dans l'enseignement supérieur. On prévoit de publier.

1. des versions préliminaires de travaux originaux et de monographies
2. des cours spéciaux portant sur un domaine nouveau ou sur des aspects nouveaux de domaines classiques
3. des rapports de séminaires
4. des conférences faites lors de congrès ou de colloques

En outre il est prévu de publier dans cette série, si la demande le justifie, des rapports de séminaires et des cours multicopiés ailleurs mais déjà épuisés.

Dans l'intérêt d'une diffusion rapide, les contributions auront souvent un caractère provisoire; le cas échéant, les démonstrations ne seront données que dans les grandes lignes. Les travaux présentés pourront également paraître ailleurs. Une réserve suffisante d'exemplaires sera toujours disponible. En permettant aux personnes intéressées d'être informées plus rapidement, les éditeurs Springer espèrent, par cette série de «prépublications», rendre d'appréciables services aux instituts de physique. Les annonces dans les revues spécialisées, les inscriptions aux catalogues et les copyrights rendront plus facile aux bibliothèques la tâche de réunir une documentation complète.

Présentation des manuscrits
Les manuscrits, étant reproduits par procédé photomécanique, doivent être soigneusement dactylographiés type assez grand. Il est recommandé d'écrire à l'encre de Chine noire les formules non dactylographiées. Les corrections nécessaires doivent être effectuées soit par collage du nouveau texte sur l'ancien soit en recouvrant les endroits à corriger par du vernis correcteur blanc. Les illustrations; en dimension originale, préparées pour reproduction sont à insérer dans le texte. S'il s'avère nécessaire d'écrire de nouveau le manuscrit, soit complètement, soit en partie, la maison d'édition se déclare prête à verser à l'auteur, lors de la parution du volume, le montant des frais correspondants. Les auteurs recoivent 50 exemplaires gratuits.

Pour obtenir une reproduction optimale il est désirable que le texte dactylographié sur une page ne dépasse pas 26,5 cm en hauteur et 18 cm en largeur. **Sur demande la maison d'edition met à la disposition des auteurs du papier spécialement préparé.**

Les manuscrits en anglais, allemand ou français peuvent être adressés à Springer-Verlag, 6900 Heidelberg, Postfach 1780.